Adaptive and Iterative Signal Processing in Communications

Adaptive signal processing (ASP) and iterative signal processing (ISP) are important techniques in improving the performance of receivers in communication systems. Using examples from practical transceiver designs, this book describes the fundamental theory and practical aspects of both methods, providing a link between the two where possible. The book is divided into three parts: the first two parts deal with ASP and ISP, respectively, each in the context of receiver design over intersymbol interference (ISI) channels. In the third part, the applications of ASP and ISP to receiver design in other interference-limited channels, including CDMA and MIMO, are considered; the author attempts to illustrate in this section how the two techniques can be used to solve problems in channels that have inherent uncertainty. With illustrations and worked examples, this text will be suitable for graduate students and researchers in electrical engineering, as well as for practitioners in the telecommunications industry. Further resources for this title are available online at www.cambridge.org/9780521864862.

PROFESSOR JINHO CHOI received his Ph.D. in electrical engineering from the Korea Advanced Institute of Science and Technology (KAIST) in 1994. He currently holds a Chair position in the Institute of Advanced Telecommunications (IAT), the University of Wales Swansea, UK. He is a senior member of the IEEE and an associate editor of *IEEE Transactions on Vehicular Technology* and an editor of the *Journal of Communications and Networks*. In 1999, he was awarded the Best Paper Award for Signal Processing from EURASIP.

Adaptive and Iterative Signal Processing in Communications

JINHO CHOI

CAMBRIDGE
UNIVERSITY PRESS

CAMBRIDGE UNIVERSITY PRESS
Cambridge, New York, Melbourne, Madrid, Cape Town, Singapore, São Paulo

Cambridge University Press
The Edinburgh Building, Cambridge CB2 2RU, UK

Published in the United States of America by Cambridge University Press, New York

www.cambridge.org
Information on this title: www.cambridge.org/9780521864862

© Cambridge University Press 2006

First published 2006

Printed in the United Kingdom at the University Press, Cambridge

A catalog record for this publication is available from the British Library

ISBN-13 978-0-521-86486-2 hardback
ISBN-10 0-521-86486-0 hardback

Contents

Figures

Tables

Preface

Various signal processing techniques are actively used in communication systems to improve the performance. In particular, adaptive signal processing has a strong impact on communications. For example, various adaptive algorithms are applied to the channel equalization and interference rejection. Adaptive equalizers and interference cancellers can effectively mitigate interference and adapt to time-varying channel environments.

Even though iterative signal processing is not as advanced as adaptive signal processing, it plays a significant role in improving the performance of receivers, which may be limited by interfering signals. In addition, the estimation error of certain channel parameters, for example the channel impulse response, can degrade the performance. An improvement in interference cancelation or a better estimate of channel parameters may be available due to iterative signal processing. After each iteration, more information about interfering signals or channel parameters is available. Then, the interference cancelation is more precise and the channel parameters can be estimated more accurately. This results in an improvement in performance for each iteration.

It would be beneficial if we could study adaptive and iterative signal processing with respect to communications. There are a number of excellent books on adaptive signal processing and communication systems, though it is difficult to find a single book that covers both topics in detail. Furthermore, as iterative signal processing is less advanced, I have been unable to find a book that balances the subjects of signal processing and its applications in communication. My desire to locate such a book increased when I took a postgraduate course entitled "Adaptive Signal Processing in Telecommunications." This provided me with the motivation to write this book, in which I attempt to introduce adaptive and iterative signal processing along with their applications in communications.

This book can be divided into three parts. In Part I, we introduce intersymbol interference (ISI) channels and adaptive signal processing techniques for ISI channels. The ISI channel is a typical interference-limited channel, and its performance is limited by the ISI. There are a number of methods used to mitigate the ISI to improve the performance. The reader will learn how adaptive signal processing techniques can be used successfully to mitigate the ISI.

In Part II, two different key methods for iterative signal processing are introduced. One is based on the expectation-maximization (EM) algorithm and the other is based on the turbo-principle. The EM algorithm was introduced to solve the maximum likelihood (ML) estimation problem. The EM algorithm is an iterative algorithm that can find ML estimates numerically. Since the EM algorithm is numerically stable and improves the likelihood through iterations, it has been extensively studied in statistics. In statistical signal process-ing areas, the EM algorithm is regarded as a standard approach for parameter estimation problems. As the channel estimation problem is a parameter estimation problem, it is natural to apply the EM algorithm.

The turbo-principle was quite a suprising idea when it appeared for turbo decoding. The performance of a simple channel code can approach a limit with the turbo decoding algorithm of reasonably low complexity. Once the turbo-principle was understood, it was widely adopted to solve difference problems including the channel equalization problem. Based on the turbo-principle, turbo equalizers were employed suppress the ISI effectively through iterations.

In Part III, we introduce different interference-limited channels. Code division multiple access (CDMA) systems suffer from multiuser interference (MUI). Therefore, the perfor-mance of CDMA is limited by MUI and can be improved by sucessfully mitigating it. Multiple input multiple output (MIMO) channels are also interference-limited since multi-ple transmit antennas transmit signals simultaneously and the transmitted signals from the other antennas become interfering signals. For both CDMA and MIMO channels, adaptive and iterative signal processing techniques are used to mitigate interfering signals effectively and estimate channels more precisely.

I would like to thank many people for supporting this work: J. Ritcey (University of Washington), K. Wong and W. Zhang (University of Waterloo), H. Jamali (University of Tehran), and F. Chan (University of New South Wales), who helped by providing constructive comments. All these colleagues assisted with proofreading, especially K. Wong and F. Chan. Needless to say the responsibility for the remaining errors, typos, unclear passages, and weaknesses is mine.

Special thanks go to C.-C. Lim (University of Adelaide), Y.-C. Liang (Institute for Infocomm Research), S. Choi (Hanyang University), X. Shen (University of Waterloo), and F. Adachi (Tohoku University) for encouragement and long-term friendship.

I also want to thank Assistant Editor A. Littlewood and Publishing Director P. Meyler at Cambridge University Press.

Finally, I would like to offer my very special thanks to my family, Kila, Seji, and Wooji, for their support, encouragement, and love.

Symbols

$\mathbf{b}$	symbol vector
b_l	(binary) data symbol transmitted at time l
$\{b_m\}$	symbol sequence
$f(\cdot)$	generic expression for a pdf
$\mathbf{h}$	CIR
$\mathbf{h}_l$	CIR at time l
$\Im(x)$	imaginary part of a complex number x
$\Re(x)$	real part of a complex number x
$\mathbf{R_y}$	covariance matrix of $\mathbf{y}$
$\lceil x \rceil$	smallest integer that is greater than or equal to x
$\lfloor x \rfloor$	largest integer that is smaller than or equal to x
$\mathbf{y}$	received signal vector
y_l	received signal at time l
$\{y_m\}$	received signal sequence

Abbreviations

AWGN	additive white Gaussian noise
BCJR algorithm	Bahl–Cocke–Jelinek–Raviv algorithm
BER	bit error rate
CAI	co-antenna interference
cdf	cumulative density function
CDMA	code division multiple access
CIR	channel impulse response
CLT	central limit theorem
CP	cyclic prefix
CRB	Cramer–Rao bound
CSI	channel state information
DFE	decision feedback equalization (or equalizer)
EGC	equal gain combining
EM	expectation-maximization
EXIT	extrinsic information transfer
FBF	feedback filter
FFF	feedforward filter
FIR	finite impulse response
IDD	iterative detector and decoder
iid	independent identically distributed
ISI	intersymbol interference
KF	Kalman filter
LE	linear equalization (or equalizer)
LLR	log likelihood ratio
LMS	least mean square
LRT	likelihood ratio test
LS	least squares
MAI	multiple access interference
MAP	maximum *a posteriori* probability
MIMO	multiple input multiple output
ML	maximum likelihood
MLSD	maximum likelihood sequence detection
MMSE	minimum mean square error
MRC	maximal ratio combining

MSE	mean square error
MUI	multiple user interference
OFDM	orthogonal frequency division multiplexing
PAPR	peak-to-average-power ratio
pdf	probability density function
PSP	per-survivor processing
RLS	recursive least square
SC	soft (interference) cancelation
SD	selection diversity
SINR	signal-to-interference-plus-noise ratio
SNR	signal-to-noise ratio
VA	Viterbi algorithm
ZF	zero-forcing

1 Introduction

Signal processing and communications are closely related; indeed, various signal processing techniques are adopted for communications at both transmitters and receivers. In particular, the role of signal processing is very important in receiver design. For example, signal processing techniques are applied to carrier and clock synchronization, channel equalization, channel estimation, interference rejection, etc. It is our aim in this book to introduce adaptive and iterative signal processing techniques for receiver design in interference-limited environments.

1.1 Communications in interference-limited environments

Generally, the performance of communication systems is limited by the interference, of which there are various sources. For example, multipaths of a radio channel cause inter-symbol interference (ISI), as shown in Fig. 1.1. The received signal becomes a sum of delayed transmitted signals with different attenuation factors. Although the background noise is negligible, the ISI can degrade the performance because the received signal is distorted by the ISI.

In a multiuser system, such as the one shown in Fig. 1.2, the other users' signals become interfering signals. Thus, it is important to alleviate interfering signals to achieve a satisfactory performance.

Throughout this book, we introduce a few communication systems under interference-limited environments. For each communication system, adaptive and/or iterative signal processing methods are discussed to overcome the interference.

We will briefly review some approaches for the interference mitigation in each interference-limited channel below.

1.1.1 ISI channels

A nonideal dispersive communication channel can introduce the ISI, and the receiver has to mitigate the ISI to achieve a satisfactory performance. Since a dispersive channel is seen as a linear filter, a linear filtering approach can be used to equalize a dispersive channel. The resulting approach is called the linear equalization. A block diagram for a communication system with a linear equalizer is shown in Fig. 1.3, in which $H(z)$ and $G(z)$ denote the $\mathcal{Z}$-transforms of an ISI channel and a linear equalizer, respectively. The transmitted signal,

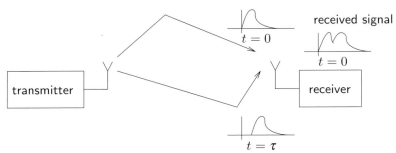

Figure 1.1. Radio channel with multipaths. In this illustration, the propagation delay of the upper path is zero, while the propagation delay of the lower path is $\tau > 0$. The received signal is a superposition of the multipath signals.

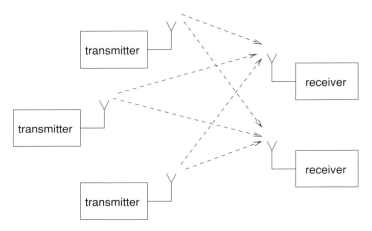

Figure 1.2. Multiuser system.

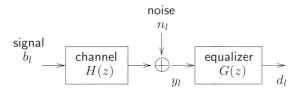

Figure 1.3. Dispersive channel and linear equalizer.

noise, channel output, and equalizer output at discrete time l are denoted by b_l, n_l, y_l, and d_l, respectively. The role of the linear equalizer is to equalize the frequency response of the channel. The zero-forcing (ZF) linear equalizer has the inverse frequency response of the channel so that the overall frequency response becomes flat and there is no ISI. That is, if $G(z) = 1/H(z)$ and the noise is negligible, we can expect that $d_l = b_l$. However, this intuitive approach often fails. If the frequency response of an ISI channel is small at some frequencies, the inverse of the frequency response, $1/H(z)$, can be large at those frequencies. The noise components corresponding to those frequencies can be enhanced, resulting in a low signal-to-noise ratio (SNR) after the equalization. To avoid this problem, other approaches can be considered.

The minimum mean square error (MMSE) filtering is well established in statistical signal processing. A linear filter is designed to minimize the difference between a desired signal and the output of the filter so that the output of the filter can be used as an estimate of the desired signal. The desired signal is a random signal and the input signal to the filter is generally correlated with the desired signal. Generally, the mean square error (MSE) is chosen to optimize the filter since the MMSE criterion is mathematically tractable.

The MMSE equalization is based on the MMSE filtering. In the MMSE equalizer, the error consists of the ISI as well as the noise. Since the MSE is to be minimized, the noise is not enhanced. The MMSE equalizer generally outperforms the ZF equalizer.

Extension of the linear equalization is possible by incorporating the cancelation of (detected) past symbols. This results in the decision feedback equalization. The main difference between this and the linear equalizer is that the decision feedback equalizer (DFE) is nonlinear. However, under certain assumptions, the DFE can be considered as a linear equalizer. The DFE consists of a feedback filter and a feedforward filter. The role of the feedforward filter is to combine dispersed signals, while the role of the feedback filter is to suppress ISI components. Even though this approach looks simple, it is very insightful, and the idea is applicable to other systems suffering from interfering signals.

A different approach, based on the sequence detection, is available that allows effective detection of signals over an ISI channel. The data sequence that maximizes the likelihood function is chosen using the maximum likelihood (ML) criterion. This approach is called ML sequence detection (MLSD) and the solution sequence is called the ML sequence. An exhaustive search, in which the likelihoods for all the possible data sequences are obtained, is carried out to find the ML sequence. Clearly, this approach becomes prohibitive because the number of candidate sequences grows exponentially with the length of the data sequences. For example, there are 2^L candidate data sequences for a binary sequence of length L.

Fortunately, there are computationally efficient algorithms for MLSD. The Viterbi algorithm (VA) is an example. The VA can find the ML sequence, and its complexity grows linearly with the length of data sequence.

Both sequence detection and channel equalization require channel estimation; this is crucial because the channel estimation error can degrade the performance. The channel estimation and data detection are closely coupled. Usually, joint channel estimation and data detection approaches can outperform decoupled approaches (in which channel estimation and data detection are carried out separately).

1.1.2 CDMA channels

For multiuser communications, there are various multiple access schemes. A multiple access scheme allows multiple users to share a common communication channel, for example the same frequency band. In frequency division multiple access (FDMA), a frequency band is divided into multiple subbands so that each user can use one of the subbands without the interference from the other users in the same system. In time division multiple access (TDMA), multiple users share the same frequency band. However, they use different time

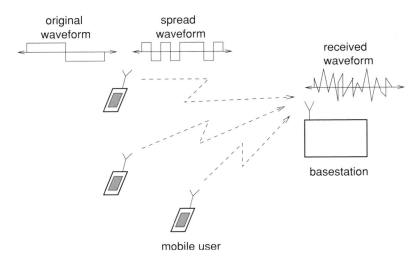

Figure 1.4. Uplink channels with spread signals.

slots to avoid multiple access interference (MAI). Code division multiple access (CDMA) is another multiple access scheme. In CDMA, users can transmit signals simultaneously over the same frequency band during the same time interval. However, to distinguish each user from the others, users' signals are spread by different spreading codes. Figure 1.4 shows the uplink channels[†] of CDMA systems, where the mobile users transmit spread signals simultaneously. CDMA is chosen for some cellular systems as it has a number of advantages over the others.

CDMA becomes interference-limited if spreading codes are not orthogonal. Since there are multiple users in a CDMA system, the signals from the other users become the interfering signals or the MAI. As the MAI limits the performance, it is desirable to mitigate the MAI to achieve a good performance. Linear multiuser detection and interference cancelation are introduced to deal effectively with the MAI. The main idea behind linear multiuser detection is similar to that behind linear equalization; that is, a linear transform is used to recover users' signals. There is also a strong relationship between interference cancelation and decision feedback equalization. If signals are detected, they are canceled to allow better detection of the as yet undetected signals.

Both ZF and MMSE criteria can be employed to design linear multiuser detectors. In addition to the linear multiuser detector and the interference canceler, the ML multiuser detector is also available. The ML multiuser detector is analogous to the MLSD. Owing to the ML criterion, all the users' signals are simultaneously detected. However, a straightforward implementation requires an exhaustive search.

1.1.3 MIMO channels

To increase the channel capacity of wireless communications, multiple antennas were introduced. Since there are multiple transmit and multiple receive antennas, the channel is called

---

[†]   The uplink channel is the name given to a channel from a user to a basestation.

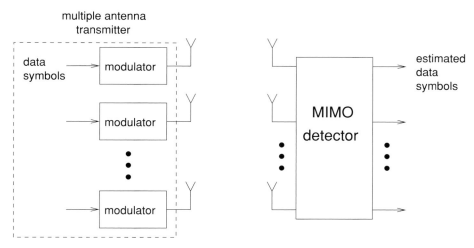

Figure 1.5. Block diagram for MIMO systems.

the multiple input multiple output (MIMO) channel. It is shown that the MIMO channel capacity increases linearly with the minimum of the numbers of transmit and receive antennas. This is an important finding, because MIMO systems can be used to provide high data rate services over wireless channels.

Various transmission schemes are studied, including a simple scheme in which each antenna transmits different data sequences. If a receiver is able to detect all the different data sequences, the data rate increases linearly with the number of transmit antennas. A receiver equipped with multiple receive antennas, as shown in Fig. 1.5, can detect multiple signals simultaneously and the resulting detection is called the MIMO detection. An optimal MIMO detector is the ML detector. Finding the ML solution vector that maximizes the likelihood function may require an exhaustive search; since the number of the candidate vectors grows exponentially with the number of transmit antennas, the ML detector becomes impractical. Thus, suboptimal, but less complex, detection algorithms are required. A combination of linear filtering and cancelation can lead to computationally efficient MIMO detection algorithms.

1.2 Issues in receiver design

There are a number of factors to consider in receiver design, and we focus on two particular issues in this book. The first is the relationship between data detection and channel estimation, and the second is the computational complexity.

1.2.1 Data detection and channel estimation

In practice, the receiver performs two major tasks: channel estimation and data detection. Even though channel estimation and data detection are tightly coupled, they are often dealt with separately. In the conventional approach, the channel estimation is carried out first to obtain a channel estimate. Then, the data detection is performed with the estimated channel. In fact, the channel estimation and data detection can be carried out jointly to achieve better performance.

There are various methods employed in data detection and channel estimation. In general, the following three groups of methods can be considered.

  (i) Group 1: joint channel estimation and data detection methods;
 (ii) Group 2: data detection oriented methods;
(iii) Group 3: channel estimation oriented methods.

For an ISI channel, a method in Group 1 attempts to find the estimates of the channel impulse response and data sequence simultaneously under a certain performance criterion, for example the maximum likelihood (ML) criterion.

Group 2 methods focus on the data detection, and the channel estimation is often implicitly carried out within the data detection process. For example, if each candidate data sequence has its own channel estimate, an exhaustive search can find an optimal data sequence (and its corresponding channel estimate) under a certain performance criterion. To avoid an exhaustive search, less complex suboptimal approaches could be investigated. In particular for an ISI channel, the trellis structure is exploited to reduce the complexity of the search. Some Group 2 methods apply to joint channel estimation and data detection (i.e., some Group 2 methods can be seen as Group 1 methods).

Group 3 methods are mainly investigated using statistical signal processing techniques, because the channel estimation problem can be seen as a parameter estimation problem. The main idea of Group 3 methods is based on the conventional approach for receiver design. As mentioned earlier, the conventional receiver has two separate steps: channel estimation and then data detection. If the channel is ideally estimated, the ensuing data detection will not experience any performance degradation. In one Group 3 method, the data detection can be carried out implicitly as part of the channel estimation.

It is difficult to compare these three groups in terms of performance and complexity. Each group has advantages and disadvantages over the others. However, throughout this book we mainly focus on Group 2 and 3 methods as they are suitable for adaptive and iterative receivers.

1.2.2 Complexity

The complexity, especially the computational complexity, is of extreme importance in receiver design. Fundamentally, it is possible to design an ideal (or optimal) receiver if an exhaustive search is affordable. However, in general, an exhaustive search becomes impractical because the complexity, which grows exponentially with the length of data sequences and/or number of users, is too high. A high complexity can result in undesirable results, such as long processing delays and high power consumption.

Adaptive and iterative signal processing techniques play a key role in deriving computationally efficient algorithms for receivers. The resulting receivers are adaptive or iterative.

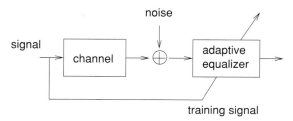

Figure 1.6. Adaptive equalization.

1.3 Adaptive signal processing

Adaptive signal processing has a wide range of applications, including its use in communication systems. The main idea of adaptive signal processing is learning.

For a channel equalizer, optimal equalization (coefficient) vectors can be found in terms of the channel impulse response (CIR) of an ISI channel. In practice, since the CIR is unknown, it has to be estimated. Then, with the estimated CIR as the true CIR, the corresponding optimal equalization vectors can be found. However, this approach has drawbacks. If the CIR varies, a new CIR should be estimated and the equalization vectors should be updated accordingly. In addition, due to the estimation error, the optimality would not be guaranteed.

Adaptive algorithms can find the equalization vectors without explicit knowledge of the channel. In general, an adaptive equalizer can find an equalization vector using a training sequence, as shown in Fig. 1.6. If adaptive algorithms are capable of tracking a time-varying CIR, the equalization vector can be continuously updated according to the channel variation.

Adaptive algorithms are also applicable to the channel estimation for an ISI channel. Adaptive channel estimation can be carried out in conjunction with the sequence detection. This results in an adaptive joint channel estimation and data detection.

For CDMA systems, adaptive algorithms are applied to the multiuser detection as the spreading codes of the other users may not be known. Adaptive linear multiuser detectors learn the statistical properties of interfering signals to suppress interfering signals.

In CDMA systems, statistical properties of the MAI can vary when there are incoming or outgoing users. Thus, the statistical properties of interfering signals are time-variant. In this case, the tracking ability of adaptive algorithms becomes essential in order to adjust the coefficients of a multiuser detector. Adaptive receivers are also important in tracking the variation of fading channels.

1.4 Iterative signal processing

There are various forms of iterative signal processing for communication systems. Iterative signal processing is an attractive idea as the iterations are expected to produce better performance. Generally, iterative signal processing techniques are off-line algorithms, while

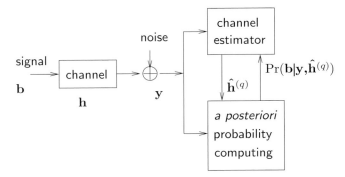

Figure 1.7. Iterative channel estimation approach.

adaptive signal processing techniques are on-line algorithms. In this book, we focus on two different methods for iterative signal processing: one is based on the expectation-maximization (EM) algorithm, and other is based on the turbo principle.

The EM algorithm is an iterative method used to solve various maximum likelihood (ML) estimation problems. If the likelihood function is highly nonlinear, a closed-form solution may not be available. In this case, numerical approaches can be used to find the ML solution. The main advantage of the EM algorithm over other numerical approaches is that it is numerically stable with each EM iteration increasing the likelihood. The EM algorithm has been widely used in a number of signal processing areas. The channel estimation is a typical example in which the EM algorithm is applicable.

Different approaches could be used to apply the EM algorithm to the channel estimation. Figure 1.7 shows a block diagram of an iterative channel estimation approach based on the EM algorithm approach, where $\mathbf{b}$, $\mathbf{h}$, and $\mathbf{y}$ denote the transmitted signal vector, the channel vector, and the received signal vector, respectively. The channel vector $\mathbf{h}$ may be estimated from $\mathbf{y}$. In general, channel estimation becomes difficult if $\mathbf{b}$ is unknown; in this case it is known as blind channel estimation. The iterative channel estimation provides a sequence of channel estimates through iterations. Since $\mathbf{b}$ is not available, we attempt to find the *a posteriori* probability of $\mathbf{b}$ from the received signal vector, $\mathbf{y}$, and, say, the qth estimate of $\mathbf{h}$, $\hat{\mathbf{h}}^{(q)}$, for the next estimation to obtain $\hat{\mathbf{h}}^{(q+1)}$, which is the $(q+1)$th estimate of $\mathbf{h}$. Through iterations, if the *a posteriori* probability of $\mathbf{b}$ becomes more reliable, a better estimate of $\mathbf{h}$ is expected. According to the principle of the EM algorithm, the channel estimate converges to the ML estimate if an initial estimate is sufficiently close to the ML estimate.

The iterative channel estimation approach is generic. It is applicable to ISI, CDMA, and MIMO channels. In addition, the channel estimator in EM-based channel estimation can also be modified as long as it can provide good performance. The main requirement of the channel estimator in Fig. 1.7 is that it should exploit the *a posteriori* probability of $\mathbf{b}$. There are other channel estimators which meet this requirement.

The major problem with the iterative channel estimation approach in Fig. 1.7 is that the complexity of finding the *a posteriori* probability of $\mathbf{b}$ can be too great. A straightforward implementation of the calculation of the *a posteriori* probability of $\mathbf{b}$ has the complexity

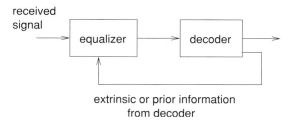

received
signal

extrinsic or prior information
from decoder

Figure 1.8. Turbo equalizer within the iterative receiver.

growing exponentially with the length of **b**. There are computationally efficient approaches that can find the *a posteriori* probability of **b** depending on the structure of the channels. For example, the trellis structure of an ISI channel can be exploited to find the *a posteriori* probability of **b** with a lower complexity.

The turbo principle was originally introduced to decode turbo codes, and it led to the proposal of an iterative decoding algorithm that provided an (almost) ideal performance. The turbo principle is also applicable to channel equalization for coded signals. The main idea behind the turbo equalizer is to exchange extrinsic information between the channel equalizer and channel decoder. Since more reliable prior information from a channel decoder is available, the ISI can be effectively suppressed in the turbo equalizer through iterations. An equalizer within an iterative receiver, as shown in Fig. 1.8, should take advantage of the prior information from the channel decoder.

The extrinsic information is regarded as prior information and is used to detect or decode data symbols together with the received signal. Therefore, the maximum *a posteriori* probability (MAP) principle is suitable for both equalizer and decoder. However, there can be other approaches to incorporating prior information. In particular, for channel equalization, a cancelation based approach can be employed. Using prior information of data symbols, a cancelation method can effectively mitigate the ISI with a low complexity. Thus, the MAP equalizer can be replaced by a cancelation based equalizer such as a DFE at the expense of minor performance degradation.

In CDMA systems, the turbo principle can be applied in the design of iterative receivers. For coded signals, an iterative receiver consists of a multiuser detector and a group of channel decoders. Using decoded signals, a better interference cancelation is expected if an interference cancelation based multiuser detector is used. As the output of the multiuser detector becomes more reliable, the resulting decoding would provide more reliable outputs. Through iterations, an (almost) ideal performance can be achieved.

The iterative receiver has several drawbacks. Generally, it has a long processing delay as several iterations are required. In addition, since iterative receivers are based on off-line processing, a number of memory elements are required to store a block of received signal samples as well as soft information of decoded bit sequences. Even though the iterative receiver has these drawbacks, it will be widely employed as it can provide almost ideal performance. In addition, given that computing power will increase and the cost of memory elements will become cheaper, there is no doubt that the iterative receiver will become popular in the future.

1.5 Outline of the book

The book is divided into three parts as follows.

- Part I: Chapters 2, 3, and 4 belong to Part I, in which the ISI channel is considered. To alleviate the ISI, various approaches are discussed with adaptive signal processing techniques.

  Chapter 2 mainly concentrates on channel equalizers. The sequence detection is an alternative approach to deal with the ISI problem. The adaptive channel estimation is introduced in conjunction with sequence detection in Chapter 3. In Chapter 4, we discuss sequence detection and channel estimation for fading channels, a brief introduction of which is also presented in Chapter 4.

- Part II: Iterative signal processing techniques and iterative receivers that include a channel decoder are introduced in Part II, which consists of Chapters 5, 6, and 7.

  We discuss the EM algorithm and introduce an iterative channel estimation method based on the EM algorithm in Chapter 5. An optimal symbol detection method is studied in conjunction with the iterative channel estimation. In Chapter 6, an iterative receiver is introduced for signal detection/decoding over ISI channels based on the turbo principle. We briefly introduce information theory and channel coding to provide a background for understanding of the iterative receiver. Two different approaches to the design of the iterative receiver for random channels are discussed in Chapter 7.

- Part III: Applications of adaptive and iterative signal processing techniques to other interference-limited communication systems are studied in Part III (Chapters 8, 9, 10, and 11).

  In Chapter 8, CDMA systems are introduced along with multiuser detection. Adaptive linear multiuser detectors are also discussed. The iterative receiver for CDMA systems is the major topic of Chapter 9.

  MIMO channels with multiple antennas are studied in Chapter 10 using diversity techniques. Various MIMO detection algorithms that are suitable for the iterative receiver are also presented.

  Orthogonal frequency division multiplexing (OFDM) is introduced in Chapter 11. OFDM is not interference-limited in an ideal situation. However, OFDM is introduced in this book as it is becoming important for wireless communications and an iterative receiver is included to improve the performance. The code diversity in OFDM systems and the iterative receiver based on the EM algorithm are discussed.

1.6 Limitations of the book

In this book, we do not attempt to introduce a comprehensive guide for communication systems. Rather, we aim at introducing adaptive and iterative signal processing with examples in communication systems. Thus, we do not address several important topics of communication systems, including signal design, modulation schemes, synchronization, and so on. In addition, we confine ourselves to the binary phase shift keying (BPSK) signaling to simplify the problems discussed in the book. At the end of each chapter, we include a summary, and we refer readers to references that cover topics not considered in this book.

I ISI channels and adaptive signal processing

2 Channel equalization for dispersive channels

In this chapter, we study the intersymbol interference channel. Intersymbol interference (ISI) is a self-noise introduced by a dispersive channel. Since the ISI channel can significantly degrade the performance of communication systems, we need to mitigate the ISI. One of the methods that can achieve this is channel equalization.

There are two different structures for channel equalization. Linear equalization is based on a linear filtering approach. Since an ISI channel can be seen as a linear filter, a linear equalizer at the receiver equalizes the frequency response of the ISI channel to mitigate the ISI. The other approach is called decision feedback equalization, in which linear filtering and cancelation are used to alleviate the ISI. In addition to these approaches, which are based on the structure of the equalizer, the performance criterion is important in the optimization of an equalizer. Two of the most popular criteria will be introduced in this chapter.

We also study adaptive equalizers with fundamental properties of adaptive algorithms. For a given ISI channel, an adaptive equalizer can adjust itself to achieve the optimum performance. Therefore, adaptive equalizers are important practically.

In this chapter, we assume that the reader is familiar with the fundamental concepts of signal processing, including convolution, linear filtering, and the sampling theorem. In addition, a working knowledge of probability and random processes is required.

2.1 ISI channels and the equalization problem

We will derive a model for the ISI channel in the discrete-time domain since most receiver operations are generally carried out with digital circuits or digital signal processors after sampling.

2.1.1 Discrete time model for ISI channels

An overall communication system over a dispersive linear channel is depicted in Fig. 2.1. A data sequence, denoted by $\{b_m\}$, is to be transmitted as an analog waveform that is generated by the impulse modulator and transmitter filter. The output of the impulse modulator is given

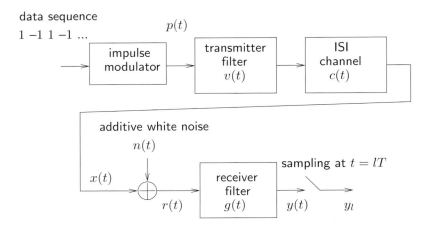

Figure 2.1. Communications through a bandlimited channel.

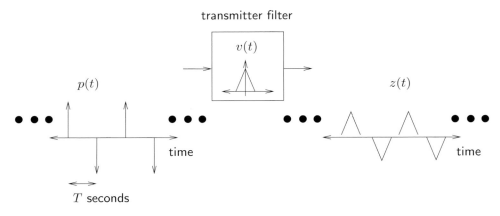

Figure 2.2. Examples of waveforms of $p(t)$ and $z(t)$.

by

$$p(t) = \sum_{l=0}^{L-1} b_l \delta(t - lT),$$

where L is the length of the data sequence, T stands for the symbol duration, and $\delta(t)$ denotes the Dirac delta function. If $v(t)$ denotes the impulse response of the transmitter filter, the transmitted signal that is the output of the transmitter filter is given by

$$z(t) = \sum_{l=0}^{L-1} b_l v(t - lT).$$

The waveforms of $p(t)$ and $z(t)$ are illustrated in Fig. 2.2.

Let $c(t)$ be the channel impulse response (CIR) and let $g(t)$ be the impulse response of the receiver filter; see Fig. 2.1. Furthermore, let $\bar{v}(t) = v(t) * c(t)$, where $*$ denotes the convolution. Then, $\bar{v}(t)$ can be considered as a modified transmitter filter that includes the

CIR. The received signal after receiver filtering is given by

$$y(t) = \int g(t-\tau)(x(\tau)+n(\tau))\,d\tau$$

$$= \int g(t-\tau)\left(\sum_m \bar{v}(\tau - mT)b_m\right)d\tau + \int g(t-\tau)n(\tau)\,d\tau$$

$$= \sum_m b_m \int g(t-\tau)\bar{v}(\tau-mT)\,d\tau + \int g(t-\tau)n(\tau)\,d\tau,$$

where $x(t) = c(t) * z(t) = v(t) * p(t)$ is the channel output without noise and $n(t)$ is a zero-mean white Gaussian noise with double-sided spectral density $N_0/2$. This means that the spectral density of $n(t)$ is given by $S_n(f) = N_0/2, \forall f$, where f denotes the frequency. The corresponding autocorrelation function is given by $R_n(\tau) = E[n(t)n(t-\tau)] = (N_0/2)\delta(\tau)$.

The sampled signal of $y(t)$ at every T seconds (we assume a critical sampling) is given by

$$y_l = y(lT)$$

$$= \sum_m b_m \underbrace{\int g(lT-\tau)\bar{v}(\tau-mT)\,d\tau}_{=g(t)*\bar{v}(t)|_{t=(l-m)T}} + n_l$$

$$= \sum_m b_m h_{l-m} + n_l$$

$$= b_l * h_l + n_l,$$

where $n_l = \int g(lT-\tau)n(\tau)\,d\tau$ and $h_{l-m} = g(t)*\bar{v}(t)|_{t=(l-m)T}$. Here, n_l is the sampled noise and $\{h_m\}$ is the discrete-time CIR.

Example 2.1.1 This example considers how the discrete-time CIR may be obtained after sampling.

Let

$$c(t) = \begin{cases} 1/\sqrt{T}, & 0 \le t < 1.5T; \\ 0, & \text{otherwise} \end{cases}$$

and

$$g(t) = \begin{cases} 1/\sqrt{T}, & 0 \le t < T; \\ 0, & \text{otherwise.} \end{cases}$$

In addition, assume that $v(t) = \delta(t)$.

With $\bar{v}(t) = c(t) * v(t) = c(t)$, the discrete-time CIR, $\{h_m\}$, can be found:

$$h_l = \int g(lT-\tau)\bar{v}(\tau)\,d\tau$$

$$= \int_0^{1.5T} (1/\sqrt{T})g(lT-\tau)\,d\tau$$

$$= \begin{cases} 0, & \text{if } l \le 0; \\ 1, & \text{if } l = 1; \\ 0.5, & \text{if } l = 2; \\ 0, & \text{if } l \ge 3. \end{cases}$$

The sampled noise n_l is given by

$$n_l = \int g(lT - \tau)n(\tau)\,d\tau$$

$$= \int g(\tau)n(lT - \tau)\,d\tau$$

$$= \frac{1}{\sqrt{T}}\int_0^T n(lT - \tau)\,d\tau.$$

It can be shown that $E[n_l] = 0$. The variance of n_l can be found as follows:

$$E[|n_l|^2] = \frac{1}{T}E\left[\int_0^T n(lT - \tau)\,d\tau \int_0^T n(lT - \tau')\,d\tau'\right]$$

$$= \frac{1}{T}\int_0^T \int_0^T E[n(lT - \tau)n(lT - \tau')]\,d\tau\,d\tau'$$

$$= \frac{1}{T}\int_0^T \frac{N_0}{2}\,d\tau$$

$$= \frac{N_0}{2}.$$

It can also be shown that $E[n_l n_m] = 0$ if $l \neq m$.

Generally, we may make the following observations.

- If the receiver filter $g(t)$ is the Nyquist filter, then the sampled noise sequence, n_l, is a white noise sequence. For a normalized receiver filter $g(t)$, we can assume that $E[n_l] = 0$ and $E[n_m n_l] = (N_0/2)\delta_{m,l}$, where $\delta_{m,l}$ stands for the Kronecker delta. The Kronecker delta is defined as follows:

$$\delta_{m,l} = \begin{cases} 1, & \text{if } m = l; \\ 0, & \text{otherwise.} \end{cases}$$

- The CIR, $\bar{v}(t)$, may not satisfy the Nyquist condition even if the transmitter filter, $v(t)$, does. Therefore, $h_{l-m} \neq \delta_{m,l}$.

If $v(t)$ and $c(t)$ have finite-length supports, $\{h_m\}$ is a finite time sequence, i.e. $h_l = 0$ for $l > M_u$ and $l < M_l$, where $M_l < M_u$ are integers. Hence, after introducing a time delay, the discrete-time CIR, $\{h_m\}$, can be seen as a causal impulse response of a linear system (i.e. the discrete-time channel). That is, $h_l = 0$ for $l < 0$ and $l \geq P$, where P is the length of the CIR. Consequently, the received signal in the discrete-time domain is given by

$$y_l = \sum_{p=0}^{P-1} h_p b_{l-p} + n_l. \tag{2.1}$$

The corresponding ISI channel in the discrete-time domain is shown in Fig. 2.3. If P is finite, the ISI channel (excluding the noise) can be seen as a finite impulse response (FIR) filter.

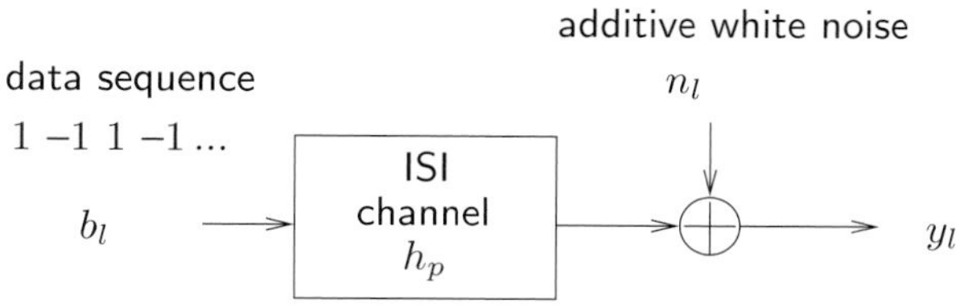

Figure 2.3. Discrete-time domain ISI channel.

2.1.2   The ISI problem

The ISI is a self-noise. According to Eq. (2.1), the received signal is a sum of P transmitted data symbols with different time delays. Hence, although the noise is zero, there exist interfering signals. Suppose that $s_l = b_{l-\bar{m}}$ is the desired signal, where $\bar{m}$ ($\bar{m} \geq 0$) is a delay. Then, Eq. (2.1) can be rewritten as follows:

$$y_l = h_{\bar{m}} \underbrace{b_{l-\bar{m}}}_{\text{desired signal}} + \underbrace{\sum_{p \neq \bar{m}} h_p b_{l-p}}_{\text{ISI terms}} + n_l.$$

Due to the ISI terms, the desired signal, $s_l = b_{l-\bar{m}}$, cannot be clearly observed from the received signal. Therefore, it is necessary to eliminate the ISI terms to extract the desired signal.

2.2   Linear equalizers

From Eq. (2.1), an ISI channel can be considered as a linear filter. The channel equalization can employ a linear filtering approach to mitigate the ISI.

2.2.1   Zero-forcing equalizer

Both structure and performance criteria are important in equalizer design. In order to discuss the structure of an equalizer, a frequency domain approach based on the $\mathcal{Z}$-transform becomes useful.

Consider the $\mathcal{Z}$-transform of the received signal in the absence of noise:

$$Y(z) = \mathcal{Z}(y_l)$$
$$= H(z)B(z), \tag{2.2}$$

where $H(z)$ and $B(z)$ are the $\mathcal{Z}$-transforms of h_p and b_l, respectively. The role of a linear equalizer (LE) is to equalize the effect of the channel in the frequency domain. Hence, we can use a linear filter to equalize the channel as in Fig. 2.4 to maintain the following relationship:

$$G(z)H(z) = z^{-\bar{m}}, \qquad \forall z = e^{j\omega}, \quad -\pi \leq \omega < \pi, \tag{2.3}$$

where $G(z)$ is the $\mathcal{Z}$-transform of the impulse response of an LE. The impulse response and the output of an LE are denoted by $\{g_m\}$ and d_l, respectively. If $G(z)H(z) = z^{-\bar{m}}$, we

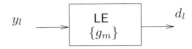

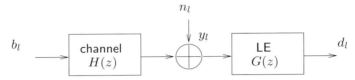

Figure 2.4. Linear equalizer.

have $d_l = s_l = b_{l-\bar{m}}$, ignoring the noise. The LE whose $\mathcal{Z}$-transform is $G(z) = z^{-\bar{m}}/H(z)$ is called the zero-forcing (ZF) equalizer, because the ISI is forced to be zero.

Although a ZF equalizer can completely remove the ISI, there are some drawbacks, for example the ZF equalizer can enhance the noise. To see this, consider the noise variance after ZF equalization. Let $\{g_m\}$ denote the impulse response of a ZF equalizer, $G(z) = z^{-\bar{m}}/H(z)$. The noise after ZF equalization can be written as follows:

$$\eta_l = \sum_m g_m n_{l-m}.$$

Using Parseval's equality, the variance of η_l can be found as follows:

$$\sigma_\eta^2 = E[|\eta_l|^2]$$

$$= E\left[\left(\sum_m g_m n_{l-m}\right)\left(\sum_q g_q n_{l-q}\right)\right]$$

$$= \sum_m \sum_q g_m g_q \underbrace{E[n_{l-m} n_{l-q}]}_{=(N_0/2)\delta_{m,q}}$$

$$= \frac{N_0}{2} \sum_m |g_m|^2$$

$$= \frac{N_0}{2} \frac{1}{2\pi} \int_{-\pi}^{\pi} \frac{1}{|H(e^{j\omega})|^2} \, d\omega. \tag{2.4}$$

This shows that if $H(e^{j\omega})$ has nulls (in the frequency response), the variance of the noise can be infinity.

There is another disadvantage of the ZF equalization. A ZF equalizer has an infinite impulse response for finite-length channels. For example, let $h_0 = 1$ and $h_1 = a$ with $P = 2$ and $\bar{m} = 0$. The impulse response of the ZF equalizer that is given by

$$G(z) = \frac{1}{H(z)}$$

$$= \frac{1}{1 + az^{-1}}$$

$$= 1 - az^{-1} + a^2 z^{-2} - a^3 z^{-3} + a^4 z^{-4} - \cdots$$

has infinite length.

2.2.2 Minimum mean square error linear equalizer

There exists another LE that can overcome the major problems of the ZF equalization. Using the minimum mean square error (MMSE) criterion, a better LE can be constructed.

In the ZF equalization, a deterministic approach is used to equalize an ISI channel. On the other hand, in the MMSE approach, a statistical method is used. It is assumed that the transmitted sequence $\{b_m\}$ is a random sequence. The role of the MMSE equalizer is to minimize the error between the output of the equalizer and the desired signal. Since the error is also a random variable, the mean square error (MSE) or the mean absolute error can be used to indicate an average performance.

Suppose that the MSE between the output of a causal LE, d_l, and the desired signal, $s_l = b_{l-\bar{m}}$, is to be minimized. We assume that $\{b_m\}$ is independent identically distributed (iid). In addition, assume that $E[b_l] = 0$ and $E[|b_l|^2] = \sigma_b^2$. The MSE is given by

$$\begin{aligned}
\text{MSE} &= E[|u_l|^2] \\
&= E[|s_l - d_l|^2] \\
&= E\left[\left|s_l - \sum_{m=0}^{\infty} g_m y_{l-m}\right|^2\right],
\end{aligned} \tag{2.5}$$

where $u_l = s_l - d_l$ is the error. Since the MSE includes the noise as well as the ISI, the minimization of the MSE can result in a better performance as the noise would not be enhanced.

Suppose that an LE has an FIR of length M. Define

$$\begin{aligned}
\mathbf{y}_l &= [\,y_l \quad y_{l-1} \quad \cdots \quad y_{l-M+1}]^{\text{T}}, \\
\mathbf{g} &= [g_0 \quad g_1 \quad \cdots \quad g_{M-1}]^{\text{T}}.
\end{aligned}$$

Then, the MSE in Eq. (2.5) is rewritten as follows:

$$\begin{aligned}
\text{MSE}(\mathbf{g}) &= E[|s_l - d_l|^2] \\
&= E\left[|s_l - \mathbf{y}_l^{\text{T}}\mathbf{g}|^2\right].
\end{aligned} \tag{2.6}$$

If $\mathbf{g}$ is optimal in terms of the MMSE criterion, according to the orthogonality principle, the error, $u_l = s_l - d_l$, should be uncorrelated with $\mathbf{y}_l$. This implies that

$$E[u_l \mathbf{y}_l] = \mathbf{0}. \tag{2.7}$$

It follows that

$$\begin{aligned}
\mathbf{0} &= E\left[\mathbf{y}_l(s_l - \mathbf{y}_l^{\text{T}}\mathbf{g})\right] \\
&= E[\mathbf{y}_l s_l] - E[\mathbf{y}_l \mathbf{y}_l^{\text{T}}]\mathbf{g}.
\end{aligned} \tag{2.8}$$

Thus, we can have

$$\mathbf{g}_{\text{mmse}} = \mathbf{R}_{\mathbf{y}}^{-1}\mathbf{r}_{y,s}, \tag{2.9}$$

where

$$\mathbf{R_y} = E[\mathbf{y}_l \mathbf{y}_l^\mathsf{T}],$$
$$\mathbf{r}_{y,s} = E[\mathbf{y}_l s_l].$$

Furthermore, the MMSE is given by

$$
\begin{aligned}
\text{MMSE} &= E\big[\big|s_l - \mathbf{y}_l^\mathsf{T}\mathbf{g}_{\text{mmse}}\big|^2\big] \\
&= E\big[s_l\big(s_l - \mathbf{y}_l^\mathsf{T}\mathbf{g}_{\text{mmse}}\big)\big] \\
&= \sigma_b^2 - E\big[s_l\mathbf{y}_l^\mathsf{T}\big]\mathbf{g}_{\text{mmse}} \\
&= \sigma_b^2 - \mathbf{r}_{y,s}^\mathsf{T}\mathbf{R_y}^{-1}\mathbf{r}_{y,s}.
\end{aligned}
\tag{2.10}
$$

A general expression is often required for $\mathbf{R_y}$. The covariance matrix of $\mathbf{y}_l$ is a Toeplitz matrix[†] as follows:

$$[\mathbf{R_y}]_{m,k} = E[y_{l-m+1}y_{l-k+1}] = r_y(k-m), \quad m,k = 1,2,\ldots,M.$$

This is a consequence because y_l is a wide-sense stationary random process. From Eq. (2.1), we can show that

$$
\begin{aligned}
[\mathbf{R_y}]_{m,k} &= E[y_{l-m+1}y_{l-k+1}] \\
&= E\left[\left(\sum_{p=0}^{P-1} h_p b_{l-m+1-p} + n_{l-m+1}\right)\left(\sum_{p'=0}^{P-1} h_{p'}b_{l-k+1-p'} + n_{l-k+1}\right)\right] \\
&= \sum_p \sum_{p'} h_p h_{p'} E[b_{l-m+1-p}b_{l-k+1-p'}] + \frac{N_0}{2}\delta_{m,k} \\
&= \sigma_b^2 \sum_p h_p h_{m-k+p} + \frac{N_0}{2}\delta_{m,k}.
\end{aligned}
\tag{2.11}
$$

As shown in Eq. (2.11), $[\mathbf{R_y}]_{m,k} = r_y(k-m)$ is a function of $k-m$. In addition, $\mathbf{R_y}$ is symmetric, i.e. $[\mathbf{R_y}]_{m,k} = [\mathbf{R_y}]_{k,m}$. Thus, we only need to find $[\mathbf{R_y}]_{m,k}$ for $m \geq k$. If $m \geq k$, we have

$$[\mathbf{R_y}]_{m,k} = \sigma_b^2 \sum_{p=0}^{P-1-(m-k)} h_p h_{m-k+p} + \frac{N_0}{2}\delta_{m,k}.
\tag{2.12}$$

From this, we can easily find the covariance matrix $\mathbf{R_y}$.

Suppose that $0 \leq \bar{m} \leq M-1$. To compute the correlation vector $\mathbf{r}_{y,s}$, we need to find

$$
\begin{aligned}
[\mathbf{r}_{y,s}]_m &= E[s_l y_{l-m+1}] \\
&= E\left[b_{l-\bar{m}}\left(\sum_{p=0}^{P-1} h_p b_{l-m+1-p} + n_{l-m+1}\right)\right] \\
&= \begin{cases} \sigma_b^2 h_{\bar{m}-m+1}, & \text{if } 0 \leq \bar{m}-m+1 \leq P-1; \\ 0, & \text{otherwise.} \end{cases}
\end{aligned}
$$

[†]   A matrix $\mathbf{A}$ is called a Toeplitz matrix if $[\mathbf{A}]_{m,n} = \alpha_{m-n}$.

For example, if $\bar{m} = 0$, we have

$$\mathbf{r}_{y,s} = \left[(\sigma_b^2 h_0) \ 0 \ \cdots \ 0 \right]^{\mathrm{T}}. \tag{2.13}$$

Equations (2.12) and (2.13) help to derive the MMSE LE coefficients.

Example 2.2.1 In this example, we find the MMSE LE that minimizes the MSE cost function MSE $= E[|s_l - d_l|^2]$ when $h_0 = 1$ and $h_1 = a$ and $s_l = b_l$. We assume that $M = 3$ for an LE.

First of all, we need to find $E[y_l y_{l-m}]$:

$$
\begin{aligned}
E[y_l y_{l-m}] &= E[(b_l + ab_{l-1} + n_l)(b_{l-m} + ab_{l-m-1} + n_{l-m})] \\
&= E[b_l b_{l-m}] + aE[b_{l-1}b_{l-m}] + aE[b_l b_{l-m-1}] \\
&\quad + a^2 E[b_{l-1}b_{l-m-1}] + E[n_l n_{l-m}] \\
&= \sigma_b^2(\delta_m + a\delta_{m-1} + a\delta_{m+1} + a^2\delta_m) + \frac{N_0}{2}\delta_m \\
&= \begin{cases} \sigma_b^2(1+a^2) + \frac{N_0}{2}, & \text{if } m = 0; \\ \sigma_b^2 a, & \text{if } m = 1 \text{ or } -1; \\ 0, & \text{otherwise}, \end{cases}
\end{aligned}
$$

where δ_m stands for the unit pulse:

$$
\delta_m = \begin{cases} 1, & \text{if } m = 0; \\ 0, & \text{otherwise}. \end{cases}
$$

In addition, the elements of $\mathbf{r}_{y,s}$ can be found as follows:

$$
\begin{aligned}
E[b_l y_{l-m}] &= E[b_l(b_{l-m} + ab_{l-m-1} + n_{l-m})] \\
&= \sigma_b^2(\delta_m + a\delta_{m+1}) \\
&= \begin{cases} \sigma_b^2, & \text{if } m = 0; \\ a\sigma_b^2, & \text{if } m = -1; \\ 0, & \text{otherwise}. \end{cases}
\end{aligned}
$$

Hence, the vector $\mathbf{g}_{\mathrm{mmse}}$ becomes

$$
\mathbf{g}_{\mathrm{mmse}} = \begin{bmatrix} \sigma_b^2(1+a^2) + \frac{N_0}{2} & \sigma_b^2 a & 0 \\ \sigma_b^2 a & \sigma_b^2(1+a^2) + \frac{N_0}{2} & \sigma_b^2 a \\ 0 & \sigma_b^2 a & \sigma_b^2(1+a^2) + \frac{N_0}{2} \end{bmatrix}^{-1} \begin{bmatrix} \sigma_b^2 \\ 0 \\ 0 \end{bmatrix}.
$$

The orthogonality principle can be explained by a geometrical approach with the MSE cost function in Eq. (2.6). Note that the MSE cost function is a quadratic function of $\mathbf{g}$. Hence, the minimum can be obtained when the first order derivative of the MSE cost function with respect to $\mathbf{g}$ is zero. That is,

$$
\left. \frac{\partial \mathrm{MSE}(\mathbf{g})}{\partial \mathbf{g}} \right|_{\mathbf{g}_{\mathrm{mmse}}} = \mathbf{0}.
$$

Since

$$\frac{\partial \mathrm{MSE}(\mathbf{g})}{\partial \mathbf{g}} = \frac{\partial}{\partial \mathbf{g}} E\left[\left(s_l - \mathbf{y}_l^{\mathrm{T}} \mathbf{g}\right)^2\right]$$
$$= -E\left[2\mathbf{y}_l\left(s_l - \mathbf{y}_l^{\mathrm{T}} \mathbf{g}\right)\right],$$

the condition that the first order derivative is zero implies that

$$E\left[\mathbf{y}_l\left(s_l - \mathbf{y}_l^{\mathrm{T}} \mathbf{g}\right)\right] = E[\mathbf{y}_l u_l] = 0$$

or

$$E[u_l y_{l-m}] = 0, \quad m = 0, 1, \ldots, M-1.$$

This is the same condition found from the orthogonality principle.

Although a finite-length MMSE LE is used in practical situations, it is interesting to see the MMSE solution for an infinite-length MMSE LE. The $\mathcal{Z}$-transform of an infinite-length MMSE LE is given by

$$\begin{aligned} G(z) &= \frac{\sigma_b^2 H^*(z) z^{-\bar{m}}}{\sigma_b^2 |H(z)|^2 + \frac{N_0}{2}} \\ &= \frac{H^*(z) z^{-\bar{m}}}{|H(z)|^2 + \frac{1}{\gamma}}, \end{aligned} \tag{2.14}$$

where $\gamma = 2\sigma_b^2/N_0$. From this, an implementation of the MMSE LE can be considered by using the cascade of two filters. The first filter is the matched filter to the CIR $\{h_p\}$ with a delay $\bar{m}$, whose impulse response is $H^*(z) z^{-\bar{m}}$, and the second filter is an infinite-tap filter whose impulse response is the inverse of $|H(z)|^2 + (1/\gamma)$.

To see the performance of the MMSE LE, we consider the following two CIRs of length $P = 5$:

$$(\text{Channel A}) \quad \mathbf{h} = [0.227 \ \ 0.460 \ \ 0.688 \ \ 0.460 \ \ 0.227]^{\mathrm{T}},$$
$$(\text{Channel B}) \quad \mathbf{h} = [0.688 \ \ 0.460 \ \ 0.460 \ \ 0.227 \ \ 0.227]^{\mathrm{T}},$$

where $\mathbf{h} = [h_0 \ h_1 \ \cdots \ h_{P-1}]^{\mathrm{T}}$. For both Channel A and Channel B, we have $\|\mathbf{h}\|^2 = 1$. The optimal MMSE LEs are found for each channel with $\bar{m} = 2$. The frequency responses of Channel A and Channel B are shown in Fig. 2.5. The zeros of Channel A are at $-0.5257 \pm j0.8951$ and $-0.4875 \pm j0.8305$. It can be easily verified that the zeros of Channel A are close to the unit circle. Thus, as shown in Fig. 2.5, there are frequency nulls. On the other hand, Channel B does not have frequency nulls.

The decision error probability (referred to as the bit error rate, BER) after the channel equalization is obtained by simulations and is shown in Fig. 2.6. The signal to noise ratio (SNR) is defined as follows:

$$\mathrm{SNR} = \frac{\|\mathbf{h}\|^2}{N_0} = \frac{1}{N_0}.$$

It is shown that the performance of the MMSE LE varies significantly depending on the ISI channel. The performance is mainly affected by spectral properties of the ISI channel. If

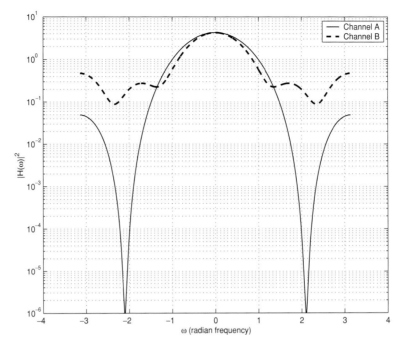

Figure 2.5. Frequency responses of two channels.

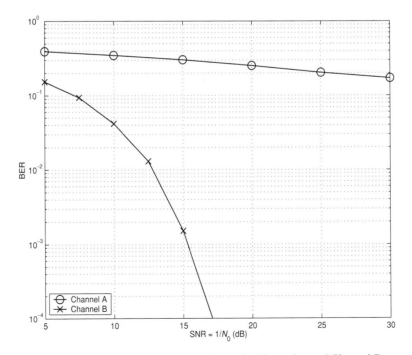

Figure 2.6. Bit error rate performance of the MMSE LE for Channel A and Channel B.

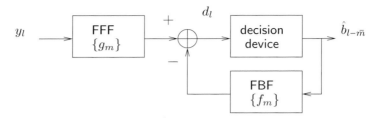

Figure 2.7. Structure of the DFE.

there are frequency nulls, as shown in Eq. (2.14), the corresponding frequency responses of the MMSE LE become zero to suppress the noise since there is no useful signal. Generally, if there are more frequency nulls, more corresponding frequency responses of the MMSE LE become zero and this results in a worse performance of Channel A, as shown in Fig. 2.6.

2.3   Decision feedback equalizers

Depending on the channel spectrum, the performance of the LE varies. In particular, if there are frequency nulls, the performance of the LE would not be satisfactory even though the SNR is high. To overcome this difficulty, the decision feedback equalizer (DFE) can be used.

The DFE consists of the following three components: a feedforward filter (FFF), a feedback filter (FBF), and a decision device. A block diagram for the DFE is depicted in Fig. 2.7. In this section, we introduce a DFE with finite numbers of taps of FFF and FBF. The two criteria, ZF and MMSE, will be applied to derive the filter coefficients.

2.3.1   Zero-forcing DFE

Suppose that $\{g_m\}$ is the impulse response of the FFF and that the length of the impulse response is M. The convolution of the FFF's impulse response $\{g_m\}$ and the channel impulse response $\{h_p\}$ becomes

$$c_l = \sum_{m=0}^{M-1} g_m h_{l-m}. \tag{2.15}$$

Using matrix notation, we can show that

$$\begin{aligned} \mathbf{c} &= [c_0 \ c_1 \ \cdots \ c_{M+P-2}] \\ &= \mathbf{Hg} \\ &= \underbrace{\begin{bmatrix} h_0 & 0 & \cdots & 0 \\ h_1 & h_0 & \cdots & 0 \\ \vdots & \vdots & \ddots & \vdots \\ 0 & 0 & \cdots & h_{P-1} \end{bmatrix}}_{(M+P-1)\times M} \mathbf{g}, \end{aligned} \tag{2.16}$$

where $\mathbf{H}$ is the channel filtering matrix and $\mathbf{g}$ is the equalization vector for FFF given by

$$\mathbf{g} = [g_0 \; g_1 \; \cdots \; g_{M-1}].$$

Note that the length of $\{c_m\}$ is $N = M + P - 1$. The output of the FFF is given by

$$q_l = g_l * (h_l * b_l + n_l)$$
$$= \sum_{m=0}^{N-1} c_m b_{l-m} + \sum_{m=0}^{M-1} g_m n_{l-m}.$$

Consider the signal part only and assume that the DFE is to estimate $b_{l-\bar{m}}$ at time l, where $\bar{m}$ is a decision delay. The signal part can be rewritten as follows:

$$\sum_{m=0}^{N-1} c_m b_{l-m} = \sum_{m=0}^{\bar{m}} c_m b_{l-m} + \sum_{m=\bar{m}+1}^{N-1} c_m b_{l-m}.$$

Assume that the decisions on the past symbols,

$$b_{l-\bar{m}-1}, b_{l-\bar{m}-2}, \ldots, b_{l-N+1},$$

are available. Then, the ISI due to these past symbols can be eliminated by cancelation, which is carried out by the FBF, and the output of the DFE is given by

$$d_l = q_l - \sum_{m=\bar{m}+1}^{N-1} f_m \hat{b}_{l-m}$$
$$= \sum_{m=0}^{\bar{m}} c_m b_{l-m} + \sum_{m=\bar{m}+1}^{N-1} (c_m b_{l-m} - f_m \hat{b}_{l-m}) + \sum_{m=0}^{M-1} g_m n_{l-m}, \quad (2.17)$$

where $\{f_m\}$ is the impulse response of the FBF and $\hat{b}_{l-m}$ denotes the detected symbol of b_{l-m}. Note that the delay of the FBF is introduced for the causality since the decision is made after an $\bar{m}$-symbol delay. If the decisions are correct ($\hat{b}_{l-m} = b_{l-m}$) and $f_m = c_m$, for $m = \bar{m} + 1, \bar{m} + 2, \ldots, N - 1$, then the output of the DFE becomes

$$d_l = q_l - \sum_{m=\bar{m}+1}^{N-1} c_m b_{l-m}$$
$$= \sum_{m=0}^{\bar{m}} c_m b_{l-m} + \sum_{m=0}^{M-1} g_m n_{l-m}. \quad (2.18)$$

Since the DFE is to estimate $b_{l-\bar{m}}$, it is expected that $d_l = b_{l-\bar{m}}$ without the noise when the FFF is properly designed. From this, we can find the ZF condition for the DFE.

According to Eq. (2.18), the ZF condition for the FFF becomes

$$c_m = \begin{cases} 1, & \text{if } m = \bar{m}; \\ 0, & \text{if } m = 0, 1, \ldots, \bar{m} - 1. \end{cases} \quad (2.19)$$

Hence, from Eq. (2.16), by taking the first $(\bar{m}+1)$ elements of $\mathbf{c}$, i.e. c_0 to $c_{\bar{m}}$, the ZF condition is given by

$$
\begin{bmatrix} 0 \\ 0 \\ \vdots \\ 1 \end{bmatrix} = \begin{bmatrix} c_0 \\ c_1 \\ \vdots \\ c_{\bar{m}} \end{bmatrix} = \underbrace{\begin{bmatrix} h_0 & 0 & \cdots & 0 \\ h_1 & h_0 & \cdots & 0 \\ \vdots & \vdots & \ddots & \vdots \\ 0 & 0 & \cdots & h_{\bar{m}+1-M} \end{bmatrix}}_{(\bar{m}+1)\times M} \begin{bmatrix} g_0 \\ g_1 \\ \vdots \\ g_{M-1} \end{bmatrix}.
\tag{2.20}
$$

In particular, when $\bar{m} = M - 1$, the FFF, $\mathbf{g}$, with the ZF condition can be obtained through a matrix inversion as follows:

$$
\begin{bmatrix} g_0 \\ g_1 \\ \vdots \\ g_{M-1} \end{bmatrix} = \begin{bmatrix} h_0 & 0 & \cdots & 0 \\ h_1 & h_0 & \cdots & 0 \\ \vdots & \vdots & \ddots & \vdots \\ 0 & 0 & \cdots & h_0 \end{bmatrix}^{-1} \begin{bmatrix} 0 \\ 0 \\ \vdots \\ 1 \end{bmatrix}.
$$

It is necessary that $\bar{m}$ is smaller than or equal to $M - 1$ so that there exists a vector $\mathbf{g}$ that solves Eq. (2.20).

The ISI terms due to c_m, $\bar{m} + 1 \le m \le N - 1$, are called the postcursors, while the ISI terms due to c_m, $0 \le m \le \bar{m} - 1$, are called the precursors. The postcursors are canceled by the FBF and the precursors are suppressed by the FFF. As shown in Eq. (2.20), the operation of the FFF can be seen as a linear transform. Note that, since the FBF performs cancelation and a decision device (which is a nonlinear function in general) is involved, the DFE is not an LE.

Example 2.3.1 This example shows how the filter coefficients of the ZF DFE can be obtained with an ISI channel of $h_0 = 0.5$, $h_1 = 1$, and $h_2 = 0.5$, where $P = 3$. The ZF DFE with $\bar{m} + 1 = M = 3$ can be found as follows.

Since $N = M + P - 1$, we have $N = 5$. The FFF vector is denoted by $\mathbf{g} = [g_0 \ g_1 \ g_2]^T$ and the FBF vector is denoted by $\mathbf{f} = [f_3 \ f_4]^T$. The coefficients of the FFF are given by

$$
\begin{bmatrix} g_0 \\ g_1 \\ g_2 \end{bmatrix} = \begin{bmatrix} 0.5 & 0 & 0 \\ 1 & 0.5 & 0 \\ 0.5 & 1 & 0.5 \end{bmatrix}^{-1} \begin{bmatrix} 0 \\ 0 \\ 1 \end{bmatrix} = \begin{bmatrix} 0 \\ 0 \\ 2 \end{bmatrix}.
\tag{2.21}
$$

To find the coefficients of the FBF, we need to compute

$$
\begin{bmatrix} c_0 \\ c_1 \\ c_2 \\ c_3 \\ c_4 \end{bmatrix} = \begin{bmatrix} 0.5 & 0 & 0 \\ 1 & 0.5 & 0 \\ 0.5 & 1 & 0.5 \\ 0 & 0.5 & 1 \\ 0 & 0 & 0.5 \end{bmatrix} \begin{bmatrix} g_0 \\ g_1 \\ g_2 \end{bmatrix} = \begin{bmatrix} 0 \\ 0 \\ 1 \\ 2 \\ 1 \end{bmatrix}.
\tag{2.22}
$$

This gives $f_3 = 2$ and $f_4 = 1$.

If the FBF is not used, the FFF solely mitigates the ISI components (in this case, the DFE becomes the LE). As shown in Eq. (2.22), the ISI can be suppressed if

$$\mathbf{c} = [0 \ 0 \ 1 \ 0 \ 0]^{\mathrm{T}}.$$

However, there are more equations than variables unless M is infinity, as shown in Eq. (2.22). Thus, in general, the length of an LE should be long enough to achieve a satisfactory performance. However, when the FBF is employed, the FFF only needs to suppress the precursors. As shown in Eq. (2.21), the FFF of a finite length can suppress the precursors. From this, we can see that a DFE can provide a good performance with small numbers of taps of the FFF and FBF.

2.3.2 MMSE DFE

Since the ZF DFE only attempts to remove the ISI, the noise can be enhanced. To avoid this, it is desirable to consider the MMSE criterion.

With the equalizer output, d_l, that estimates the desired symbol $s_l = b_{l-\bar{m}}$, the MSE is given by

$$
\begin{aligned}
\mathrm{MSE} &= E[|s_l - d_l|^2] \\
&= E[|b_{l-\bar{m}} - d_l|^2] \\
&= E\left[\left| b_{l-\bar{m}} - \left(\sum_{m=0}^{M-1} g_m y_{l-m} - \sum_{m=\bar{m}+1}^{N-1} f_m \hat{b}_{l-m} \right) \right|^2\right].
\end{aligned}
\tag{2.23}
$$

Let

$$\mathbf{f} = [f_{\bar{m}+1} \quad f_{\bar{m}+2} \quad \cdots \quad f_{N-1}]^{\mathrm{T}},$$
$$\hat{\mathbf{b}}_{l-\bar{m}-1} = [\hat{b}_{l-\bar{m}-1} \quad \hat{b}_{l-\bar{m}-2} \quad \cdots \quad \hat{b}_{l-N+1}]^{\mathrm{T}}.$$

Then, it follows that

$$
\begin{aligned}
\mathrm{MSE} &= E\left[\left| b_{l-\bar{m}} - \left(\mathbf{y}_l^{\mathrm{T}} \mathbf{g} - \hat{\mathbf{b}}_{l-\bar{m}-1}^{\mathrm{T}} \mathbf{f} \right) \right|^2\right] \\
&= E\left[\left| b_{l-\bar{m}} - [\mathbf{y}_l^{\mathrm{T}} \ \hat{\mathbf{b}}_{l-\bar{m}-1}^{\mathrm{T}}] \begin{bmatrix} \mathbf{g} \\ -\mathbf{f} \end{bmatrix} \right|^2\right].
\end{aligned}
\tag{2.24}
$$

Using the orthogonality principle, we can find the optimal vectors for $\mathbf{g}$ and $\mathbf{f}$ for the MMSE DFE as follows:

$$
\begin{bmatrix} \mathbf{g} \\ -\mathbf{f} \end{bmatrix}_{\mathrm{mmse}} = \begin{bmatrix} E[\mathbf{y}_l \mathbf{y}_l^{\mathrm{T}}] & E[\mathbf{y}_l \hat{\mathbf{b}}_{l-\bar{m}-1}^{\mathrm{T}}] \\ E[\hat{\mathbf{b}}_{l-\bar{m}-1} \mathbf{y}_l^{\mathrm{T}}] & E[\hat{\mathbf{b}}_{l-\bar{m}-1} \hat{\mathbf{b}}_{l-\bar{m}-1}^{\mathrm{T}}] \end{bmatrix}^{-1} \begin{bmatrix} E[\mathbf{y}_l b_{l-\bar{m}}] \\ E[\hat{\mathbf{b}}_{l-\bar{m}-1} b_{l-\bar{m}}] \end{bmatrix}.
\tag{2.25}
$$

In general, it is not easy to find the solution for the MMSE DFE in Eq. (2.25), because $\hat{\mathbf{b}}_{l-\bar{m}-1}$ is a function of $\mathbf{f}$ and $\mathbf{g}$. Consequently, the solution in Eq. (2.25) becomes highly nonlinear and cannot easily be solved. In practice, however, we assume that the past decided symbols are correct, i.e.

$$\hat{b}_{l-\bar{m}-1} = b_{l-\bar{m}-1}, \ldots, \hat{b}_{l-N+1} = b_{l-N+1}.$$

In this case, the solution given in Eq. (2.25) can be readily obtained.

The performance of the DFE depends on decision errors. Suppose that $\hat{b}_{l-\bar{m}-1} \neq b_{l-\bar{m}-1}$ due to the decision errors of past symbols. As the cancelation through the FBF is erroneous, it is more likely to incur a decision error in detecting the current desired symbol, $s_l = b_{l-\bar{m}}$. Hence, the decision error can propagate to the detection for the following symbols. This phenomenon is called the error propagation and it degrades the performance of the DFE when the SNR is low or moderate.

Example 2.3.2   In this example, we find the MMSE DFE with an ISI channel which has the impulse response $\{h_0, h_1\} = \{1, 0.5\}$ with $P = 2$ and $N_0/2 = 0.5$. We assume that $M = 2$. The MMSE DFE will be found when (i) $\bar{m} = 0$ and (ii) $\bar{m} = 1$.

(i) Since $\bar{m} = 0$ and $M = 2$, we have $N = 3$ and

$$\mathbf{g} = [g_0 \ \ g_1]^T \text{ and } \mathbf{f} = [f_1 \ \ f_2 \ \ f_3]^T.$$

The MSE cost function becomes

$$\text{MSE} = E[|g_0 y_l + g_1 y_{l-1} - f_1 \hat{b}_{l-1} - f_2 \hat{b}_{l-2} - f_3 \hat{b}_{l-3} - b_l|^2].$$

We need to find $E[\mathbf{y}_l \mathbf{y}_l^T]$, $E[\mathbf{y}_l \hat{b}_{l-1}]$, $E[\mathbf{y}_l \hat{b}_{l-2}]$, and $E[\hat{b}_{l-m} \hat{b}_{l-p}]$, $p, m = 1, 2, 3$, for $\mathbf{y}_l = [y_l \ \ y_{l-1}]^T$ and $\hat{b}_{l-m} = b_{l-m}$, $m = 1, 2, 3$. Since

$$y_l = b_l + 0.5 b_{l-1} + n_l,$$

we can show that

$$E[\mathbf{y}_l \mathbf{y}_l^T] = \begin{bmatrix} 1.75 & 0.5 \\ 0.5 & 1.75 \end{bmatrix}, \quad E[\mathbf{y}_l \hat{b}_{l-1}] = \begin{bmatrix} 0.5 \\ 1 \end{bmatrix}, \quad \text{and } E[\mathbf{y}_l \hat{b}_{l-2}] = \begin{bmatrix} 0 \\ 0.5 \end{bmatrix},$$

and $E[\mathbf{y}_l \hat{b}_{l-3}] = \mathbf{0}$. In addition, we have $E[\hat{b}_{l-m} \hat{b}_{l-m}] = E[b_{l-m}^2] = 1$, $m = 1, 2, 3$, $E[\mathbf{y}_l b_l] = [1 \ 0]^T$, and $E[\hat{b}_{l-m} b_l] = 0$, $m = 1, 2, 3$. According to Eq. (2.25), we can show that

$$\begin{bmatrix} \mathbf{g} \\ -\mathbf{f} \end{bmatrix}_{\text{mmse}} = \begin{bmatrix} 1.75 & 0.5 & 0.5 & 0 & 0 \\ 0.5 & 1.75 & 1 & 0.5 & 0 \\ 0.5 & 1 & 1 & 0 & 0 \\ 0 & 0.5 & 0 & 1 & 0 \\ 0 & 0 & 0 & 0 & 1 \end{bmatrix}^{-1} \begin{bmatrix} 1 \\ 0 \\ 0 \\ 0 \\ 0 \end{bmatrix}.$$

From this, we have

$$\mathbf{g} = [(2/3) \ \ 0]^T \text{ and } \mathbf{f} = [(1/3) \ \ 0 \ \ 0]^T.$$

The MMSE is $1/3$.

(ii) We have $\mathbf{f} = [f_2 \ \ f_3]$. The MSE cost function is given by

$$\text{MSE} = E[|g_0 y_l + g_1 y_{l-1} - f_2 \hat{b}_{l-2} - f_3 \hat{b}_{l-3} - b_{l-1}|^2].$$

After applying the same approach used above, we have

$$\mathbf{g} = [0.1053 \ \ 0.6316]^T \text{ and } \mathbf{f} = [f_2 \ \ f_3]^T = [0.3158 \ \ 0]^T.$$

In this case, the MMSE is 0.3158. This shows that the MMSE depends on $\bar{m}$.

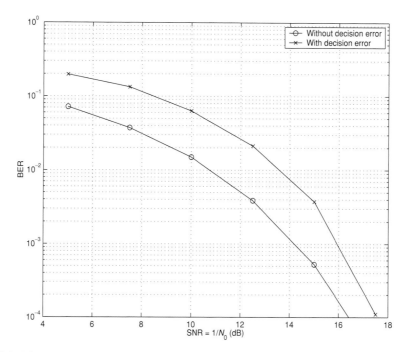

Figure 2.8. Bit error rate performance of the MMSE DFE with and without decision errors.

Figure 2.8 shows simulation results for the MMSE DFE with Channel A. It is assumed that $M = 5$ and $\bar{m} = 4$. To see the impact of error propagation, simulation results obtained without decision errors are also presented. As the SNR increases, there would be less decision errors and the performance of the MMSE DFE with decision errors can approach that without decision errors. As shown in Fig. 2.8, there is about a 3 dB SNR gap at a BER of 10^{-2}, and a 1 dB SNR gap at a BER of 10^{-4}.

Comparing this with the performance of the MMSE LE shown in Fig. 2.6, it is shown that the MMSE DFE significantly outperforms and is robust against frequency nulls.

It is possible to find a simpler derivation of the MMSE DFE. Using $y_l = \sum_{p=0}^{P-1} h_p b_{l-p} + n_l$, the output of the FFF is given by

$$
\begin{aligned}
\sum_{m=0}^{M-1} g_m y_{l-m} &= \sum_{m=0}^{M-1} g_m \sum_{p=0}^{P-1} h_p b_{l-m-p} + \sum_{m=0}^{M-1} g_m n_{l-m} \\
&= \sum_{m=0}^{M-1} g_m \sum_{k=m}^{m+P-1} h_{k-m} b_{l-k} + \sum_{m=0}^{M-1} g_m n_{l-m} \quad (\text{letting } k = m + p) \\
&= \sum_{k=0}^{N-1} \left(\sum_{m=0}^{M-1} g_m h_{k-m} \right) b_{l-k} + \sum_{m=0}^{M-1} g_m n_{l-m},
\end{aligned}
$$

where we assume that $h_k = 0$ for $k < 0$ and $k \geq P$. Then, the MSE cost function in Eq. (2.23) is rewritten as follows:

$$
\text{MSE} = E\left[\left| \sum_{k=0}^{N-1} \underbrace{\left(\sum_{m=0}^{M-1} g_m h_{k-m}\right) b_{l-k}}_{=c_k} + \sum_{m=0}^{M-1} g_m n_{l-m} - \sum_{m=\bar{m}+1}^{N-1} f_m \hat{b}_{l-m} - b_{l-\bar{m}} \right|^2 \right]. \quad (2.26)
$$

If $\hat{b}_{l-m} = b_{l-m}$ for $m = \bar{m}+1, \bar{m}+2, \ldots, N-1$, we can find the f_m's of the FBF to minimize the MSE from Eq. (2.26) as follows:

$$
f_m = c_m, \quad m = \bar{m}+1, \bar{m}+2, \ldots, N-1.
$$

This is identical to the FFF of the ZF DFE. Then, the MSE cost function is given by

$$
\text{MSE} = E\left[\left| \sum_{k=0}^{\bar{m}} \underbrace{\left(\sum_{m=0}^{M-1} g_m h_{k-m}\right) b_{l-k}}_{=c_k} + \sum_{m=0}^{M-1} g_m n_{l-m} - b_{l-\bar{m}} \right|^2 \right]. \quad (2.27)
$$

Let $\tilde{\mathbf{c}}_1 = [c_0 \ c_1 \ \cdots \ c_{\bar{m}}]^\text{T}$. Then, we can show that

$$
\tilde{\mathbf{c}}_1 = \tilde{\mathbf{H}}_1 \mathbf{g}, \quad (2.28)
$$

where $\tilde{\mathbf{H}}_1$ is a Toeplitz matrix of size $(\bar{m}+1) \times M$ with the first column vector

$$
\mathbf{h}_1 = \begin{cases} \left[h_0 \ h_1 \ \cdots \ h_{P-1} \ \underbrace{0 \cdots 0}_{(\bar{m}+1-P) \text{ times}} \right]^\text{T}, & \text{if } \bar{m} > P-1; \\ [h_0 \ h_1 \ \cdots \ h_{\bar{m}}]^\text{T}, & \text{if } \bar{m} \leq P-1, \end{cases}
$$

and the first row vector

$$
[h_0 \ \underbrace{0 \ \cdots \ 0}_{(M-1) \text{ zeros}}].
$$

Using Eq. (2.28), the MSE cost function is given by

$$
\text{MSE} = E\left[\left| \tilde{\mathbf{c}}_1^\text{T} \tilde{\mathbf{b}}_l - b_{l-\bar{m}} + \sum_{m=0}^{M-1} g_m n_{l-m} \right|^2 \right]
$$

$$
= \tilde{\mathbf{c}}_1^\text{T} E[\tilde{\mathbf{b}}_l \tilde{\mathbf{b}}_l^\text{T}] \tilde{\mathbf{c}}_1 - 2\tilde{\mathbf{c}}_1^\text{T} E[\tilde{\mathbf{b}}_l b_{l-\bar{m}}] + E[b_{l-\bar{m}}^2] + \|\mathbf{g}\|^2 \frac{N_0}{2}
$$

$$
= \mathbf{g}^\text{T} \tilde{\mathbf{H}}_1^\text{T} \tilde{\mathbf{H}}_1 \mathbf{g} - 2\mathbf{g}^\text{T} \tilde{\mathbf{H}}_1^\text{T} \tilde{\mathbf{1}}_1 + 1 + \|\mathbf{g}\|^2 \frac{N_0}{2}, \quad (2.29)
$$

where $\tilde{\mathbf{b}}_l = [b_l \ b_{l-1} \ \cdots \ b_{l-\bar{m}}]^\text{T}$ and $E[\tilde{\mathbf{b}}_l b_{l-\bar{m}}] = \tilde{\mathbf{1}}_1 = [0 \ 0 \ \cdots \ 0 \ 1]$. Taking the derivative with respect to $\mathbf{g}$, we can show that

$$
\frac{\partial}{\partial \mathbf{g}} \text{MSE} = 2\left(\tilde{\mathbf{H}}_1^\text{T} \tilde{\mathbf{H}}_1 + \frac{N_0}{2}\mathbf{I}\right)\mathbf{g} - 2\tilde{\mathbf{H}}_1^\text{T} \tilde{\mathbf{1}}_1. \quad (2.30)
$$

From this, the optimal vector $\mathbf{g}$ is obtained as follows:

$$\mathbf{g} = \left(\tilde{\mathbf{H}}_1^T \tilde{\mathbf{H}}_1 + \frac{N_0}{2} \mathbf{I} \right)^{-1} \tilde{\mathbf{H}}_1^T \tilde{\mathbf{1}}_1. \qquad (2.31)$$

Instead of using Eq. (2.25) to find the optimal FFF and FBF vectors, we can use Eq. (2.31), which has a lower complexity. In Eq. (2.25), a matrix inversion of size $(M + N - \bar{m} - 1) \times (M + N - \bar{m} - 1)$, where $0 \leq \bar{m} \leq N - 1$, is required, while a matrix inversion of size $M \times M$ is required in Eq. (2.31).

2.4 Adaptive linear equalizers

In this section, we introduce adaptive methods for channel equalization. Adaptive equalizers can be considered as practical approaches because they do not require second-order statistics of signals. Instead of second-order statistics, a training sequence is used to find the equalization vector(s) for the LE or DFE.

2.4.1 Iterative approaches

Suppose that an LE is causal and has a finite length, say M. Then, the output of the LE is given by

$$d_l = \sum_{m=0}^{M-1} g_m y_{l-m}. \qquad (2.32)$$

As shown in Section 2.2, the MMSE solution can be found using second-order statistics as in Eq. (2.9). However, if (i) second-order statistics (e.g., $\mathbf{R}_y$ and $\mathbf{r}_{y,s}$) are not known or available and/or (ii) the matrix inversion in Eq. (2.9) demands a higher computing power than the receiver's hardware can provide, alternative approaches that can overcome these difficulties should be sought. One of the practical approaches is *adaptive equalization*. Generally, an adaptive equalizer requires neither second-order statistics nor the matrix inversion.

To derive the adaptive LE, we first consider an iterative algorithm that can find the MMSE solution without the matrix inversion. From Eq. (2.5), the MSE as a function of $\mathbf{g}$ is given by

$$\begin{aligned}
\text{MSE}(\mathbf{g}) &= E \left[\left| s_l - \sum_{m=0}^{M-1} g_m y_{l-m} \right|^2 \right] \\
&= E[|s_l|^2] - 2 \sum_{m=0}^{M-1} g_m E[|s_l y_{l-m}|] + E \left[\left| \sum_{m=0}^{M-1} g_m y_{l-m} \right|^2 \right] \\
&= \sigma_b^2 - 2 \mathbf{r}_{y,s}^T \mathbf{g} + \mathbf{g}^T \mathbf{R}_y \mathbf{g}. \qquad (2.33)
\end{aligned}$$

As shown in Eq. (2.33), the MSE is a quadratic function of $\mathbf{g}$. This indicates that $\mathbf{g}$ has a unique global minimum. Obviously, this minimum can be found by taking the derivative and setting it equal to zero. This leads to Eq. (2.9). Note that we use the orthogonality principle for the derivation of Eq. (2.9) in Section 2.2, while the above derivation relies on the property of a quadratic function.

As the MSE function is a quadratic function, there exist iterative techniques to find the minimum. One of the simple techniques is the steepest descent (SD) algorithm. Suppose that $\mathbf{g}^{(0)}$ is an initial vector. The gradient vector at $\mathbf{g}^{(0)}$ is given by

$$
\begin{aligned}
\nabla_{(0)} &= \left.\frac{\partial}{\partial \mathbf{g}}\text{MSE}(\mathbf{g})\right|_{\mathbf{g}=\mathbf{g}^{(0)}} \\
&= \left.\frac{\partial}{\partial \mathbf{g}}(\sigma_b^2 - 2\mathbf{r}_{\mathbf{y},s}^{\mathsf{T}}\mathbf{g} + \mathbf{g}^{\mathsf{T}}\mathbf{R}_{\mathbf{y}}\mathbf{g})\right|_{\mathbf{g}=\mathbf{g}^{(0)}} \\
&= -2\mathbf{r}_{\mathbf{y},s} + 2\mathbf{R}_{\mathbf{y}}\mathbf{g}^{(0)},
\end{aligned} \tag{2.34}
$$

where the gradient $\frac{\partial}{\partial \mathbf{g}}\text{MSE}(\mathbf{g})$ is given by

$$
\frac{\partial}{\partial \mathbf{g}}\text{MSE}(\mathbf{g}) = \left[\frac{\partial}{\partial g_0}\text{MSE}(\mathbf{g}) \quad \frac{\partial}{\partial g_1}\text{MSE}(\mathbf{g}) \cdots \frac{\partial}{\partial g_{M-1}}\text{MSE}(\mathbf{g})\right]^{\mathsf{T}}.
$$

With the SD direction $-\nabla_{(0)}$ at $\mathbf{g}^{(0)}$ and a constant step size $\mu > 0$, the next vector $\mathbf{g}^{(1)}$ which may yield a smaller MSE than $\mathbf{g}^{(0)}$ can be given by

$$
\begin{aligned}
\mathbf{g}^{(1)} &= \mathbf{g}^{(0)} - \mu\nabla_{(0)} \\
&= \mathbf{g}^{(0)} - 2\mu\left(\mathbf{R}_{\mathbf{y}}\mathbf{g}^{(0)} - \mathbf{r}_{\mathbf{y},s}\right).
\end{aligned}
$$

Consequently, a recursion toward the minimum can be written as follows:

$$
\mathbf{g}^{(k)} = \mathbf{g}^{(k-1)} - 2\mu\left(\mathbf{R}_{\mathbf{y}}\mathbf{g}^{(k-1)} - \mathbf{r}_{\mathbf{y},s}\right), \quad k = 0, 1, \ldots, \tag{2.35}
$$

where k is the iteration index. The iteration is terminated when the SD direction becomes zero, i.e. $\nabla_{(k)} = \mathbf{0}$ for some k. This point achieves the minimum and we can find the optimal solution as follows:

$$
\begin{aligned}
0 &= \nabla_{(k)} \\
&= -2\mathbf{r}_{\mathbf{y},s} + 2\mathbf{R}_{\mathbf{y}}\mathbf{g}^{(k)} \\
\Leftrightarrow \mathbf{g}^{(k)} &= \mathbf{R}_{\mathbf{y}}^{-1}\mathbf{r}_{\mathbf{y},s}.
\end{aligned}
$$

The recursion in Eq. (2.35) is called the SD algorithm.

Prior to discussing the properties of the SD algorithm, we briefly explain how the recursion in Eq. (2.35) works with a simple example. Consider the MSE for $M = 1$:

$$
\text{MSE}(g) = \sigma_b^2 - 2r_{y,s}g + g^2 r_y,
$$

where $r_y = \mathbf{R}_{\mathbf{y}}, r_{y,s} = \mathbf{r}_{\mathbf{y},s}$, and $\mathbf{g} = g$ for $M = 1$. Clearly, the $\text{MSE}(g)$ is a quadratic function of a scalar coefficient g. The gradient can be written as follows:

$$
\frac{\mathrm{d}}{\mathrm{d}g}\text{MSE}(g) = -2r_{y,s} + 2r_y g.
$$

If $g > g_{\text{mmse}} = r_{y,s}/r_y$, we can see that $\frac{\mathrm{d}}{\mathrm{d}g}\text{MSE}(g) > 0$ and a new g should be smaller than the current g as given in Eq. (2.35) to approach g_{mmse}. An illustration of the MSE function and its derivative is shown in Fig. 2.9.

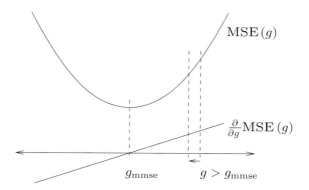

Figure 2.9. MSE function and its derivative when $M = 1$.

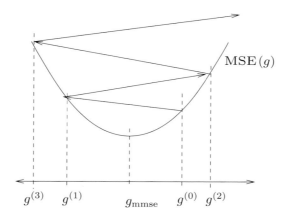

Figure 2.10. A case of divergence when μ is too large.

From Fig. 2.9, we can see that μ should not be large. If μ is too large, the next g will be smaller than g_{mmse} and may result in a higher MSE than the current MSE. In this case, the recursion diverges and never finds the minimum. This is illustrated in Fig. 2.10. On the other hand, if μ is too small, it would take too many iterations to converge. Therefore, it is important to determine the value of μ such that the recursion can converge at a fast rate.

2.4.2   Convergence analysis of the SD algorithm

In order to determine the value of μ, we need to understand the properties of the recursion in Eq. (2.35). For convenience, using the relation $\mathbf{R_y}\mathbf{g}_{\mathrm{mmse}} = \mathbf{r}_{\mathbf{y},s}$, we can modify the recursion in Eq. (2.35) with the difference vector, $\mathbf{g}^{(k)} - \mathbf{g}_{\mathrm{mmse}}$, as follows:

$$
\begin{aligned}
\tilde{\mathbf{g}}^{(k)} &= \mathbf{g}^{(k)} - \mathbf{g}_{\mathrm{mmse}} \\
&= \tilde{\mathbf{g}}^{(k-1)} - 2\mu\mathbf{R_y}\tilde{\mathbf{g}}^{(k-1)} \\
&= (\mathbf{I} - 2\mu\mathbf{R_y})\tilde{\mathbf{g}}^{(k-1)}.
\end{aligned} \tag{2.36}
$$

Equation (2.36) becomes an autonomous difference equation. We can easily rewrite Eq. (2.36) as follows:

$$\tilde{\mathbf{g}}^{(k)} = (\mathbf{I} - 2\mu\mathbf{R_y})^k \tilde{\mathbf{g}}^{(0)}. \qquad (2.37)$$

The analysis of Eq. (2.37) can be made easier if the difference equation is decoupled. To this end, the eigendecomposition of $\mathbf{R_y}$ is required.

The eigendecomposition of $\mathbf{R_y}$ is given by

$$\mathbf{R_y} = \mathbf{E}\mathbf{\Lambda}\mathbf{E}^\mathsf{T}, \qquad (2.38)$$

where

$$\mathbf{E} = [\mathbf{e}_1 \ \ \mathbf{e}_2 \ \ \cdots \ \ \mathbf{e}_M]$$

and

$$\mathbf{\Lambda} = \mathrm{Diag}(\lambda_1, \lambda_2, \dots, \lambda_M)$$

$$= \begin{bmatrix} \lambda_1 & 0 & \cdots & 0 \\ 0 & \lambda_2 & \cdots & 0 \\ \vdots & \vdots & \ddots & \vdots \\ 0 & 0 & \cdots & \lambda_M \end{bmatrix}.$$

Here, λ_m and $\mathbf{e}_m$ are the mth largest eigenvalue and its corresponding eigenvector of $\mathbf{R_y}$, respectively. Note that the covariance matrix $\mathbf{R_y}$ is symmetric and positive semidefinite. Thus, the eigenvalues are real and nonnegative. Using the orthogonality of eigenvectors, we can show that

$$\mathbf{E}\mathbf{E}^\mathsf{T} = \mathbf{E}^\mathsf{T}\mathbf{E} = \mathbf{I}. \qquad (2.39)$$

Note that $\mathbf{E}$ is a unitary matrix. Using Eqs (2.38) and (2.39), it follows that

$$\mathbf{I} - 2\mu\mathbf{R_y} = \mathbf{E}\mathbf{E}^\mathsf{T} - 2\mu\mathbf{E}\mathbf{\Lambda}\mathbf{E}^\mathsf{T}$$
$$= \mathbf{E}\left(\mathbf{I} - 2\mu\mathbf{\Lambda}\right)\mathbf{E}^\mathsf{T}$$

and

$$(\mathbf{I} - 2\mu\mathbf{R_y})^k = \mathbf{E}\left(\mathbf{I} - 2\mu\mathbf{\Lambda}\right)^k \mathbf{E}^\mathsf{T}.$$

For diagonalization, pre-multiplying $\mathbf{E}^\mathsf{T}$ to Eq. (2.37) yields

$$\mathbf{E}^\mathsf{T}\tilde{\mathbf{g}}^{(k)} = (\mathbf{I} - 2\mu\mathbf{\Lambda})^k \, \mathbf{E}^\mathsf{T}\tilde{\mathbf{g}}^{(0)}.$$

Let $\tilde{\mathbf{q}}^{(k)} = \mathbf{E}^\mathsf{T}\tilde{\mathbf{g}}^{(k)}$. Then, it follows that

$$\tilde{\mathbf{q}}^{(k)} = (\mathbf{I} - 2\mu\mathbf{\Lambda})^k \, \tilde{\mathbf{q}}^{(0)} \qquad (2.40)$$

or

$$\tilde{q}_m^{(k)} = (1 - 2\mu\lambda_m)^k \, \tilde{q}_m^{(0)}, \ \ m = 1, 2, \dots, M, \qquad (2.41)$$

where $\tilde{q}_m^{(k)}$ represents the mth element of $\tilde{\mathbf{q}}_m^{(k)}$ and is called the mth eigenmode.

From Eq. (2.41), we can find the following property:

$$\lim_{k \to \infty} \tilde{q}_m^{(k)} = 0, \quad \text{if } |1 - 2\mu\lambda_m| < 1. \tag{2.42}$$

This yields the following convergence condition:

$$0 < \mu < \frac{1}{\lambda_m}, \quad \forall m. \tag{2.43}$$

Note that $\lambda_m \geq 0$, $\forall m$. If μ satisfies Eq. (2.43), we can show that $\tilde{\mathbf{q}}^{(k)} \to \mathbf{0}$ or $\mathbf{g}^{(k)} \to \mathbf{g}_{\text{mmse}}$.

Let $\lambda_{\max}$ be the maximum eigenvalue. Then we can find a sufficient condition for convergence as follows:

$$0 < \mu < \frac{1}{\lambda_{\max}}. \tag{2.44}$$

Since $\lambda_{\max} \leq \text{Tr}(\mathbf{R_y}) = \sum_{m=1}^{M} \lambda_m$, another sufficient condition can be found as

$$0 < \mu < \frac{1}{\text{Tr}(\mathbf{R_y})}. \tag{2.45}$$

The condition in Eq. (2.45) can be more practical because μ is directly determined by $\mathbf{R_y}$ without finding the largest eigenvalue.

There are several approaches that may be taken to find the rate of convergence. One is to consider the squared norm of $\tilde{\mathbf{q}}^{(k)}$:

$$\left\| \tilde{\mathbf{q}}^{(k)} \right\|^2 = \left\| \mathbf{E}^{\mathsf{T}} \tilde{\mathbf{g}}^{(k)} \right\|^2 = \left\| \tilde{\mathbf{g}}^{(k)} \right\|^2.$$

From Eq. (2.41) it follows that

$$\begin{aligned}
\left| \tilde{q}_m^{(k)} \right|^2 &= (1 - 2\mu\lambda_m)^2 \left| \tilde{q}_m^{(k-1)} \right|^2 \\
&= \left(1 - 4\mu\lambda_m + 4\mu^2\lambda_m^2 \right) \left| \tilde{q}_m^{(k-1)} \right|^2, \quad m = 1, 2, \ldots, M.
\end{aligned} \tag{2.46}$$

Since $\mu > 0$ and $\lambda_m \geq 0$, we can show that

$$1 - 4\mu\lambda_m + 4\mu^2\lambda_m^2 \leq 1 - 4\mu\lambda_{\min} + 4\mu^2\lambda_{\max}^2,$$

where $\lambda_{\min}$ stands for the minimum eigenvalue of $\mathbf{R_y}$. Let

$$\beta_{\text{upp}} = 1 - 4\mu\lambda_{\min} + 4\mu^2\lambda_{\max}^2.$$

This gives an upper bound sequence as follows:

$$\left| \tilde{q}_m^{(k)} \right|^2 \leq \beta_{\text{upp}}^k \left| \tilde{q}_m^{(0)} \right|^2. \tag{2.47}$$

Hence, the value of β_{upp} can be minimized for a faster convergence rate. Since β is a function of μ, the value of μ that minimizes β_{upp} is given by

$$\mu_{\text{opt}} = \frac{1}{2} \frac{\lambda_{\min}}{\lambda_{\max}^2} \tag{2.48}$$

and the minimum of β_{upp} is given by

$$\bar{\beta}_{\text{upp}} = 1 - \left(\frac{\lambda_{\min}}{\lambda_{\max}} \right)^2. \tag{2.49}$$

Substituting Eq. (2.49) into Eq. (2.47), we have

$$
|\tilde{q}_m^{(k)}|^2 \le \left[1 - \left(\frac{\lambda_{\min}}{\lambda_{\max}} \right)^2 \right]^k |\tilde{q}_m^{(0)}|^2. \tag{2.50}
$$

Furthermore, since $\|\tilde{\mathbf{q}}^{(k)}\|^2 = \sum_{m=0}^{M-1} |\tilde{q}_m^{(k)}|^2$, we can show that

$$
\|\tilde{\mathbf{q}}^{(k)}\|^2 \le \left[1 - \left(\frac{\lambda_{\min}}{\lambda_{\max}} \right)^2 \right]^k \|\tilde{\mathbf{q}}^{(k-1)}\|^2. \tag{2.51}
$$

This shows that as the ratio, $\lambda_{\max}/\lambda_{\min}$, which is called the eigenspread or condition number of $\mathbf{R}_y$, decreases, a faster rate of convergence can be achieved. Note that the rate of convergence depends on the value of μ, and the result in Eq. (2.51) is valid when the value of μ in Eq. (2.48) is used.

The transient behavior of the SD algorithm is closely related to the rate of convergence. As the rate of convergence increases, the transient time becomes shorter. From Eq. (2.41), we can represent a (geometric) decaying curve by an exponential envelope as follows:

$$
\begin{aligned}
\tilde{q}_m^{(k)} &= (1 - 2\mu\lambda_m)^k \tilde{q}_m^{(0)} \\
&= e^{-(k/\tau_m)} \tilde{q}_m^{(0)},
\end{aligned} \tag{2.52}
$$

where τ_m is the time constant for the mth element. By a Taylor expansion, we have

$$
e^{-1/\tau_m} = 1 - \frac{1}{\tau_m} + \frac{1}{2!\tau_m^2} - \frac{1}{3!\tau_m^3} + \cdots,
$$

and (a large) τ_m can be approximately obtained as follows:

$$
\begin{aligned}
1 - 2\mu\lambda_m &= e^{-\frac{1}{\tau_m}} \\
&\simeq 1 - \frac{1}{\tau_m} \\
\Rightarrow \tau_m &\simeq \frac{1}{2\mu\lambda_m}.
\end{aligned}
$$

If we let $\mu = 1/\lambda_{\max}$ from Eq. (2.44), the time constant is given and bounded as follows:

$$
\tau_m \simeq \frac{\lambda_{\max}}{2\lambda_m} \le \frac{\lambda_{\max}}{2\lambda_{\min}}. \tag{2.53}
$$

Since the rate of convergence is decided by the slowest eigenmode, the largest time constant $\tau_{\max} = \lambda_{\max}/2\lambda_{\min}$ decides the overall rate of convergence. The rate of convergence becomes faster as $\tau_{\max}$ becomes smaller. Thus, the smaller the eigenspread $\lambda_{\max}/\lambda_{\min}$ of $\mathbf{R}_y$, the faster the convergence. This is the same result as in Eq. (2.51). Consequently, we can see that the eigenspread of $\mathbf{R}_y$ plays a key role in deciding the rate of convergence of the SD algorithm.

Since the eigenspread of $\mathbf{R}_y$ depends on the CIR, $\{h_p\}$, it is interesting to characterize ISI channels for the rate of convergence. The covariance matrix $\mathbf{R}_y$ given in Eq. (2.11) is a Toeplitz matrix. It is known that the eigenspread of a Toeplitz matrix $\mathbf{R}_y$ is related to the power spectral density of the stationary process y_l or the Fourier transform of the autocorrelation $r_y(m) = E[y_l y_{l-m}]$ (Gray, 2006). It can be shown that

$$\lim_{M \to \infty} \frac{\lambda_{\max}}{\lambda_{\min}} = \frac{\max_{0 \le \omega < 2\pi} S_y(\omega)}{\min_{0 \le \omega < 2\pi} S_y(\omega)},$$

where $S_y(\omega)$ is the power spectral density given by

$$S_y(\omega) = \sum_m r_y(m) e^{-jm\omega}$$

$$= H(e^{j\omega}) H^*(e^{j\omega}) + \frac{N_0}{2}, \quad 0 \le \omega < 2\pi.$$

From this, we can see that the eigenspread of $\mathbf{R_y}$ can be smaller if the power spectral density becomes more flat. On the other hand, if the power spectral density has nulls (i.e. $S_y(\omega) = 0$ for some $\omega \in [0, 2\pi)$), the eigenspread becomes infinite. Hence, the rate of convergence of the SD algorithm can be very slow if the ISI channel has nulls and the noise spectral density $N_0/2$ is sufficiently low (or the SNR is high) since the eigenspread is large.

Example 2.4.1 Consider an ISI channel with the following CIR:

$$h_0 = a, \; h_1 = 1, \; h_2 = a,$$

where $P = 3$. The $\mathcal{Z}$-transform of the CIR, $H(z)$, is given by

$$H(z) = z^{-1}(1 + a(z + z^{-1})).$$

We can show that

$$H(e^{j\omega}) H^*(e^{j\omega}) = |1 + 2a \cos \omega|^2.$$

For ω that satisfies

$$\cos \omega = -\frac{1}{2a},$$

$H(e^{j\omega}) H^*(e^{j\omega})$ has a frequency null. From this, we can show that there exist nulls if $|a| \ge 1/2$. For example, if $a = 2/3$, a null happens at $\omega = \pm 2.4189 = \pm 0.7699\pi$ (radian). However, if $|a| < 1/2$, there is no frequency null.

2.4.3 Least mean square algorithm

The SD algorithm can overcome the second difficulty of the MMSE approach (i.e. the matrix inversion can be avoided). However, the first difficulty, i.e. the need for second-order statistics, has not been overcome yet. Using an approximation in the SD algorithm, this difficulty can be resolved and it results in the least mean square (LMS) algorithm.

Recall the MSE in Eq. (2.33):

$$\text{MSE}(\mathbf{g}) = E\left[\left| s_l - \sum_{m=0}^{M-1} g_m y_{l-m} \right|^2 \right]$$

$$= E[|u_l|^2], \tag{2.54}$$

where $u_l = s_l - \sum_{m=0}^{M-1} g_m y_{l-m}$. The SD algorithm can be represented by the following recursion:

$$\mathbf{g}^{(l)} = \mathbf{g}^{(l-1)} - \mu \frac{\partial}{\partial \mathbf{g}} \text{MSE}(\mathbf{g}) \Big|_{\mathbf{g} = \mathbf{g}^{(l-1)}}$$
$$= \mathbf{g}^{(l-1)} + 2\mu E[\mathbf{y}_l u_l] \Big|_{\mathbf{g} = \mathbf{g}^{(l-1)}}, \qquad (2.55)$$

where $\mathbf{y}_l = [y_l \ y_{l-1} \ \cdots \ y_{l-M+1}]^{\text{T}}$. If $E[\mathbf{y}_l u_l]$, is replaced by $\mathbf{y}_l u_l$, without the expectation, we can obtain the LMS algorithm. We can see that $\mathbf{y}_l u_l$ is an *estimate* of $E[\mathbf{y}_l u_l]$. A generalization is possible using a sample average (e.g. $(1/Q) \sum_{q=l-Q+1}^{l} \mathbf{y}_q u_q$ for a $Q > 0$) to replace $\mathbf{y}_l u_l$.

The LMS algorithm is as follows:

$$\mathbf{g}^{(l)} = \mathbf{g}^{(l-1)} + 2\mu (\mathbf{y}_l u_l) \Big|_{\mathbf{g} = \mathbf{g}^{(l-1)}}$$
$$= \mathbf{g}^{(l-1)} + 2\mu \mathbf{y}_l e_l, \qquad (2.56)$$

where

$$e_l = u_l \Big|_{\mathbf{g} = \mathbf{g}^{(l-1)}}$$
$$= s_l - \sum_{m=0}^{M-1} g_m^{(l-1)} y_{l-m}$$
$$= s_l - \mathbf{y}_l^{\text{T}} \mathbf{g}^{(l-1)}.$$

The recursion for the LMS algorithm in Eq. (2.56) does not require second-order statistics of the signals. Therefore, it overcomes the first difficulty of the MMSE approach.

The vector $\mathbf{g}^{(l)}$ is a random vector as shown in Eq. (2.56). Hence, convergence properties of the LMS algorithm are not deterministic and are more involved. In general, the LMS algorithm has the same convergence properties as the SD algorithm in terms of mean sense. In the Appendix to this chapter, a second-order analysis of the LMS algorithm is addressed in detail.

2.4.4   Least squares approach and the RLS algorithm

To find the MMSE equalization vector $\mathbf{g}$, second-order statistics of signals are used. To avoid the need for second-order statistics, an equalization vector can be found from actual signals.

Consider a sum of squared errors (SSE) as follows:

$$\text{SSE} = \sum_{l=0}^{N-1} \left| s_l - \sum_{m=0}^{M-1} g_m y_{l-m} \right|^2$$
$$= \sum_{l=0}^{N-1} |s_l - \mathbf{y}_l^{\text{T}} \mathbf{g}|^2, \qquad (2.57)$$

where N is the number of samples. The least squares (LS) approach finds the optimal solution of $\mathbf{g}$ that minimizes the SSE in Eq. (2.57). Let $\mathbf{s} = [s_0 \ s_1 \ \cdots \ s_{N-1}]^{\text{T}}$ and

$$\mathbf{Y} = [\mathbf{y}_0 \ \mathbf{y}_1 \ \cdots \ \mathbf{y}_{N-1}].$$

Then, the SSE is written as follows:

$$\mathrm{SSE} = \|\mathbf{s} - \mathbf{Y}^{\mathrm{T}}\mathbf{g}\|^2.$$

By taking the derivative and setting it equal to zero, we can show that

$$\begin{aligned} \mathbf{0} &= \nabla_{\mathbf{g}}\mathrm{SSE} \\ &= \mathbf{Y}(\mathbf{s} - \mathbf{Y}^{\mathrm{T}}\mathbf{g}). \end{aligned} \tag{2.58}$$

This leads to the optimal vector $\mathbf{g}$ minimizing the SSE, called the LS solution, which is given by

$$\mathbf{g}_{\mathrm{ls}} = (\mathbf{Y}\mathbf{Y}^{\mathrm{T}})^{-1}\mathbf{Y}\mathbf{s}. \tag{2.59}$$

If the inverse of $\mathbf{Y}\mathbf{Y}^{\mathrm{T}}$ does not exist, it can be replaced by the pseudo-inverse.

The LS approach is an off-line approach to find an equalization vector. It would be desirable to derive an on-line algorithm based on the LS approach.

The recursive LS (RLS) algorithm is an on-line algorithm employed to perform the LS estimation. To derive the RLS, we need to introduce the forgetting factor, λ. From Eq. (2.57), the exponential weighted SSE up to time l is given by

$$\mathrm{SSE}_l = \sum_{k=0}^{l} \lambda^{l-k}\big|s_k - \mathbf{y}_k^{\mathrm{T}}\mathbf{g}\big|^2, \tag{2.60}$$

where $0 < \lambda < 1$. From Eq. (2.60), we can show that

$$\mathrm{SSE}_l = \lambda \times \mathrm{SSE}_{l-1} + \big|s_l - \mathbf{y}_l^{\mathrm{T}}\mathbf{g}\big|^2.$$

To find the optimal solution that minimizes SSE_l at time l, the same approach which is performed in Eq. (2.58) can be used. After taking the derivative and setting it equal to zero on the right hand side, we can show that

$$\begin{aligned} 0 &= \nabla_{\mathbf{g}}\mathrm{SSE}_l \\ &= \nabla_{\mathbf{g}}\big(\lambda \times \mathrm{SSE}_{l-1} + \big|s_l - \mathbf{y}_l^{\mathrm{T}}\mathbf{g}\big|^2\big) \\ &= \left(\lambda\left(\sum_{k=0}^{l-1}\lambda^{l-1-k}\mathbf{y}_k\mathbf{y}_k^{\mathrm{T}}\right) + \mathbf{y}_l\mathbf{y}_l^{\mathrm{T}}\right)\mathbf{g}^{(l)} - \lambda\left(\sum_{k=0}^{l-1}\lambda^{l-1-k}\mathbf{y}_k s_k\right) - \mathbf{y}_l s_l, \end{aligned} \tag{2.61}$$

where $\mathbf{g}^{(l)}$ denotes the solution vector $\mathbf{g}$ that minimizes SSE_l. Define

$$\mathbf{\Sigma}_l = \sum_{k=0}^{l} \lambda^{l-k}\mathbf{y}_k\mathbf{y}_k^{\mathrm{T}}$$

and

$$\mathbf{r}_l = \sum_{k=0}^{l} \lambda^{l-k}\mathbf{y}_k s_k.$$

We can readily show that

$$\mathbf{\Sigma}_l = \lambda\mathbf{\Sigma}_{l-1} + \mathbf{y}_l\mathbf{y}_l^{\mathrm{T}} \text{ and } \mathbf{r}_l = \lambda\mathbf{r}_{l-1} + \mathbf{y}_l s_l.$$

Using these, Eq. (2.61) can be rewritten as follows:

$$\left(\lambda \Sigma_{l-1} + \mathbf{y}_l \mathbf{y}_l^T\right) \mathbf{g}^{(l)} = \left(\lambda \mathbf{r}_{l-1} + \mathbf{y}_l s_l\right). \tag{2.62}$$

Since the LS solution can be directly found from Eq. (2.60) to be

$$\Sigma_l \mathbf{g}^{(l)} = \mathbf{r}_l,$$

Eq. (2.62) does not seem useful. However, using the matrix inversion lemma or Woodbury's identity, which is given by

$$\left(\mathbf{A} + \gamma^2 \mathbf{u} \mathbf{u}^T\right)^{-1} = \mathbf{A}^{-1} - \frac{\gamma^2}{1 + \gamma^2 \mathbf{u}^T \mathbf{A}^{-1} \mathbf{u}} \mathbf{A}^{-1} \mathbf{u} \mathbf{u}^T \mathbf{A}^{-1}, \tag{2.63}$$

where $\mathbf{A}$ is a full-rank Hermitian matrix, Eq. (2.62) can lead to a computationally efficient algorithm.

Substituting Eq. (2.63) into Eq. (2.62) yields

$$\begin{aligned}
\mathbf{g}^{(l)} &= \Sigma_l^{-1} \mathbf{r}_l \\
&= \left(\lambda \Sigma_{l-1} + \mathbf{y}_l \mathbf{y}_l^T\right)^{-1} \left(\lambda \mathbf{r}_{l-1} + \mathbf{y}_l s_l\right) \\
&= \Sigma_{l-1}^{-1} \mathbf{r}_{l-1} - \lambda \Omega_l \mathbf{r}_{l-1} + \lambda^{-1} \Sigma_{l-1}^{-1} \mathbf{y}_l s_l - \Omega_l \mathbf{y}_l s_l,
\end{aligned} \tag{2.64}$$

where

$$\Omega_l = \frac{\lambda^{-2}}{1 + \lambda^{-1} \mathbf{y}_l^T \Sigma_{l-1}^{-1} \mathbf{y}_l} \Sigma_{l-1}^{-1} \mathbf{y}_l \mathbf{y}_l^T \Sigma_{l-1}^{-1}.$$

For further simplification, we define

$$\begin{aligned}
\beta_l &= \frac{\lambda^{-1}}{1 + \lambda^{-1} \mathbf{y}_l^T \Sigma_{l-1}^{-1} \mathbf{y}_l}, \\
\mathbf{m}_l &= \beta_l \Sigma_{l-1}^{-1} \mathbf{y}_l.
\end{aligned} \tag{2.65}$$

Then, Ω_l is rewritten as

$$\Omega_l = (\lambda \beta_l)^{-1} \mathbf{m}_l \mathbf{m}_l^T.$$

Substituting this into Eq. (2.64), we have

$$\begin{aligned}
\mathbf{g}^{(l)} &= \mathbf{g}^{(l-1)} - \mathbf{m}_l \mathbf{y}_l^T \Sigma_{l-1}^{-1} \mathbf{r}_{l-1} + (\lambda \beta_l)^{-1} \mathbf{m}_l s_l - (\lambda \beta_l)^{-1} \mathbf{m}_l \mathbf{m}_l^T \mathbf{y}_l s_l \\
&= \mathbf{g}^{(l-1)} - \mathbf{m}_l \mathbf{y}_l^T \mathbf{g}^{(l-1)} + (\lambda \beta_l)^{-1} \mathbf{m}_l \left(1 - \mathbf{m}_l^T \mathbf{y}_l\right) s_l \\
&= \mathbf{g}^{(l-1)} + \mathbf{m}_l \left((\lambda \beta_l)^{-1} \left(1 - \mathbf{m}_l^T \mathbf{y}_l\right) s_l - \mathbf{y}_l^T \mathbf{g}^{(l-1)}\right).
\end{aligned} \tag{2.66}$$

Using the definitions of β_l and $\mathbf{m}_l$, we can show that

$$1 - \mathbf{m}_l^T \mathbf{y}_l = 1 - \frac{\lambda^{-1} \mathbf{y}_l \Sigma_{l-1}^{-1} \mathbf{y}_l}{1 + \lambda^{-1} \mathbf{y}_l \Sigma_{l-1}^{-1} \mathbf{y}_l} = \lambda \beta_l.$$

This finally simplifies the recursion as follows:

$$\mathbf{g}^{(l)} = \mathbf{g}^{(l-1)} + \mathbf{m}_l \left(s_l - \mathbf{y}_l^T \mathbf{g}^{(l-1)}\right). \tag{2.67}$$

This recursion is called the RLS algorithm since it solves the exponentially weighted SSE recursively. The only difference from the LMS algorithm is the updating vector. We can summarize this as follows:

$$\text{(LMS)} \quad \mathbf{g}^{(l)} = \mathbf{g}^{(l-1)} + 2\mu \mathbf{y}_l \left(s_l - \mathbf{y}_l^{\mathrm{T}} \mathbf{g}^{(l-1)} \right), \tag{2.68a}$$

$$\text{(RLS)} \quad \mathbf{g}^{(l)} = \mathbf{g}^{(l-1)} + \mathbf{m}_l \left(s_l - \mathbf{y}_l^{\mathrm{T}} \mathbf{g}^{(l-1)} \right)$$

$$= \mathbf{g}^{(l-1)} + \beta_l \boldsymbol{\Sigma}_{l-1}^{-1} \mathbf{y}_l \left(s_l - \mathbf{y}_l^{\mathrm{T}} \mathbf{g}^{(l-1)} \right). \tag{2.68b}$$

The updating vector is proportional to $\mathbf{y}_l$ in the LMS algorithm, and proportional to $\boldsymbol{\Sigma}_{l-1}^{-1} \mathbf{y}_l$ in the RLS algorithm.

In the RLS algorithm, we need to compute the inverse of $\boldsymbol{\Sigma}_l$. The matrix inversion requires the complexity of order $O(M^3)$. Since the matrix inversion shall be performed for every received signal sample, y_l, it is obviously prohibitive. However, the inverse of $\boldsymbol{\Sigma}_l$, $\boldsymbol{\Sigma}_l^{-1}$, can be recursively computed from the previous one, $\boldsymbol{\Sigma}_{l-1}^{-1}$. To see this, let $\boldsymbol{\Phi}_l = \boldsymbol{\Sigma}_l^{-1}$. Then, from Eq. (2.63), we have the following recursion:

$$\boldsymbol{\Phi}_l = \lambda^{-1} \boldsymbol{\Phi}_{l-1} - \boldsymbol{\Omega}_l$$

$$= \lambda^{-1} \left(\boldsymbol{\Phi}_{l-1} - \mathbf{m}_l \mathbf{y}_l^{\mathrm{T}} \boldsymbol{\Phi}_{l-1} \right). \tag{2.69}$$

This implies that we can obtain the matrix inversion of $\boldsymbol{\Sigma}_l$ recursively with the complexity of order $O(M^2)$.

Finally, the RLS algorithm can be summarized as follows:

$$\mathbf{g}^{(l)} = \mathbf{g}^{(l-1)} + \beta_l \boldsymbol{\Phi}_{l-1} \mathbf{y}_l \left(s_l - \mathbf{y}_l^{\mathrm{T}} \mathbf{g}^{(l-1)} \right), \tag{2.70a}$$

$$\beta_l = \frac{\lambda^{-1}}{1 + \lambda^{-1} \mathbf{y}_l^{\mathrm{T}} \boldsymbol{\Phi}_{l-1} \mathbf{y}_l}, \tag{2.70b}$$

$$\mathbf{m}_l = \beta_l \boldsymbol{\Phi}_{l-1} \mathbf{y}_l, \tag{2.70c}$$

$$\boldsymbol{\Phi}_l = \lambda^{-1} \left(\boldsymbol{\Phi}_{l-1} - \mathbf{m}_l \mathbf{y}_l^{\mathrm{T}} \boldsymbol{\Phi}_{l-1} \right). \tag{2.70d}$$

The initial matrix, $\boldsymbol{\Phi}_0$, is given by

$$\boldsymbol{\Phi}_0 = \phi_0 \mathbf{I},$$

where $\phi_0 > 0$.

2.5 Adaptive decision feedback equalizers

When the training sequence is available, we can always use the correct symbols in the DFE. From Eq. (2.24), let

$$\mathbf{a} = [\mathbf{g}^{\mathrm{T}} \ -\mathbf{f}^{\mathrm{T}}]^{\mathrm{T}} \quad \text{and} \quad \mathbf{x}_l = \left[\mathbf{y}_l^{\mathrm{T}} \ \hat{\mathbf{b}}_{l-\tilde{m}-1}^{\mathrm{T}} \right]^{\mathrm{T}} = \left[\mathbf{y}_l^{\mathrm{T}} \ \mathbf{b}_{l-\tilde{m}-1}^{\mathrm{T}} \right]^{\mathrm{T}}. \tag{2.71}$$

Then, the MSE for the DFE is rewritten as follows:

$$\text{MSE} = E\left[\left| s_l - \mathbf{x}_l^{\mathrm{T}} \mathbf{a} \right|^2 \right]$$

$$= E\left[\left| b_{l-\tilde{m}} - \mathbf{x}_l^{\mathrm{T}} \mathbf{a} \right|^2 \right]. \tag{2.72}$$

The adaptive DFE finds the equalization vector $\mathbf{a}$ from the $\mathbf{x}_l$'s and $s_l = b_{l-\bar{m}}$'s. Clearly, both LMS and RLS algorithms can be used for the adaptive DFE.

2.6   Summary and notes

We discussed channel equalization in this chapter. We also considered adaptive algorithms, such as the LMS and RLS algorithms, for channel equalization.

Adaptive equalization for digital communication was introduced by Lucky (1965, 1966). The DFE was proposed by Austin (1967). The reader is referred to Gitlin, Hayes, and Weinstein (1992), Lucky, Salz, and Weldon (1968), and Proakis (1995) for detailed accounts of adaptive equalization.

Besides the LMS and RLS algorithms, other adaptive algorithms can be applied to the channel equalization (Proakis, 1995). There are available excellent books on adaptive signal processing including Haykin (1996), Honig and Messerschmitt (1984), Solo and Kong (1995), and Widrow and Stearns (1985). In this chapter, we do not discuss blind algorithms for the channel equalization. A blind algorithm, such as the constant modulus algorithm (CMA), can train an equalizer without a training sequence; see Ding and Li (2001) for details.

2.7   Appendix to Chapter 2: Convergence of the LMS algorithm

For a convergence analysis, the reader is referred to Honig and Messerschmitt (1984) and Solo and Kong (1995). In this appendix, we follow the approach in Honig and Messerschmitt (1984).

We need to make a few assumptions for a convergence analysis.

(A1) The received signal vector $\mathbf{y}_l$ is uncorrelated with its past vectors,

$$\{\mathbf{y}_{l-1}, \mathbf{y}_{l-2}, \ldots, \mathbf{y}_0\}.$$

(A2) There exists the optimal solution of $\mathbf{g}$, $\mathbf{g}_{\mathrm{mmse}}$.

Recall the LMS algorithm:

$$\mathbf{g}^{(l)} = \mathbf{g}^{(l-1)} + 2\mu \mathbf{y}_l \left(s_l - \mathbf{y}_l^{\mathrm{T}} \mathbf{g}^{(l-1)}\right);$$

after defining $\tilde{\mathbf{g}}^{(l)} = \mathbf{g}^{(l)} - \mathbf{g}_{\mathrm{mmse}}$ (it is assumed that $\mathbf{g}_{\mathrm{mmse}}$ exists; see (A2)), we obtain

$$\tilde{\mathbf{g}}^{(l)} = \tilde{\mathbf{g}}^{(l-1)} + 2\mu \mathbf{y}_l \left(s_l - \mathbf{y}_l^{\mathrm{T}} \mathbf{g}_{\mathrm{mmse}} - \mathbf{y}_l^{\mathrm{T}} \tilde{\mathbf{g}}^{(l-1)}\right). \tag{2.73}$$

Let

$$\epsilon_l = s_l - \mathbf{y}_l^{\mathrm{T}} \mathbf{g}_{\mathrm{mmse}}.$$

Then, Eq. (2.73) is rewritten as follows:

$$\tilde{\mathbf{g}}^{(l)} = \tilde{\mathbf{g}}^{(l-1)} + 2\mu \mathbf{y}_l \left(\epsilon_l - \mathbf{y}_l^{\mathrm{T}} \tilde{\mathbf{g}}^{(l-1)}\right). \tag{2.74}$$

Firstly, consider the first-order analysis. From Eq. (2.74), we have

$$E[\tilde{\mathbf{g}}^{(l)}] = E[\tilde{\mathbf{g}}^{(l-1)}] + 2\mu \left(E[\mathbf{y}_l \epsilon_l] - E[\mathbf{y}_l \mathbf{y}_l^{\mathrm{T}} \tilde{\mathbf{g}}^{(l-1)}]\right). \tag{2.75}$$

Due to the orthogonality principle, $E[\mathbf{y}_l \epsilon_l] = \mathbf{0}$. Since $\tilde{\mathbf{g}}^{(l-1)}$ is a function of the past received signal vectors $\{\mathbf{y}_{l-1}, \mathbf{y}_{l-2}, \ldots\}$, we can see that $\mathbf{y}_l$ and $\tilde{\mathbf{g}}^{(l-1)}$ are uncorrelated according to (A1). Then, it follows that

$$E[\mathbf{y}_l \mathbf{y}_l^{\mathrm{T}} \tilde{\mathbf{g}}^{(l-1)}] = E[\mathbf{y}_l \mathbf{y}_l^{\mathrm{T}}] E[\tilde{\mathbf{g}}^{(l-1)}], \tag{2.76}$$

and from Eq. (2.75) we can immediately show that

$$E[\tilde{\mathbf{g}}^{(l)}] = (\mathbf{I} - 2\mu \mathbf{R}_{\mathbf{y}}) E[\tilde{\mathbf{g}}^{(l-1)}]. \tag{2.77}$$

Therefore, we can see that the first-order analysis is the same as the convergence analysis of the SD algorithm. That is, the first-order convergence is guaranteed if

$$0 < \mu < \frac{1}{\lambda_{\max}}.$$

The second-order analysis is not straightforward as fourth-order statistics of $\mathbf{y}_l$ are required. Hence, some approximations are required unless $\mathbf{y}_l$ are Gaussian random vectors; see Solo and Kong (1995) for a detailed account. First of all, we need to note that $E[\tilde{\mathbf{g}}^{(l)}]$ becomes zero as $l \to \infty$. In addition, it is useful to consider $\tilde{\mathbf{q}}^{(l)} = \mathbf{E}^{\mathrm{T}} \tilde{\mathbf{g}}^{(l)}$ rather than $\tilde{\mathbf{g}}^{(l)}$. Once we find $E[\tilde{\mathbf{q}}^{(l)}(\tilde{\mathbf{q}}^{(l)})^{\mathrm{T}}]$, we can obtain $E[\tilde{\mathbf{g}}^{(l)}(\tilde{\mathbf{g}}^{(l)})^{\mathrm{T}}]$ by using the following relationship:

$$E[\tilde{\mathbf{g}}^{(l)}(\tilde{\mathbf{g}}^{(l)})^{\mathrm{T}}] = \mathbf{E} E[\tilde{\mathbf{q}}^{(l)}(\tilde{\mathbf{q}}^{(l)})^{\mathrm{T}}] \mathbf{E}^{\mathrm{T}}.$$

Hence, we directly consider the second-order moment $\mathbf{\Gamma}_l = E[\tilde{\mathbf{q}}^{(l)}(\tilde{\mathbf{q}}^{(l)})^{\mathrm{T}}]$. From Eq. (2.74), we have

$$\tilde{\mathbf{q}}^{(l)} = \tilde{\mathbf{q}}^{(l-1)} + 2\mu \mathbf{z}_l \left(s_l - \mathbf{z}_l^{\mathrm{T}} \mathbf{q}_{\mathrm{mmse}} - \mathbf{z}_l^{\mathrm{T}} \tilde{\mathbf{q}}^{(l-1)}\right),$$

where $\mathbf{z}_l = \mathbf{E}^{\mathrm{T}} \mathbf{y}_l$. It is easy to show that

$$E[\mathbf{z}_l \mathbf{z}_l^{\mathrm{T}}] = \mathbf{\Lambda}.$$

Since $s_l - \mathbf{y}_l^{\mathrm{T}} \mathbf{g}_{\mathrm{mmse}} = s_l - \mathbf{z}_l^{\mathrm{T}} \mathbf{q}_{\mathrm{mmse}} = \epsilon_l$, it follows that

$$\mathbf{\Gamma}_l = \mathbf{\Gamma}_{l-1} + 2\mu E\left[\mathbf{z}_l \left(\epsilon_l - \mathbf{z}_l^{\mathrm{T}} \tilde{\mathbf{q}}^{(l-1)}\right)(\tilde{\mathbf{q}}^{(l-1)})^{\mathrm{T}}\right]$$
$$+ 2\mu E\left[\tilde{\mathbf{q}}^{(l-1)}\left(\epsilon_l - \mathbf{z}_l^{\mathrm{T}} \tilde{\mathbf{q}}^{(l-1)}\right)\mathbf{z}_l^{\mathrm{T}}\right] + 4\mu^2 E\left[\mathbf{z}_l \mathbf{z}_l^{\mathrm{T}}\left(\epsilon_l - \mathbf{z}_l^{\mathrm{T}} \tilde{\mathbf{q}}^{(l-1)}\right)^2\right].$$

Furthermore, since $E[\mathbf{z}_l \epsilon_l] = \mathbf{0}$ and $E[\tilde{\mathbf{q}}_l] \to \mathbf{0}$ as $l \to \infty$, we have

$$\mathbf{\Gamma}_l = \mathbf{\Gamma}_{l-1} - 2\mu \mathbf{\Lambda} \mathbf{\Gamma}_{l-1} - 2\mu \mathbf{\Gamma}_{l-1} \mathbf{\Lambda} + 4\mu^2 \left(\mathbf{\Lambda} \sigma_\epsilon^2 + E\left[\mathbf{z}_l \mathbf{z}_l^{\mathrm{T}} \left(\mathbf{z}_l^{\mathrm{T}} \tilde{\mathbf{q}}^{(l-1)}\right)^2\right]\right), \tag{2.78}$$

where $\sigma_\epsilon^2 = E[\epsilon_l^2]$ is the MMSE. As shown in Eq. (2.78), fourth-order statistics of $\mathbf{z}_l$ are required to go further.

For convenience, let $\mathbf{z} = \mathbf{z}_l$ and $\tilde{\mathbf{q}} = \tilde{\mathbf{q}}^{(l-1)}$. The matrix which is given by

$$E[\mathbf{z}\mathbf{z}^{\mathrm{T}}(\mathbf{z}^{\mathrm{T}} \tilde{\mathbf{q}})^2] = E[\mathbf{z}\mathbf{z}^{\mathrm{T}} \tilde{\mathbf{q}} \tilde{\mathbf{q}}^{\mathrm{T}} \mathbf{z}\mathbf{z}^{\mathrm{T}}] \tag{2.79}$$

has the (k, m)th element as follows:

$$E[z_k(\mathbf{z}^\mathsf{T}\tilde{\mathbf{q}})(\tilde{\mathbf{q}}^\mathsf{T}\mathbf{z})z_m] = E\left[z_k\left(\sum_i z_i\tilde{q}_i\right)\left(\sum_j z_j\tilde{q}_j\right)z_m\right]$$

$$= \sum_i \sum_j E[z_k z_i\tilde{q}_i z_j\tilde{q}_j z_m]$$

$$= \sum_i \sum_j E[\tilde{q}_i\tilde{q}_j]E[z_k z_i z_j z_m]. \qquad (2.80)$$

We can approximate that the matrix $\mathbf{\Gamma}_l$ is diagonal. This means that $E[\tilde{q}_i\tilde{q}_j] = 0$ if $i \neq j$. Then, we have

$$E[z_k(\mathbf{z}^\mathsf{T}\tilde{\mathbf{q}})(\tilde{\mathbf{q}}^\mathsf{T}\mathbf{z})z_m] \simeq \sum_i E[(\tilde{q}_i)^2]E[z_k z_m(z_i)^2]. \qquad (2.81)$$

Furthermore, if we use the approximation $E[z_k z_m(z_i)^2] \simeq E[z_k z_m]E[(z_i)^2] = E[(z_k)^2] \times E[(z_i)^2]\delta_{k,m}$ because $E[\mathbf{zz}^\mathsf{T}] = \mathbf{\Lambda}$ is diagonal, we have

$$E[z_k(\mathbf{z}^\mathsf{T}\tilde{\mathbf{q}})(\tilde{\mathbf{q}}^\mathsf{T}\mathbf{z})z_m] \simeq \sum_i E[(\tilde{q}_i)^2]E[(z_i)^2]E[(z_k)^2]\delta_{k,m}$$

$$= \sum_i E[(\tilde{q}_i)^2]\lambda_i\lambda_k\delta_{k,m}. \qquad (2.82)$$

This approximation leads to the result that the matrix $E[\mathbf{z}_l\mathbf{z}_l^\mathsf{T}(\mathbf{z}_l^\mathsf{T}\tilde{\mathbf{q}}^{(l-1)})^2]$ is diagonal and is given by

$$E[\mathbf{z}_l\mathbf{z}_l^\mathsf{T}(\mathbf{z}_l^\mathsf{T}\tilde{\mathbf{q}}^{(l-1)})^2] \simeq \mathbf{\Lambda}\left(\sum_i E[(\tilde{q}_i)^2]\mathbf{\Lambda}\right)$$

$$= \mathbf{\Lambda}\,\mathrm{Tr}(\mathbf{\Gamma}_{l-1}\mathbf{\Lambda}). \qquad (2.83)$$

Hence, the recursion in Eq. (2.78) is approximated as follows:

$$\mathbf{\Gamma}_l \simeq \mathbf{\Gamma}_{l-1} - 2\mu\mathbf{\Lambda}\mathbf{\Gamma}_{l-1} - 2\mu\mathbf{\Gamma}_{l-1}\mathbf{\Lambda} + 4\mu^2\left(\mathbf{\Lambda}\sigma_\epsilon^2 + \mathbf{\Lambda}\,\mathrm{Tr}(\mathbf{\Gamma}_{l-1}\mathbf{\Lambda})\right). \qquad (2.84)$$

Let $\mathbf{v}$ and $\mathbf{d}_l$ be the vectors whose elements are the diagonal elements of $\mathbf{\Lambda}$ and $\mathbf{\Gamma}_l$, respectively. Then, it follows that

$$\mathbf{d}_l \simeq \mathbf{d}_{l-1} - 4\mu\mathbf{\Lambda}\mathbf{d}_{l-1} + 4\mu^2\left(\sigma_\epsilon^2\mathbf{v} + \mathbf{v}(\mathbf{v}^\mathsf{T}\mathbf{d}_{l-1})\right)$$

$$= \left(\mathbf{I} - 4\mu\mathbf{\Lambda} + 4\mu^2\mathbf{vv}^\mathsf{T}\right)\mathbf{d}_{l-1} + 4\mu^2\sigma_\epsilon^2\mathbf{v}. \qquad (2.85)$$

Let

$$\mathbf{A} = \left(\mathbf{I} - 4\mu\mathbf{\Lambda} + 4\mu^2\mathbf{vv}^\mathsf{T}\right).$$

The convergence of Eq. (2.85) depends on the properties of matrix $\mathbf{A}$. If the absolute values of the eigenvalues of $\mathbf{A}$ are less than unity, the recursion in Eq. (2.85) will be bounded.

It is known that a symmetric matrix whose elements are nonnegative has eigenvalues whose absolute values are smaller than unity if the sum of any row (or column) vector is less than unity. Thus, if $\sum_k[\mathbf{A}]_{k,m} < 1$, Eq. (2.85) converges. The condition for second-order convergence becomes

$$\sum_k [\mathbf{A}]_{k,m} = 1 - 4\mu\lambda_m + 4\mu^2\lambda_m\left(\sum_k \lambda_k\right) < 1 \tag{2.86}$$

or

$$\mu < \frac{1}{\mathrm{Tr}(\mathbf{\Lambda})}. \tag{2.87}$$

Clearly, this is a stricter condition than that for first-order convergence.

The steady state MSE is one of the most important parameters in the understanding of the performance of adaptive algorithms. The MSE is given by

$$\begin{aligned}
\mathrm{MSE}_l &= E\left[\left|s_l - \mathbf{y}_l^{\mathrm{T}}\mathbf{g}^{(l)}\right|^2\right] \\
&= E\left[\left|s_l - \mathbf{y}_l^{\mathrm{T}}(\tilde{\mathbf{g}}^{(l)} + \mathbf{g}_{\mathrm{mmse}})\right|^2\right] \\
&= E\left[\left|\epsilon_l - \mathbf{y}_l^{\mathrm{T}}\tilde{\mathbf{g}}^{(l)}\right|^2\right] \\
&= E\left[\left|\epsilon_l - \mathbf{z}_l^{\mathrm{T}}\tilde{\mathbf{q}}^{(l)}\right|^2\right] \\
&= \sigma_\epsilon^2 + E\left[\left|\mathbf{z}_l^{\mathrm{T}}\tilde{\mathbf{q}}^{(l)}\right|^2\right]. \tag{2.88}
\end{aligned}$$

The difference between the actual MSE and the MMSE is called the *excess MSE*, given by

$$\begin{aligned}
\sigma_{\mathrm{ex},l}^2 &= E\left[\left|\mathbf{z}_l^{\mathrm{T}}\tilde{\mathbf{q}}^{(l)}\right|^2\right] \\
&\simeq \mathrm{Tr}(\mathbf{\Lambda}\mathbf{\Gamma}_l). \tag{2.89}
\end{aligned}$$

From Eq. (2.84), after the steady state, we can write

$$\begin{aligned}
\mathbf{\Gamma} &\simeq \mathbf{\Gamma} - 4\mu\mathbf{\Lambda}\mathbf{\Gamma} + 4\mu^2\left(\mathbf{\Lambda}\sigma_\epsilon^2 + \mathbf{\Lambda}\sigma_{\mathrm{ex}}^2\right) \\
&= \mathbf{\Gamma} - 4\mu\mathbf{\Lambda}\mathbf{\Gamma} + 4\mu^2\mathbf{\Lambda}\left(\sigma_\epsilon^2 + \sigma_{\mathrm{ex}}^2\right), \tag{2.90}
\end{aligned}$$

where $\mathbf{\Gamma} = \lim_{l\to\infty}\mathbf{\Gamma}_l$ and $\sigma_{\mathrm{ex}}^2 = \lim_{l\to\infty}\sigma_{\mathrm{ex},l}^2$. It follows that

$$\mathbf{\Gamma} = \mu\left(\sigma_\epsilon^2 + \sigma_{\mathrm{ex}}^2\right)\mathbf{I}. \tag{2.91}$$

From Eq. (2.89), we have

$$\begin{aligned}
\sigma_{\mathrm{ex}}^2 &= \mathrm{Tr}(\mathbf{\Lambda}\mathbf{\Gamma}) \\
&= \mu\left(\sum_m \lambda_m\right)\left(\sigma_\epsilon^2 + \sigma_{\mathrm{ex}}^2\right).
\end{aligned}$$

Finally, we have

$$\sigma_{\mathrm{ex}}^2 = \frac{\mu\left(\sum_m \lambda_m\right)}{1 - \mu\left(\sum_m \lambda_m\right)}\sigma_\epsilon^2. \tag{2.92}$$

3 Sequence detection with adaptive channel estimation

An equalizer is one of the devices that can reduce or possibly eliminate the ISI caused by a dispersive channel. After channel equalization, symbol-by-symbol detection is considered, though there are alternative methods of detecting symbols. We can attempt to detect the symbol sequence itself rather than individual symbols by taking the ISI into account. This approach is called sequence detection. In this chapter, we introduce a sequence detection approach based on the maximum likelihood (ML) criterion called the ML sequence detection (MLSD).

Since the complexity of the sequence detection grows exponentially with the length of sequence when an exhaustive search is employed, computationally efficient algorithms are required to solve the sequence detection. A computationally efficient algorithm is also studied in this chapter.

For the sequence detection, the CIR should be known. We study the channel estimation problem with adaptive algorithms. It is possible to include adaptive channel estimation within a data sequence detection method. There are various approaches for joint adaptive channel estimation and data sequence detection. A well known computationally efficient method will be discussed in this chapter.

3.1 MLSD and the Viterbi algorithm

We will discuss the sequence detection problem over ISI channels under the ML criterion in this section. A computationally efficient algorithm for the sequence detection is also discussed.

3.1.1 ISI channels

Recall the received signal over a dispersive channel:

$$y_l = \sum_{p=0}^{P-1} h_p b_{l-p} + n_l, \quad l = 0, 1, \ldots, L-1, \tag{3.1}$$

where $\{h_p\}$ is the finite impulse response of the channel, $\{b_m\}$ is the transmitted data symbol sequence, L is the length of the data sequence, and $\{n_m\}$ is the additive white Gaussian

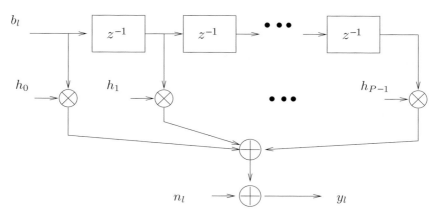

Figure 3.1. FIR channel with additive white Gaussian noise.

noise sequence with $E[n_l] = 0$ and $E[n_l^2] = N_0/2$. Generally, the output of a dispersive channel of finite impulse response depends on the current data symbol as well as past data symbols. As shown in Eq. (3.1), the received signal or the output of a dispersive channel is dependent on (1) the current input symbol and (2) the channel state decided by the last $(P - 1)$ symbols. Figure 3.1 shows an FIR channel in which we can see that the channel output y_l is dependent on the current data symbol b_l and the last $(P - 1)$ data symbols $\{b_{l-1}, b_{l-2}, \ldots, b_{l-P+1}\}$.

3.1.2 Symbol detection without ISI

It would be helpful to review signal detection theory to understand sequence detection; see Whalen (1971) for a detailed account of signal detection theory. Consider a simple example with $P = 1$ (i.e. there is no ISI). The received signal is given by

$$y_l = h_0 b_l + n_l.$$

The received signal y_l is a random variable conditioned on the transmitted data symbol b_l. Since n_l is a Gaussian random variable, y_l becomes a conditional Gaussian random variable. The conditional mean and variance of y_l given b_l are

$$E[y_l|b_l] = h_0 b_l,$$
$$Var(y_l|b_l) = N_0/2,$$

respectively, where the conditional variance given b_l is obtained as follows:

$$\begin{aligned} Var(y_l|b_l) &= E[(y_l - E[y_l|b_l])^2|b_l] \\ &= E[(y_l)^2|b_l] - (E[y_l|b_l])^2 \\ &= (h_0 b_l)^2 + \frac{N_0}{2} - (h_0 b_l)^2 \\ &= \frac{N_0}{2}. \end{aligned}$$

Thus, the conditional pdf of y_l given b_l is given by

$$f(y_l|b_l) = \frac{1}{\sqrt{\pi N_0}} \exp\left(-\frac{1}{N_0}|y_l - h_0 b_l|^2\right). \tag{3.2}$$

The conditional pdf becomes the likelihood function of b_l given y_l. With the received signal y_l, we can decide whether $b_l = 1$ or $b_l = -1$ using the likelihood function as follows:

$$b_l = \begin{cases} +1, & \text{if } f(y_l|b_l = 1) \geq f(y_l|b_l = -1) \\ -1, & \text{if } f(y_l|b_l = 1) < f(y_l|b_l = -1). \end{cases} \tag{3.3}$$

This decision rule is called the ML detection rule as the symbol that maximizes the likelihood function is chosen. Substituting Eq. (3.2) into Eq. (3.3), the ML decision rule is further simplified as follows:

$$b_l = \begin{cases} +1, & \text{if } 4h_0 y_l / N_0 \geq 0 \\ -1, & \text{if } 4h_0 y_l / N_0 < 0. \end{cases}$$

This approach for the symbol detection can be extended to the sequence detection when $P > 1$.

There can be errors in decision in the ML detection. Suppose that $b_l = 1$ and the ML detection provides a decision of $b_l = -1$, which is an incorrect decision. In this case, the probability of decision error can be written as follows:

$$\begin{aligned} \Pr(\text{error}|b_l = 1) &= \Pr(f(y_l|b_l = 1) < f(y_l|b_l = -1) \mid b_l = 1) \\ &= \Pr\left(\frac{f(y_l|b_l = 1)}{f(y_l|b_l = -1)} < 1 \mid b_l = 1\right) \\ &= \Pr\left(\log \frac{f(y_l|b_l = 1)}{f(y_l|b_l = -1)} < 0 \mid b_l = 1\right). \end{aligned} \tag{3.4}$$

Since

$$\begin{aligned} \log \frac{f(y_l|b_l = 1)}{f(y_l|b_l = -1)} &= -\frac{1}{N_0}\left(|y_l - h_0|^2 - |y_l + h_0|^2\right) \\ &= \frac{4h_0 y_l}{N_0}, \end{aligned}$$

it follows that

$$\begin{aligned} \Pr(\text{error}|b_l = 1) &= \Pr\left(\frac{4h_0 y_l}{N_0} < 0 | b_l = 1\right) \\ &= \Pr\left(\frac{4h_0}{N_0}(h_0 + n_l) < 0\right) \\ &= \Pr(-n_l > h_0) \end{aligned}$$

for $h_0 > 0$. (For convenience, we will assume that $h_0 > 0$ for the rest of this subsection.) It is convenient to normalize $-n_l$ to have unit variance. Let $\bar{n}_l = -n_l/\sqrt{N_0/2}$. Then, the probability of decision error becomes

$$\Pr(\text{error}|b_l = 1) = \Pr\left(\bar{n}_l > h_0/\sqrt{N_0/2}\right)$$
$$= Q\left(\frac{h_0}{\sqrt{N_0/2}}\right)$$
$$= Q\left(\sqrt{\frac{2|h_0|^2}{N_0}}\right), \tag{3.5}$$

where $Q(x)$ is the Q-function given by

$$Q(x) = \int_x^\infty \frac{1}{\sqrt{2\pi}} \exp\left(-\frac{z^2}{2}\right) dz.$$

It is well known that the Q-function is upper bounded as

$$Q(x) < e^{-x^2/2}.$$

This upper bound is called the Chernoff bound.

Due to the symmetry, the probability of decision error when $b_l = -1$ is transmitted becomes $Q\left(\sqrt{2|h_0|^2/N_0}\right)$. Hence, if b_l is equally likely (i.e. $\Pr(b_l = +1) = \Pr(b_l = -1) = 1/2$), the probability of error or bit error rate (BER) becomes

$$P_b = Q\left(\sqrt{\frac{2|h_0|^2}{N_0}}\right). \tag{3.6}$$

Letting $\text{SNR} = \|\mathbf{h}\|^2/N_0 = |h_0|^2/N_0$, the BER versus SNR curve is shown in Fig. 3.2.

3.1.3 MLSD using the Viterbi algorithm

As mentioned earlier, it is possible to extend the ML detection approach for the sequence detection. For the ML-based sequence detection, we need to derive the likelihood function of an entire symbol sequence, $\{b_m\}$, given a received signal sequence $\{y_m\}$. To this end, from Eq. (3.1), consider the conditional pdf of y_l given symbol sequence $\{b_m\}$:

$$f(y_l|\{b_m\}) = f(y_l|b_l, b_{l-1}, \ldots, b_{l-P+1})$$
$$= \frac{1}{\sqrt{\pi N_0}} \exp\left(-\frac{1}{N_0}\left|y_l - \sum_{p=0}^{P-1} h_p b_{l-p}\right|^2\right), \tag{3.7}$$

in which we can see that y_l dependent on $\{b_l, b_{l-1}, \ldots, b_{l-P+1}\}$, not an entire data symbol sequence, $\{b_m\}$. This is an important observation to make so that one can avoid an exhaustive search for the sequence detection.

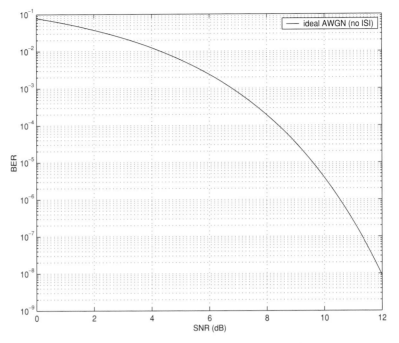

Figure 3.2. Bit error rate versus SNR for the AWGN channel without ISI.

Given $\{b_m\}$, the y_l's are independent of each other. Thus, the conditional pdf of $\{y_m\}$ is given by

$$f(\{y_m\}|\{b_m\}) = \prod_{l=0}^{L-1} f(y_l|b_l, b_{l-1}, \ldots, b_{l-P+1})$$

$$= C \exp\left(-\frac{1}{N_0} \sum_{l=0}^{L-1} \left| y_l - \sum_{p=0}^{P-1} h_p b_{l-p} \right|^2 \right), \qquad (3.8)$$

where C is a normalizing constant. Given $\{y_m\}$, $f(\{y_m\}|\{b_m\})$ becomes the likelihood function of the symbol sequence, $\{b_m\}$. From Eq. (3.8), the maximization of the likelihood function is equivalent to the minimization of the following cost function of $\{b_m\}$:

$$V(\{b_m\}) = \sum_{l=0}^{L-1} \left| y_l - \sum_{p=0}^{P-1} h_p b_{l-p} \right|^2. \qquad (3.9)$$

The optimal symbol sequence under the ML criterion, called the ML sequence, can be obtained as follows:

$$\{b_m\}_{\text{mlsd}} = \arg\max_{\{b_m\}} f(\{y_m\}|\{b_m\})$$

$$= \arg\min_{\{b_m\}} V(\{b_m\}). \qquad (3.10)$$

Table 3.1. *The costs and corresponding sequences*

Symbol sequences, $\{b_0, b_1, \ldots, b_3\}$	Costs
$\{+1, +1, +1, +1\}$	15.75
$\{-1, +1, +1, +1\}$	27.75
$\{+1, -1, +1, +1\}$	7.75
$\{-1, -1, +1, +1\}$	15.75
$\{+1, +1, -1, +1\}$	33.75
$\{-1, +1, -1, +1\}$	45.75
$\{+1, -1, -1, +1\}$	21.75
$\{-1, -1, -1, +1\}$	29.75
$\{+1, +1, +1, -1\}$	21.75
$\{-1, +1, +1, -1\}$	33.75
$\{+1, -1, +1, -1\}$	13.75
$\{-1, -1, +1, -1\}$	21.75
$\{+1, +1, -1, -1\}$	35.75
$\{-1, +1, -1, -1\}$	47.75
$\{+1, -1, -1, -1\}$	23.75
$\{-1, -1, -1, -1\}$	31.75

This is the MLSD. Since the number of the candidate symbol sequences becomes 2^L for BPSK signaling (i.e. $b_l \in \{-1, +1\}$), an exhaustive search to find the ML sequence of the smallest cost requires the complexity to grow exponentially with the length of symbol sequence, L.

Example 3.1.1   It is straightforward to derive Eq. (3.7). The received signal y_l in Eq. (3.1) is a random variable. When $\{b_l, b_{l-1}, \ldots, b_{l-P+1}\}$ are given, y_l is a Gaussian random variable with mean and variance as follows:

$$\mu = E[y_l] = \sum_{p=0}^{P-1} h_p b_{l-p};$$

$$\sigma^2 = Var(y_l) = Var(n_l) = \frac{N_0}{2}.$$

It immediately follows that

$$f(y_l | b_l, b_{l-1}, \ldots, b_{l-P+1}) = \frac{1}{\sqrt{2\pi\sigma^2}} \exp\left(-\frac{1}{2\sigma^2}|y_l - \mu|^2\right)$$

$$= \frac{1}{\sqrt{\pi N_0}} \exp\left(-\frac{1}{N_0}\left|y_l - \sum_{p=0}^{P-1} h_p b_{l-p}\right|^2\right).$$

Example 3.1.2   Let $\{y_0, y_1, \ldots, y_3\} = \{2, -1, 4, 1\}$. Suppose that the CIR is $\{h_0, h_1\} = \{1, -0.5\}$, where $P = 2$, and $b_{-1} = 0$. The ML sequence can be found through an exhaustive search. Table 3.1 shows the costs for each possible sequence for $\{b_0, b_1, \ldots, b_3\}$ (there are $2^4 = 16$ possible symbol sequences). Clearly, the ML sequence of the smallest cost is given by $\{b_0, b_1, b_2, b_3\} = \{+1, -1, +1, +1\}$.

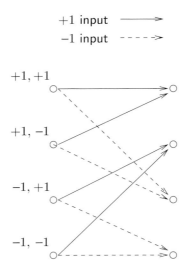

+1 input   ⟶

−1 input   - - - -➤

+1, +1

+1, −1

−1, +1

−1, −1

Figure 3.3. Trellis diagram for an ISI channel when $P = 3$.

As shown in Example 3.1.2, the complexity of the MLSD is prohibitively high when L is large if an exhaustive search is used to find the optimal symbol sequence. However, there exist some computationally efficient algorithms, including the Viterbi algorithm (VA). The VA uses the fact that y_l is dependent on $\{b_l, b_{l-1}, \ldots, b_{l-P+1}\}$, not an entire data symbol sequence.

Ignoring the noise in Eq. (3.1), the output of the channel can be broken down into two terms as follows:

$$y_l = \sum_{p=0}^{P-1} h_p b_{l-p}$$

$$= h_0 b_l + \sum_{p=1}^{P-1} h_p b_{l-p},$$

where the first term, $h_0 b_l$, represents the component decided by the current data symbol and the second term, $\sum_{p=1}^{P-1} h_p b_{l-p}$, represents the component decided by the channel state (or the memory of the channel), which depends on the last $(P - 1)$ data symbols $\{b_{l-1}, \ldots, b_{l-P+1}\}$. Assuming that $b_l \in \{-1, +1\}$ (i.e. BPSK signaling), there are 2^{P-1} different channel states. Denote the channel state at time l by $\mathcal{S}_l = [b_{l-1}, b_{l-2}, \ldots, b_{l-P+1}]$. In addition, define the set of states as $\bar{\mathbf{S}} = \{s_1, s_2, \ldots, s_{2^{P-1}}\}$, where the s_k's represent the 2^{P-1} channel states. Clearly, we have $\mathcal{S}_l \in \bar{\mathbf{S}}$. A channel state, say s_k, becomes a binary vector whose elements are 1 or -1. For example, if $P = 3$, $\bar{\mathbf{S}} = \{s_1, s_2, s_3, s_4\} = \{[1, 1], [1, -1], [-1, 1], [-1, -1]\}$.

Since the channel state depends on the input data symbol, the state transition occurs for each new input and it can be represented along the trellis diagram. For example, Fig. 3.3 shows a trellis diagram when $P = 3$. The next state, $\mathcal{S}_{l+1}$, becomes $[b_l, b_{l-1}] = [1, 1]$ if the current input data symbol b_l is 1 and the current channel state $S_l = [b_{l-1}\ b_{l-2}]$ is $[1, 1]$ or $[1, -1]$.

We make a few observations as follows:

- the number of states is 2^{P-1};
- there are two branches leaving and entering each state (as there are two possible values of a data symbol, $+1$ and -1);
- any data sequence can be shown as a path along the trellis diagram.

The VA finds the ML optimal sequence along the trellis diagram for the entire data sequence. For convenience, define

$$V_l = \sum_{m=0}^{l} \left| y_m - \sum_{p=0}^{P-1} h_p b_{m-p} \right|^2$$
$$= V_{l-1} + M_l, \qquad l = 0, 1, \ldots, L - 1, \tag{3.11}$$

where

$$M_l = \left| y_l - \sum_{p=0}^{P-1} h_p b_{l-p} \right|^2. \tag{3.12}$$

Clearly, V_{L-1} is the total cost from Eq. (3.9) and V_l is the accumulated cost up to time l. As shown in Eq. (3.11), the accumulated cost at time l can be broken down into two terms: the first term (V_{l-1}) is the accumulated cost at time $l - 1$ and the second term (M_l) is the branch metric which can have a different value for each branch from time $l - 1$ to time l. The branch metric can be seen as a function of the current state and the current input given y_l:

$$M_l = \mathcal{M}(b_l, S_l; y_l)$$
$$= \left| y_l - \sum_{p=0}^{P-1} h_p b_{l-p} \right|^2$$
$$= |y_l - h_0 b_l - g(S_l)|^2, \tag{3.13}$$

where $g(S_l) = \sum_{p=1}^{P-1} h_p b_{l-p}$. The main idea of the VA is to determine the survival path and the competing path for each state at each time.

Consider a state, say s_k, at time l. There are two paths entering s_k and each path may have a different cost at time l. We assume that the cost of Path 1 is smaller than that of Path 2 and we denote by $V_l(\text{Path 1})$ and $V_l(\text{Path 2})$ the costs of Path 1 and Path 2 at time l, respectively. As the two paths enter the same state, they must have the same path, which minimizes the cost from time l to the final time $L - 1$. If we denote by V_l^{L-1} the cost associated with the optimal path (i.e. the path of the smallest cost), which starts from state s_k at time l and terminates at time $L - 1$, the total costs of Path 1 and Path 2 to the end are as follows:

$$V_{L-1}(\text{Path 1}) = V_l(\text{Path 1}) + V_l^{L-1},$$
$$V_{L-1}(\text{Path 2}) = V_l(\text{Path 2}) + V_l^{L-1}. \tag{3.14}$$

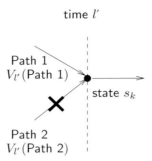

time l'

Path 1
$V_{l'}$ (Path 1)

state s_k

Path 2
$V_{l'}$ (Path 2)

Figure 3.4. Two paths enter state s_k. The path that has the larger cost at time l will be discarded. Given that V_l(Path 1) $<$ V_l(Path 2), Path 1 is the survival path and Path 2 will be discarded.

As V_l(Path 1) $<$ V_l(Path 2), we have V_{L-1}(Path 1) $<$ V_{L-1}(Path 2). Consequently, Path 1 becomes the survival path and Path 2 becomes the competing path, which can be discarded at state s_k at time l.

Any data sequences from competing paths can be discarded because we can always find the data sequences that have lower costs from survival paths, and the ML sequence should be associated with a survival path. In Fig. 3.4, we show the two paths entering state s_k at time l. Path 2 is discarded as it is the competing path, while Path 1 is kept as it is the survival path.

This observation can save on computational complexity since all the paths do not need to be taken into account. Only the survival paths are kept along the trellis. At the final time, $L-1$, the optimal path which has the smallest cost can be chosen among 2^{P-1} survival paths (as there are 2^{P-1} states and each state has one survival path). This optimal path is the ML data sequence which minimizes the cost function in Eq. (3.9).

An example of the VA is given in Fig. 3.5. With the branch metrics in Fig. 3.5(a), the VA can find the survival paths for each state at each time. The results are shown in Fig. 3.5(b). Consider the two paths entering the first state at time $l = 2$. The cost of the upper path (corresponding to the input $+1, +1, +1$) is $2 + 4 + 4 = 10$, while the cost of the lower path (corresponding to the input $-1, +1, +1$) is $1 + 1 + 2 = 4$. Hence, the lower path is chosen as the survival path and the upper path, the competing path, is discarded.

Example 3.1.3   (Example 3.1.2 revisited) The VA can be used to find the ML sequence along the trellis diagram shown in Fig. 3.6.

Let $V_l(+1)$ and $V_l(-1)$ denote the costs of survival paths of states $+1$ and -1 at time l, respectively. At time $l = 0$, the costs are given by

$$V_0(+1) = |y_0 - h_0(+1) - h_1(0)|^2 = |2 - 1|^2 = 1$$
$$V_0(-1) = |y_0 - h_0(-1) - h_1(0)|^2 = |2 + 1|^2 = 9.$$

At time $l = 1$, the two paths enter state $+1$. The path that leaves state $+1$ and enters state $+1$ at time $l = 1$ has the following cost:

$$V_0(+1) + |y_1 - h_0(+1) - h_1(+1)|^2 = 1 + |-1.5|^2 = 3.25.$$

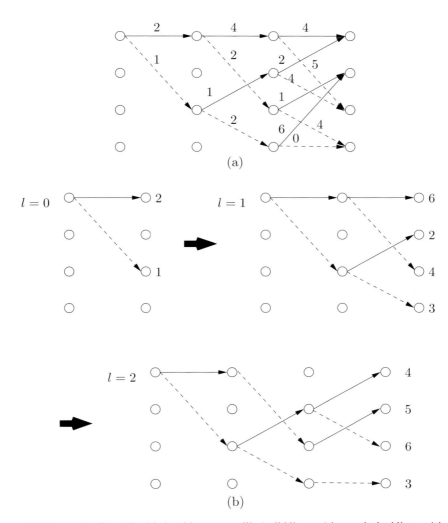

Figure 3.5. Example of the Viterbi algorithm on a trellis (solid line: $+1$ input, dashed line: -1 input). (a) Trellis with branch metrics; (b) survival paths with accumulated costs.

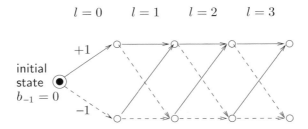

Figure 3.6. Trellis diagram with $P = 2$.

Table 3.2. *Costs and survival paths (the values in the brackets denote the previous state of the survival path)*

Cost of state	Time $l = 0$	Time $l = 1$	Time $l = 2$	Time $l = 3$
$V_l(+1)$	1	3.25 [+1]	7.5 [−1]	7.75 [+1]
$V_l(−1)$	9	1.25 [+1]	21.5 [−1]	13.75 [+1]

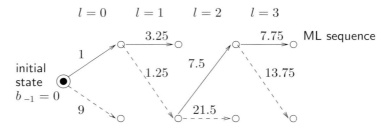

Figure 3.7. ML sequence on the trellis diagram.

The cost of the other path leaving state -1 at time $l = 0$ and entering state $+1$ at time $l = 1$ is given by

$$V_0(-1) + |y_1 - h_0(+1) - h_1(-1)|^2 = 9 + |-2.5|^2 = 15.25.$$

Thus, the survival path of state $+1$ at $l = 1$ is the path leaving state $+1$ at time $l = 0$ and entering state $+1$ at time $l = 1$, and the cost becomes $V_1(+1) = 3.25$. We can repeat this procedure to find the survival path of state -1 at $l = 1$. The results are shown in Table 3.2.

The survival path of state $+1$ at the final time $l = 3$ becomes the ML path. Hence, the ML data sequence can be found by backtracing. Taking the values in the brackets for backtracing in Table 3.2, we have

$$b_0 = +1, \ b_1 = -1, \ b_2 = +1, \ b_3 = +1.$$

This sequence is the same as that obtained by an exhaustive search in Table 3.1. Figure 3.7 shows the ML sequence on the trellis diagram.

Note that there are differences between the MLSD and equalization approaches, e.g. LE or DEF, as follows.

- The MLSD outperforms the equalization approach as the information of a received signal sequence is fully utilized. In the equalization, the symbol-by-symbol estimation is employed. Hence, the ISI is suppressed even though it contains information about adjacent symbols. On the other hand, the MLSD estimates a sequence rather than symbols by fully exploiting the information of a received signal sequence (without any suppression or elimination of information). However, this results in a long decision delay as the decision can be made after a whole sequence is received.
- The complexity of the LE or DFE is linear (in terms of the length of CIR, P), while the complexity of the MLSD grows exponentially with the length of CIR, P. Hence,

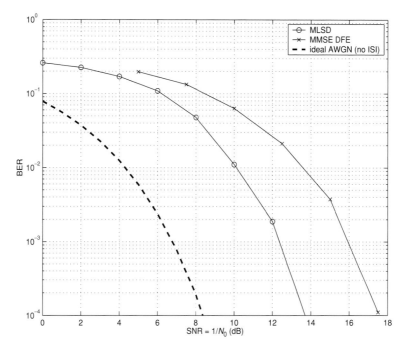

Figure 3.8. Bit error rate results of the MLSD and DFE with the ideal BER that can be obtained without ISI.

although the MLSD can outperform, the LE or DFE can still be the preferred option because of its low complexity in practice.

For an ISI channel with the following CIR:

$$\mathbf{h} = [h_0 \ h_1 \ h_2 \ h_3 \ h_4]^{\mathrm{T}} = [0.227 \ 0.460 \ 0.688 \ 0.460 \ 0.227]^{\mathrm{T}},$$

the BER results are obtained by simulations and shown in Fig. 3.8. The MLSD outperforms the MMSE DFE. At a BER of 10^{-4}, there is about 4 dB SNR gain. Note that the SNR is defined as follows:

$$\mathrm{SNR} = \frac{\|\mathbf{h}\|^2}{N_0} = \frac{1}{N_0}.$$

3.1.4 Performance analysis

In the MLSD, a sequence is decided (or detected) rather than a symbol. Hence, the performance analysis can be carried out with sequences. Let $\{b_m\}$ be the transmitted sequence and let $\{\bar{b}_m\}$ be the ML sequence. We assume that the ML sequence is not the transmitted sequence. Thus, an erroneous decision on a sequence is made. According to the ML criterion, we have the following inequality for an error event:

$$\sum_l \left| y_l - \sum_p h_p b_{l-p} \right|^2 > \sum_l \left| y_l - \sum_p h_p \bar{b}_{l-p} \right|^2.$$

The probability of this particular error event is given by

$$\Pr(\{\bar{b}_m\}|\{b_m\}) = \Pr\left(\sum_l \left|y_l - \sum_p h_p b_{l-p}\right|^2 > \sum_l \left|y_l - \sum_p h_p \bar{b}_{l-p}\right|^2\right)$$

$$= \Pr\left(\sum_l |n_l|^2 > \sum_l \left|\sum_p h_p e_{l-p} + n_l\right|^2\right)$$

$$= \Pr\left(-2\sum_l n_l u_l > \sum_l |u_l|^2\right), \tag{3.15}$$

where $e_l = b_l - \bar{b}_l$ and $u_l = \sum_p h_p e_{l-p}$. Given $\{u_m\}$, the quantity $U = -2\sum_l n_l u_l$ is a Gaussian random variable with the following mean and variance:

$$E[U] = 0,$$
$$E[U^2] = 2N_0 \sum_l |u_l|^2.$$

Using the Q-function, Eq. (3.15) is rewritten as follows:

$$\Pr(\{\bar{b}_m\}|\{b_m\}) = \Pr\left(U > \sum_l |u_l|^2\right)$$

$$= Q\left(\sqrt{\frac{\sum_l |u_l|^2}{2N_0}}\right). \tag{3.16}$$

For a particular transmitted sequence, there are a number of possible ML sequences, including the original transmitted sequence. For convenience, denote by $\mathcal{B}_{ml}(\{b_m\})$ the set of all possible ML sequences except the original transmitted sequence $\{b_m\}$. Thus, the probability of sequence error can be written as follows:

$$P_{err}(\{b_m\}) = \sum_{\{\bar{b}_m\}\in\mathcal{B}_{ml}(\{b_m\})} \Pr(\{\bar{b}_m\}|\{b_m\}). \tag{3.17}$$

Since $Q(x)$ decreases quickly as x increases, it can be shown that

$$Q(x) + Q(y) \simeq Q(x) \text{ for } 0 \ll x < y.$$

Thus, as the noise spectral density decreases or the SNR increases, the error probability can be dominated by the terms which have the minimum of $\sum_l |u_l|^2$. Let $d_{min}^2 = \min \sum_l |u_l|^2$. Then, an approximation is given by

$$P_{err}(\{b_m\}) \simeq N_{min} Q\left(\sqrt{\frac{d_{min}^2}{2N_0}}\right), \tag{3.18}$$

where N_{min} is the number of the ML sequences that have the minimum squared distance d_{min}^2. The performance of the MLSD was addressed first in Forney (1972) with the definition of the minimum squared distance d_{min}^2. In general, it is not easy to find d_{min}^2 by analytical methods. However, it is possible to carry out a numerical search to find d_{min}^2 for a given CIR.

3.2 Channel estimation

To perform the MLSD, the CIR is required. There are a number of ways of estimating the CIR. In this section, we consider several adaptive techniques used to estimate the CIR by using a training sequence.

3.2.1 MMSE approach for channel estimation

Consider the channel estimation problem as a system identification problem in which we need to estimate or identify the system by using the sets of input and output signals or their statistical properties. It is necessary to have (1) a system model and (2) a performance criterion. The system model becomes the ISI channel as follows:

$$y_l = \sum_{p=0}^{P-1} h_p b_{l-p} + n_l, \tag{3.19}$$

where the CIR $\{h_p\}$ is to be estimated. For the performance criterion, we can choose the MMSE. Then, the MMSE cost function is given by

$$\{\hat{h}_{\mathrm{mmse},p}\} = \arg \min_{\{h_p\}} E\left[\left|y_l - \sum_{p=0}^{P-1} h_p b_{l-p}\right|^2\right], \tag{3.20}$$

where the expectation is carried out with respect to $\{y_m\}$ and $\{b_m\}$. It is instructive to find the MMSE solution of the CIR. Let

$$\mathbf{h} = [h_0 \ h_1 \ \cdots \ h_{P-1}]^{\mathrm{T}},$$
$$\mathbf{b}_l = [b_l \ b_{l-1} \ \cdots \ b_{l-P+1}]^{\mathrm{T}}.$$

Then, the MSE is given by

$$\mathrm{MSE} = E\left[\left|y_l - \mathbf{b}_l^{\mathrm{T}}\mathbf{h}\right|^2\right].$$

Using the orthogonality principle, we obtain the following equation:

$$E[\mathbf{b}_l y_l] = \mathbf{R_b}\hat{\mathbf{h}}_{\mathrm{mmse}}, \tag{3.21}$$

where $\hat{\mathbf{h}}_{\mathrm{mmse}} = [\hat{h}_{\mathrm{mmse},0} \ \hat{h}_{\mathrm{mmse},1} \ \cdots \ \hat{h}_{\mathrm{mmse},P-1}]^{\mathrm{T}}$ and $\mathbf{R_b} = E[\mathbf{b}_l \mathbf{b}_l^{\mathrm{T}}]$. If $\mathbf{R_b}^{-1}$ exists, the MMSE solution becomes

$$\hat{\mathbf{h}}_{\mathrm{mmse}} = \mathbf{R_b}^{-1} E[\mathbf{b}_l y_l].$$

Since $\mathbf{b}_l$ and n_l are uncorrelated, we find

$$E[\mathbf{b}_l y_l] = E\left[\mathbf{b}_l\left(\mathbf{b}_l^{\mathrm{T}}\mathbf{h} + n_l\right)\right] = \mathbf{R_b}\mathbf{h}.$$

From Eq. (3.21), it follows that

$$\mathbf{R_b}\hat{\mathbf{h}}_{\mathrm{mmse}} = \mathbf{R_b}\mathbf{h}.$$

Hence, if $\mathbf{R_b}$ is full-rank, the MMSE solution of $\mathbf{h}$ is exactly $\mathbf{h}$. This implies that the MMSE solution is *unbiased*. In addition, it is easy to show that the MMSE is given by

$$\mathrm{MMSE} = E\left[\left|y_l - \mathbf{b}_l^{\mathrm{T}}\mathbf{h}\right|^2\right] = N_0/2. \tag{3.22}$$

3.2.2   Least mean square algorithm

When the input and output sequences, $\{b_m\}$ and $\{y_m\}$, respectively, are available (rather than their second-order statistics), adaptive techniques such as the LMS and RLS algorithms can be used to find the MMSE solution.

For example, the LMS algorithm for the channel estimation is given by

$$\mathbf{h}^{(l)} = \mathbf{h}^{(l-1)} + 2\mu\mathbf{b}_l\left(y_l - \mathbf{b}_l^{\mathsf{T}}\mathbf{h}^{(l-1)}\right), \tag{3.23}$$

where $\mathbf{h}^{(l)}$ denotes the estimated CIR vector at time l. There are some differences between the LMS algorithms for the channel estimation and the equalization.

- The desired signals are different: it is y_l in the channel estimation, b_l in the equalization.
- The data vectors are different: it is $\mathbf{b}_l$ in the channel estimation, $\mathbf{y}_l$ in the equalization.
- The weight vector in the equalization is the equalization vector. On the other hand, in the channel estimation, the weight vector itself is the channel estimate.

Adaptive techniques for the equalization and channel estimation do not differ in terms of implementation. However, there are some different aspects in terms of the performance. Let us focus on the LMS algorithm. From Eq. (3.23), the eigenspread of the covariance matrix $\mathbf{R_b} = E[\mathbf{b}_l\mathbf{b}_l^{\mathsf{T}}]$ plays a key role in determining the rate of convergence. If data symbols are independent and equally likely, we have $\mathbf{R_b} = \mathbf{I}$ and the eigenspread becomes 1. This means that the rate of convergence can be fast since a large value of μ can be chosen. However, to achieve a smaller steady state error, it is desirable to have a smaller μ.

To find the channel estimation error in Eq. (3.23), some results in Chapter 2 need to be used. From Eq. (2.91), we have

$$\mathbf{\Gamma} = \lim_{l\to\infty} E\left[\tilde{\mathbf{h}}^{(l)}\left(\tilde{\mathbf{h}}^{(l)}\right)^{\mathsf{T}}\right]$$
$$= \mu\left(\sigma_\epsilon^2 + \sigma_{\text{ex}}^2\right)\mathbf{I},$$

where $\tilde{\mathbf{h}}^{(l)} = \mathbf{h} - \mathbf{h}^{(l)}$ is the channel estimation error, σ_ϵ^2 is the actual MSE, $E[|y_l - \mathbf{b}_l^{\mathsf{T}}\mathbf{h}|^2] = N_0/2$, and σ_{ex}^2 is the excess MSE. Since $\mathbf{\Lambda} = \mathbf{I}$ (as $\mathbf{R_b} = \mathbf{I}$), from Eq. (2.92), it follows that

$$\sigma_{\text{ex}}^2 = \frac{\mu P}{1 - \mu P}\frac{N_0}{2},$$
$$\mathbf{\Gamma} = \frac{\mu}{(1 - \mu P)}\frac{N_0}{2}\mathbf{I}.$$

Since $\lim_{l\to\infty} E[\|\tilde{\mathbf{h}}^{(l)}\|^2] = \text{Tr}(\mathbf{\Gamma})$, the steady state MSE can be found as follows:

$$\lim_{l\to\infty} E\left[\|\tilde{\mathbf{h}}^{(l)}\|^2\right] = \frac{\mu P}{1 - \mu P}\frac{N_0}{2}. \tag{3.24}$$

We find that the steady state MSE of the channel estimate obtained from the LMS algorithm is independent of $\mathbf{h}$. It can also be found that $\mu < 1/P$ for the condition of the second-order convergence.

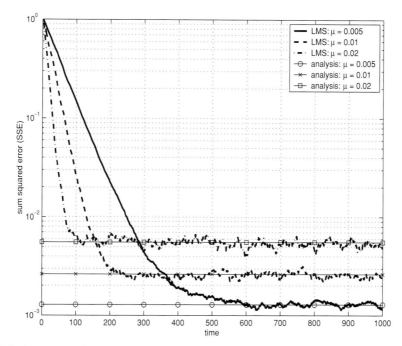

Figure 3.9. Sum squared error curves of the LMS algorithm with different values of μ. (Note that "analysis" means the steady state MSE from Eq. (3.24).)

Figure 3.9 shows simulation results of the LMS algorithm for the channel estimation with the following CIR:

$$\mathbf{h} = [0.227 \quad 0.460 \quad 0.688 \quad 0.460 \quad 0.227]^{\mathrm{T}}$$

and SNR ($= \|\mathbf{h}\|^2/N_0$) = 10 dB. The sum squared error (SSE), $\|\mathbf{h}^{(l)} - \mathbf{h}\|^2$, is taken into account to show the difference between the estimated CIR and the true CIR. Three different values of μ are considered: $\mu = 0.005$, 0.01, and 0.02. We obtain an averaged SSE from 50 realizations for each case in Fig. 3.9. The theoretical steady state MSEs from Eq. (3.24) are also shown for different values of μ. We can find a well known tradeoff between the performance (i.e. the steady state MSE) and the rate of convergence. For a lower steady state MSE, a smaller μ is desirable. However, this results in a slower rate of convergence.

Example 3.2.1 In this example, we will find the steady state MSE in Fig. 3.9. The length of CIR, P, is 5 and the SNR, $\|\mathbf{h}\|^2/N_0$, is 10 dB. Since $\|\mathbf{h}\|^2 = 1$, $N_0 = 10^{-1}$. For $\mu = 0.005$, the steady state MSE is given by

$$\begin{aligned}
\mathrm{MSE} &= \frac{\mu P}{1 - \mu P} \frac{N_0}{2} \\
&= \frac{0.025}{1 - 0.025} \times \frac{1}{20} \\
&= 0.0013.
\end{aligned}$$

For $\mu = 0.01$ and 0.02, we find the MSEs to be 0.0026 and 0.0056, respectively.

3.2.3   Least squares approach and its relation to the ML estimation

When both input and output sequences, $\{b_m\}$ and $\{y_m\}$, respectively, are available (i.e. when a training sequence is transmitted), the LS approach, which is an off-line processing, can be used to estimate the CIR as follows:

$$\{\hat{h}_{\text{ls},l}\} = \arg\min_{\{h_p\}} \sum_{l=0}^{L_{\text{train}}-1} \left| y_l - \sum_{p=0}^{P-1} h_p b_{l-p} \right|^2, \tag{3.25}$$

where L_{train} is the length of the training sequence. Since

$$\sum_{l=0}^{L_{\text{train}}-1} \left| y_l - \sum_{p=0}^{P-1} h_p b_{l-p} \right|^2 = \sum_{l=0}^{L_{\text{train}}-1} \left| y_l - \mathbf{b}_l^{\text{T}} \mathbf{h} \right|^2$$

$$= \| \mathbf{y} - \mathbf{B}^{\text{T}} \mathbf{h} \|^2, \tag{3.26}$$

the LS solution or LS estimate of CIR becomes

$$\hat{\mathbf{h}}_{\text{ls}} = (\mathbf{B}\mathbf{B}^{\text{T}})^{\dagger} \mathbf{B} \mathbf{y}. \tag{3.27}$$

where $(\ )^{\dagger}$ denotes the pseudo-inverse and $\mathbf{B}$ is a $L_{\text{train}} \times P$ Toeplitz matrix given by

$$\mathbf{B} = \begin{bmatrix} b_0 & b_{-1} & \cdots & b_{-P+1} \\ b_1 & b_0 & \cdots & b_{-P+2} \\ \vdots & \vdots & \ddots & \vdots \\ b_{L_{\text{train}}-1} & b_{L_{\text{train}}-2} & \cdots & b_{L_{\text{train}}-P} \end{bmatrix}$$

and

$$\mathbf{y} = [y_0 \ y_1 \ \cdots \ y_{L_{\text{train}}-1}]^{\text{T}}.$$

In Eq. (3.27), the pseudo-inverse becomes the inverse if the inverse exists. In fact, the LS solution is identical to the ML estimate.

The likelihood function of $\mathbf{h}$ given $\mathbf{y}$ and $\mathbf{B}$ is given by

$$f(\{y_m\}|\mathbf{B}, \mathbf{h}) = C \exp\left(-\frac{1}{N_0} \| \mathbf{y} - \mathbf{B}^{\text{T}} \mathbf{h} \|^2 \right),$$

where C is a normalizing constant. The ML estimation problem of $\mathbf{h}$ is given by

$$\hat{\mathbf{h}}_{\text{ml}} = \arg\max_{\mathbf{h}} f(\{y_m\}|\mathbf{B}, \mathbf{h})$$

$$= \arg\min_{\mathbf{h}} \| \mathbf{y} - \mathbf{B}^{\text{T}} \mathbf{h} \|^2. \tag{3.28}$$

Thus, the ML estimate coincides with the LS solution in Eq. (3.27).

We can find statistical properties of the LS or ML solution. Suppose that the inverse of $\mathbf{B}\mathbf{B}^{\text{T}}$ exists. Since $(\mathbf{B}\mathbf{B}^{\text{T}})^{-1}\mathbf{B}\mathbf{B}^{\text{T}} = \mathbf{I}$, the mean of the ML estimate of CIR becomes

$$E[\hat{\mathbf{h}}_{\text{ml}}] = E[(\mathbf{B}\mathbf{B}^{\text{T}})^{-1}\mathbf{B}\mathbf{y}]$$

$$= \mathbf{h} + E[\mathbf{e}_{\text{ml}}],$$

Table 3.3. *Values of the parameters for three different cases*

Parameters	μ	L_{train}
Case 1	0.025	40
Case 2	0.05	100
Case 3	0.025	100

where the expectation is carried out with respect to the noise and $\mathbf{e}_{\text{ml}} = (\mathbf{BB}^{\text{T}})^{-1}\mathbf{Bn}$. As $E[\mathbf{e}_{\text{ml}}] = (\mathbf{BB}^{\text{T}})^{-1}\mathbf{B}E[\mathbf{n}] = \mathbf{0}$, the ML estimate or LS solution is unbiased (i.e. $E[\hat{\mathbf{h}}_{\text{ml}}] = \mathbf{h}$). The covariance matrix of the ML estimate becomes

$$Cov(\hat{\mathbf{h}}_{\text{ml}}) = E\left[\mathbf{e}_{\text{ml}}\mathbf{e}_{\text{ml}}^{\text{T}}\right]$$
$$= \frac{N_0}{2}(\mathbf{BB}^{\text{T}})^{-1}. \tag{3.29}$$

This indicates that the performance of the LS- or ML-based approach depends on $\mathbf{B}$ or the training sequence.

The RLS algorithm can also be applied to the adaptive channel estimation with a proper forgetting factor.

3.3 MLSD with estimated channel

Once the channel estimate is obtained by either the LMS or the RLS algorithm (or using another approach), the MLSD can be carried out with the estimated channel to detect data sequences. In this section, we present simulation results and discuss a phenomenon called the error floor. We assume that the LMS algorithm is used for the channel estimation.

The performance depends on the channel estimation error. Since a smaller μ results in a smaller estimation error, provided that the convergence is achieved, a better performance is expected for a smaller μ. However, the length of training sequence or pilot sequence, L_{train}, should be long for convergence. Unfortunately, this is undesirable as the data throughput decreases.

On the other hand, if the length of a training sequence is not sufficiently long, the convergence may not be achieved, resulting in a poor performance since the estimated CIR would not be close to the true CIR. Hence, we need to optimize μ depending on the length of training sequence.

Figure 3.10 presents simulation results with the following CIR:

$$\mathbf{h} = [0.227 \ 0.460 \ 0.688 \ 0.460 \ 0.227]^{\text{T}}.$$

Assume that the length of data sequence is $L = 400$. Three cases are considered (see Table 3.3).

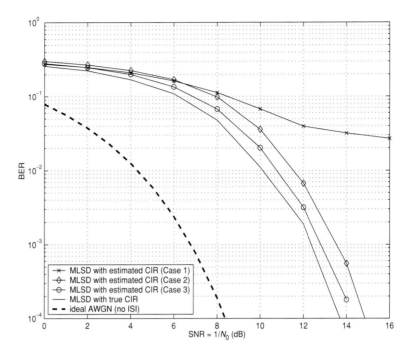

Figure 3.10. Bit error rate performance of the MLSD with the channel estimation using the LMS algorithm. Case 1: $L_{\text{train}} = 40$, $\mu = 0.025$; Case 2: $L_{\text{train}} = 100$, $\mu = 0.05$; Case 3: $L_{\text{train}} = 100$, $\mu = 0.025$.

In Case 1, the performance is not satisfactory as L_{train} is short and convergence is not achieved. Case 3 can provide a better performance than Case 2 since the value of μ is smaller while L_{train} is sufficiently long to achieve convergence in both cases.

It is important to observe an error floor in Fig. 3.10. If the channel estimation error is not small, the BER does not approach zero even though the SNR, $\|\mathbf{h}\|^2 / N_0$, goes to infinity. The BER curve for Case 3 in Fig. 3.10 has an error floor at a BER of 0.02. Since the error floor is an abnormal behavior, it is crucial to lower the error floor. In general, the receiver or MLSD needs to operate before the error floor occurs.

Error floor analysis is not simple in general. Thus, we consider a simple case. An error floor can be observed even if there is no ISI. Suppose that $P = 1$. The received signal is given by

$$y_l = h_0 b_l + n_l.$$

The channel estimate $\hat{h}_0$ is now used to detect signals. The ML decision rule becomes

$$b_l = \begin{cases} +1, & \text{if } 4\hat{h}_0 y_l / N_0 \geq 0; \\ -1, & \text{if } 4\hat{h}_0 y_l / N_0 < 0 \end{cases}$$

or

$$b_l = \begin{cases} +1, & \text{if } \hat{h}_0 y_l \geq 0; \\ -1, & \text{if } \hat{h}_0 y_l < 0. \end{cases}$$

Since $\hat{h}_0$ is different from h_0, the BER expression in Eq. (3.5) is no longer valid. There are two difference cases:

(i) $\text{Sign}(\hat{h}_0) = \text{Sign}(h_0)$;
(ii) $\text{Sign}(\hat{h}_0) \neq \text{Sign}(h_0)$.

Consider the first case. As the sign of $\hat{h}_0$ is the same as that of h_0, the ML decision rule is not changed although the channel estimate $\hat{h}_0$ is used. Hence, the same BER can be achieved. For the second case, where the sign of $\hat{h}_0$ differs from that of h_0, the BER can be written as follows:

$$P(\text{error} \mid \text{Sign}(\hat{h}_0) \neq \text{Sign}(h_0)) = \Pr(-n_l \hat{h}_0 > -1 | h_0 \hat{h}_0 |)$$
$$= 1 - Q \left(\sqrt{\frac{2|h_0|^2}{N_0}} \right).$$

Then, the average BER expression can be written as follows:

$$P_{\text{err}} = P(\text{error} \mid \text{Sign}(\hat{h}_0) = \text{Sign}(h_0)) \Pr(\text{Sign}(\hat{h}_0) = \text{Sign}(h_0))$$
$$+ P(\text{error} \mid \text{Sign}(\hat{h}_0) \neq \text{Sign}(h_0)) \Pr(\text{Sign}(\hat{h}_0) \neq \text{Sign}(h_0))$$
$$= Q \left(\sqrt{\frac{2|h_0|^2}{N_0}} \right) \Pr(\text{Sign}(\hat{h}_0) = \text{Sign}(h_0))$$
$$+ \left(1 - Q \left(\sqrt{\frac{2|h_0|^2}{N_0}} \right) \right) \Pr(\text{Sign}(\hat{h}_0) \neq \text{Sign}(h_0)). \tag{3.30}$$

Suppose that $\hat{h}_0 = h_0 + \tilde{h}_0$, where $h_0 > 0$, and the channel estimation error $\tilde{h}_0$ is a Gaussian random variable with mean zero and variance σ_h^2. Then, we can show that

$$\Pr(\text{Sign}(\hat{h}_0) \neq \text{Sign}(h_0)) = \Pr(\tilde{h}_0 < -h_0)$$
$$= Q \left(\frac{h_0}{\sigma_h} \right).$$

Finally, we obtain an expression for the average BER as follows:

$$P_{\text{err}} = Q \left(\sqrt{\frac{2|h_0|^2}{N_0}} \right) \left(1 - Q \left(\frac{h_0}{\sigma_h} \right) \right) + \left(1 - Q \left(\sqrt{\frac{2|h_0|^2}{N_0}} \right) \right) Q \left(\frac{h_0}{\sigma_h} \right). \tag{3.31}$$

It is now clear that the BER does not go to zero even though N_0 approaches zero when $\sigma_h > 0$ as

$$\lim_{N_0 \to 0} P_{\text{err}} = Q \left(\frac{h_0}{\sigma_h} \right). \tag{3.32}$$

This becomes the error floor.

An example with $h_0 = 1$ is considered in Fig. 3.11. This demonstrates that error floors can be observed even if there is no ISI. We note that the error floor can be lower as the channel estimation error becomes smaller. This can also be verified by the asymptotic BER in Eq. (3.32). Consequently, it is important to minimize the channel estimation error to lower the error floor. This conclusion is also valid for the MLSD over ISI channels.

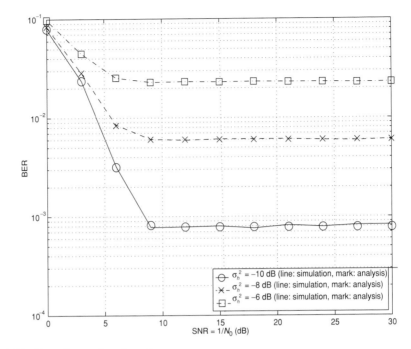

Figure 3.11. Error floors when $h_0 = 1$.

3.4   Per-survivor processing with adaptive channel estimation

In general, the performance degradation becomes negligible as the channel estimation error becomes smaller. For a smaller estimation error, a longer length of pilot symbol sequence is desirable. However, it results in the decrease of data throughput as more pilot symbols are transmitted. To avoid this problem, the use of data symbols to estimate the CIR can be considered. Blind approaches exploiting statistical and/or deterministic properties of unknown data symbols can be used to estimate CIR (Ding and Li, 2001). We can also consider semi-blind approaches that use known symbols (i.e. pilot symbols) simultaneously to estimate the CIR. Furthermore, blind or semi-blind channel estimation methods can be incorporated into the data detection. Per-survivor processing (PSP) (Raheli, Polydoros, and Tzou, 1995) is one example. Within the VA for the MLSD, the channel estimation is carried out for each survival path in the PSP. In this section, we focus on the PSP with adaptive channel estimation.

3.4.1   PSP with LMS-based channel estimation

PSP can be considered as an extended VA with the adaptive channel estimation. For the adaptive channel estimation, we assume that the LMS algorithm is used. However, the RLS algorithm is also applicable.

In PSP, each survival path has its corresponding channel estimate. To state s_k at time l ($\mathcal{S}_{l+1} = s_k$), there are two entering paths: one will be the survival path and the other will

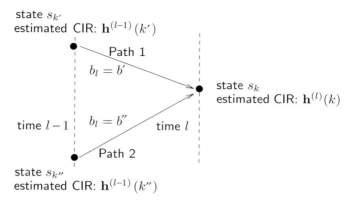

Figure 3.12. Two paths enter state k at time l. In the PSP, each state at each time has an associated estimated CIR.

be the competing path. For convenience, assume that Path 1 consists of the survival path of state $s_{k'}$ ($\mathcal{S}_l = s_{k'}$) and current symbol $b_l = b'$. In addition, assume that the second path, Path 2, consists of the survival path of state $s_{k''}$ ($\mathcal{S}_l = s_{k''}$) and current symbol $b_l = b''$. Figure 3.12 shows Path 1 and Path 2 entering state s_k at time l.

Denote by $\mathbf{h}^{(l-1)}(k')$ and $\mathbf{h}^{(l-1)}(k'')$ the CIR estimates of states $s_{k'}$ and $s_{k''}$ at time $l-1$, respectively. Then, the branch metrics for each path are given by

$$\mathcal{M}_l(k' \to k) = \left| y_l - \left(\mathbf{h}^{(l-1)}(k')\right)^{\mathrm{T}}\mathbf{b}_1 \right|^2 \text{ for Path 1,}$$
$$\mathcal{M}_l(k'' \to k) = \left| y_l - \left(\mathbf{h}^{(l-1)}(k'')\right)^{\mathrm{T}}\mathbf{b}_2 \right|^2 \text{ for Path 2,} \tag{3.33}$$

where

$$\mathbf{b}_1 = \begin{bmatrix} b' & b^l_{(l-1)}(k') & b^l_{(l-2)}(k') & \cdots & b^l_{(l-P+1)}(k') \end{bmatrix}^{\mathrm{T}},$$
$$\mathbf{b}_2 = \begin{bmatrix} b'' & b^l_{(l-1)}(k'') & b^l_{(l-2)}(k'') & \cdots & b^l_{(l-P+1)}(k'') \end{bmatrix}^{\mathrm{T}}.$$

Here, $b^l_{(l-m)}(k)$ denotes the mth symbol of the survival path of state s_k at time l. Since the last $(P-1)$ elements of $\mathbf{b}_1$ can be decided by state $\mathcal{S}_l = s_{k'}$, we can write

$$\mathbf{b}_1 = [b' \ \mathbf{b}^{\mathrm{T}}(s_{k'})]^{\mathrm{T}},$$

where $\mathbf{b}(s_k)$ represents the binary vector associated with state s_k. Similarly, $\mathbf{b}_2$ becomes

$$\mathbf{b}_2 = [b'' \ \mathbf{b}^{\mathrm{T}}(s_{k''})]^{\mathrm{T}}.$$

Note that the first $(P-1)$ symbols of $\mathbf{b}_1$ and $\mathbf{b}_2$ should be the same as Path 1 and Path 2 enters the same state, $\mathcal{S}_{l+1} = s_k$.

Once the branch metrics are found, the accumulated costs for each path can be obtained. The survival path of state s_k at time l becomes the path of the smaller cost. The channel estimate is updated for state s_k at time l using the LMS algorithm as follows:

$$\mathbf{h}^{(l)}(k) = \begin{cases} \mathbf{h}^{(l)}_1(k), & \text{if Path 1 is the survival path;} \\ \mathbf{h}^{(l)}_2(k), & \text{if Path 2 is the survival path,} \end{cases}$$

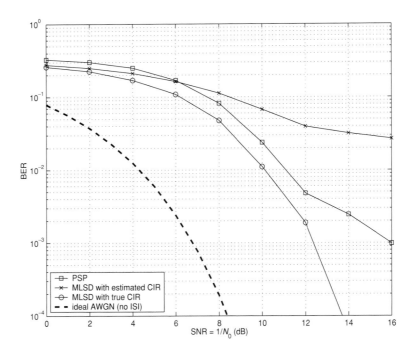

Figure 3.13. Bit error rate performance of the MLSD with estimated CIR and PSP in terms of the SNR when $\mu = 0.025$, $L = 400$, and $L_{\text{train}} = 40$.

where

$$\mathbf{h}_1^{(l)}(k) = \mathbf{h}^{(l-1)}(k') + 2\mu\mathbf{b}_1\left(y_l - (\mathbf{b}_1)^{\mathrm{T}}\,\mathbf{h}^{(l-1)}(k')\right),$$
$$\mathbf{h}_2^{(l)}(k) = \mathbf{h}^{(l-1)}(k'') + 2\mu\mathbf{b}_2\left(y_l - (\mathbf{b}_2)^{\mathrm{T}}\,\mathbf{h}^{(l-1)}(k'')\right).$$

We may want to use the posterior estimates $\mathbf{h}_1^{(l)}(k)$ and $\mathbf{h}_2^{(l)}(k)$ instead of the prior estimates $\mathbf{h}^{(l-1)}(k')$ and $\mathbf{h}^{(l-1)}(k'')$, respectively, in Eq. (3.33) as the LMS algorithm can update the CIR estimate using the current received signal y_l. However, it can be easily shown that

$$\left|y_l - \left(\mathbf{h}_1^{(l)}(k)\right)^{\mathrm{T}}\mathbf{b}_1\right|^2 = |1 - 2\mu P|^2\left|y_l - \left(\mathbf{h}^{(l-1)}(k')\right)^{\mathrm{T}}\mathbf{b}_1\right|^2$$
$$= |1 - 2\mu P|^2\mathcal{M}_l(k' \to k)$$
$$\left|y_l - \left(\mathbf{h}_2^{(l)}(k)\right)^{\mathrm{T}}\mathbf{b}_2\right|^2 = |1 - 2\mu P|^2\left|y_l - \left(\mathbf{h}^{(l-1)}(k'')\right)^{\mathrm{T}}\mathbf{b}_2\right|^2$$
$$= |1 - 2\mu P|^2\mathcal{M}_l(k'' \to k).$$

Thus, it is not necessary to use the posterior CIR estimates (i.e. updated CIR estimates).

In PSP, since each survival path has its own channel estimate, there are 2^{P-1} channel estimates at each time. This increases the complexity. However, since the data symbols are used to estimate the CIR, the channel estimation can be improved, and it results in a better performance when the pilot sequence is not long enough. Since a better channel estimate is available, the error floor can also be lower.

Figure 3.13 shows simulation results with the following CIR:

$$\mathbf{h} = [0.227\ \ 0.460\ \ 0.688\ \ 0.460\ \ 0.227]^{\mathrm{T}}.$$

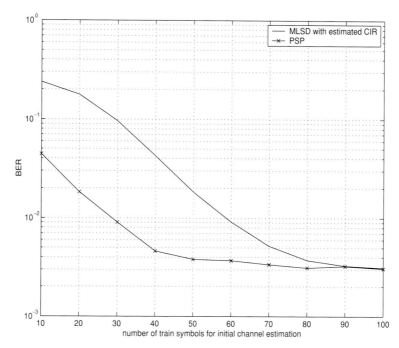

Figure 3.14. Bit error rate performance of the MLSD with estimated CIR and PSP in terms of the length of the pilot symbol sequence when $\mu = 0.025$, $L = 400$, and SNR = 12 dB.

It is assumed that the length of a training sequence is 40 and $\mu = 0.025$. This is Case 1 in Fig. 3.10 in which the LMS algorithm is not able to provide a good estimate of the CIR, because the length of training sequence is too short for it to converge. However, as shown in Fig. 3.13, PSP can provide a better performance under the same condition by utilizing data symbols for the channel estimation. In addition, we can observe that the error floor is lower.

If the training sequence is sufficiently long, the CIR can be estimated precisely and the advantage of PSP will vanish. The impact of the length of training sequence is shown in Fig. 3.14. When L_{train} is short, there is a big performance difference. However, as L_{train} increases, the difference becomes smaller.

3.4.2 MLSD for unknown CIR and PSP

Suppose that the CIR is unknown. With the ML criterion, we can formulate the following different problems for the channel estimation and/or data sequence detection.

(i) ML-1, joint ML estimation:

$$\{\hat{\mathbf{h}}_{\text{jml}}, \hat{\mathbf{b}}_{\text{jml}}\} = \arg\max_{\mathbf{h}, \mathbf{b}} f(\mathbf{y}|\mathbf{h}, \mathbf{b}),$$

where $f(\mathbf{y}|\mathbf{h}, \mathbf{b}) = \prod_{l=0}^{L-1} f(y_l|\mathbf{h}, \mathbf{b})$ is the likelihood function of $\mathbf{h}$ and $\mathbf{b}$ given $\mathbf{y}$.

(ii) ML-2, ML channel estimation:

$$\hat{\mathbf{h}}_{\text{ml}} = \arg \max_{\mathbf{h}} f(\mathbf{y}|\mathbf{h}),$$

where $f(\mathbf{y}|\mathbf{h}) = E_{\mathbf{b}}[f(\mathbf{y}|\mathbf{h}, \mathbf{b})]$ is the likelihood function of $\mathbf{h}$ given $\mathbf{y}$. Here, $E_{\mathbf{b}}[\cdot]$ stands for the expectation with respect to $\mathbf{b}$.

(iii) ML-3, MLSD for unknown CIR:

$$\hat{\mathbf{b}}_{\text{ml}} = \arg \max_{\mathbf{b}} f(\mathbf{y}|\mathbf{b}),$$

where $f(\mathbf{y}|\mathbf{b})$ is the likelihood function of $\mathbf{b}$ given $\mathbf{y}$.

The PSP is an approximation of the MLSD for unknown CIR (i.e., the ML-3 problem). This will be later discussed, after considering the other two alternative approaches.

In general, the ML-1 problem requires an exhaustive search to find the optimal solution. The problem can be rewritten as follows:

$$\{\hat{\mathbf{h}}_{\text{jml}}, \hat{\mathbf{b}}_{\text{jml}}\} = \arg \max_{\mathbf{b}} \max_{\mathbf{h}} f(\mathbf{y}|\mathbf{h}, \mathbf{b}).$$

Consider the inner problem that estimates $\mathbf{h}$ given $\mathbf{b}$ and $\mathbf{y}$:

$$\hat{\mathbf{h}}_{\text{jml}}(\mathbf{b}) = \arg \max_{\mathbf{h}} f(\mathbf{y}|\mathbf{h}, \mathbf{b}).$$

The solution is given by

$$
\begin{aligned}
\hat{\mathbf{h}}_{\text{jml}}(\mathbf{b}) &= \arg \max_{\mathbf{h}} \log f(\mathbf{y}|\mathbf{h}, \mathbf{b}) \\
&= \arg \min_{\mathbf{h}} \|\mathbf{y} - \mathbf{B}\mathbf{h}\|^2 \\
&= (\mathbf{B}^{\mathsf{T}}\mathbf{B})^{-1}\mathbf{B}^{\mathsf{T}}\mathbf{y},
\end{aligned}
\tag{3.34}
$$

where $\hat{\mathbf{h}}_{\text{jml}}(\mathbf{b})$ becomes the LS solution. Since

$$\max_{\mathbf{b}} \max_{\mathbf{h}} f(\mathbf{y}|\mathbf{h}, \mathbf{b}) = \max_{\mathbf{b}} f(\mathbf{y}|\hat{\mathbf{h}}_{\text{jml}}(\mathbf{b}), \mathbf{b}),$$

we have

$$
\begin{aligned}
\hat{\mathbf{b}}_{\text{jml}} &= \arg \max_{\mathbf{b}} f(\mathbf{y}|\hat{\mathbf{h}}_{\text{jml}}(\mathbf{b}), \mathbf{b}) \\
&= \arg \min_{\mathbf{b}} \|\mathbf{y} - \mathbf{B}\hat{\mathbf{h}}_{\text{jml}}(\mathbf{b})\|^2 \\
&= \arg \min_{\mathbf{b}} \| (\mathbf{I} - \mathbf{B}(\mathbf{B}^{\mathsf{T}}\mathbf{B})^{-1}\mathbf{B}^{\mathsf{T}}) \mathbf{y}\|^2 \\
&= \arg \min_{\mathbf{b}} \|\mathbf{P}^{\perp}(\mathbf{B})\mathbf{y}\|^2,
\end{aligned}
\tag{3.35}
$$

where $\mathbf{P}^{\perp}(\mathbf{B}) = \mathbf{I} - \mathbf{B}(\mathbf{B}^{\mathsf{T}}\mathbf{B})^{-1}\mathbf{B}^{\mathsf{T}}$ is an orthogonal projection matrix. This ML detection is also called the generalized ML (GML) detection.

Example 3.4.1   Suppose that the received signals are $\{y_0, y_1, y_2, y_3\} = \{2, -1, 4, 1\}$, where $L = 4$. It is known that $P = 2$ and $b_{-1} = 0$. The joint ML estimation is carried out with an exhaustive search (there are $2^4 = 16$ possible $\mathbf{B}$ matrices). The following two data sequences have the same smallest cost ($\min_{\mathbf{b}} \|\mathbf{P}^{\perp}(\mathbf{B})\mathbf{y}\|^2 = 4.5455$) given in Eq. (3.35):

$$\hat{\mathbf{b}}_{\mathrm{jml}} = [1 \ -1 \ 1 \ 1]^{\mathsf{T}}, \ [-1 \ 1 \ -1 \ -1]^{\mathsf{T}}.$$

The corresponding channel estimates are given by

$$\hat{\mathbf{h}}_{\mathrm{jml}} = [1.8182 \ -0.7273]^{\mathsf{T}}, \ [-1.8182 \ 0.7273]^{\mathsf{T}},$$

respectively. There exists a sign ambiguity, but if we know one of data symbols, this can be resolved.

In the ML-2 problem, we are searching for the ML channel estimate. Once the ML channel estimate is obtained, the VA can be performed for the sequence detection. We will discuss the ML-2 problem in detail in Chapter 5 using an iterative method.

The ML-3 problem is closely related to the PSP. To consider the PSP as an algorithm to solve the MLSD problem for unknown CIR (i.e. the ML-3 problem), for convenience we denote

$$\begin{aligned}
\mathbf{y}_0^l &= [y_l \quad y_{l-1} \quad \cdots \quad y_0]^{\mathsf{T}}, \\
\mathbf{b}_0^l &= [b_l \quad b_{l-1} \quad \cdots \quad b_0]^{\mathsf{T}}.
\end{aligned} \tag{3.36}$$

Using Bayes' rule, it follows that

$$\begin{aligned}
f\left(\mathbf{y}_0^{L-1} \big| \mathbf{b}_0^{L-1}\right) &\propto f\left(y_{L-1} \big| \mathbf{y}_0^{L-2}, \mathbf{b}_0^{L-1}\right) f\left(\mathbf{y}_0^{L-2} \big| \mathbf{b}_0^{L-1}\right) \\
&= f\left(y_{L-1} \big| \mathbf{y}_0^{L-2}, \mathbf{b}_0^{L-1}\right) f\left(\mathbf{y}_0^{L-2} \big| \mathbf{b}_0^{L-2}\right) \\
&= \prod_{l=0}^{L-1} f\left(y_l \big| \mathbf{y}_0^{l-1}, \mathbf{b}_0^l\right).
\end{aligned} \tag{3.37}$$

The second equality is due to the causality. Let

$$f\left(y_l \big| \mathbf{y}_0^{l-1}, \mathbf{b}_0^l\right) = f\left(y_l \big| \mathbf{y}_0^{l-1}, b_l, \mathbf{b}_0^{l-1}\right).$$

This shows that y_l depends on the previously received signals $\mathbf{y}_0^{l-1}$, the current data symbol b_l, and the past data symbols $\mathbf{b}_0^{l-1}$. Then, we can see that a state at time l can be defined by $\mathbf{b}_0^{l-1}$ and there are 2^l states. In this case, there is no trellis structure (rather we have a tree structure in which the number of states grows exponentially over time) and there is one path entering and leaving a state. Thus, it is not applicable to use the VA to reduce the complexity. For unknown CIR, according to Eq. (3.37), an exhaustive search is required to find the ML sequence. However, a forced folding (Chugg, 1998; Morley and Snyder, 1979) can be imposed to apply the VA.

The folding condition is as follows:

$$f\left(y_l \big| \mathbf{y}_0^{l-1}, \mathbf{b}_0^l\right) = f\left(y_l \big| \mathbf{y}_0^{l-1}, \mathbf{b}_{l-\bar{P}+1}^l\right), \tag{3.38}$$

where $\bar{P}$ is a positive integer. If the folding condition is satisfied, the VA can be applied with a fixed number of states ($2^{\bar{P}-1}$ states). The partial sequence $\mathbf{b}_{l-\bar{P}+1}^l$ can be characterized by a state of the last ($\bar{P} - 1$) data symbols and the current symbol b_l. If the CIR is known, the folding condition is satisfied with $\bar{P} = P$. Unfortunately, in general, the folding condition is not satisfied for unknown ISI channels. That is, there is no integer $\bar{P}$ that satisfies Eq. (3.38).

In PSP, a *forced* folding is used to apply the VA and it results in an approximation. With $\bar{P} = P$, a forced folding is considered to approximate the likelihood function as follows:

$$\log f\left(y_l \big| \mathbf{y}_0^{l-1}, \mathbf{b}_0^l\right) \simeq \log f\left(y_l \big| \mathbf{y}_0^{l-1}, \mathbf{b}_{l-P+1}^l\right). \tag{3.39}$$

Then, the information on previously received signals is converted into a channel estimate for each state for a further approximation:

$$\log f\left(y_l \big| \mathbf{y}_0^{l-1}, \mathbf{b}_{l-P+1}^l\right) \simeq \log f\left(y_l \big| \hat{\mathbf{h}}(\mathbf{y}_0^{l-1}; \mathbf{b}_{l-P+1}^{l-1}), \mathbf{b}_{l-P+1}^l\right)$$

$$\propto \left| y_l - \hat{\mathbf{h}}^{\mathrm{T}}(\mathbf{y}_0^{l-1}; \mathbf{b}_{l-P+1}^{l-1})\mathbf{b}_{l-P+1}^l \right|^2, \tag{3.40}$$

where $\hat{\mathbf{h}}(\mathbf{y}_0^{l-1}; \mathbf{b}_{l-P+1}^{l-1})$ is a CIR estimate for the state of $\mathbf{b}_{l-P+1}^{l-1}$. The CIR estimate can be obtained by using either the LMS or the RLS algorithm.

In summary, the PSP does not solve the original MLSD, i.e. the ML-3 problem. Rather, it approximates the MLSD using the forced folding in Eq. (3.39) to apply the VA in conjunction with adaptive channel estimation. We will discuss in detail the PSP in a general channel environment in Chapter 4.

3.5   Summary and notes

We studied maximum likelihood sequence detection and the Viterbi algorithm in this chapter. We showed that the MLSD is a generalization of ML detection for an ISI channel. The VA is a computationally efficient algorithm that finds the ML sequence. Channel estimation and PSP were also introduced. The PSP is an effective algorithm that can detect a data sequence when the channel is unknown.

The application of the VA to the MLSD over ISI channels was discussed first in Forney (1972). The VA was originally proposed for the decoding of convolutional codes in Viterbi (1967). It is well known that the VA is a variation of dynamic programming (Bertsekas, 1987). For the VA and its early applications to communication problems, Forney (1973) is an excellent tutorial paper; it also discusses the maximum *a posteriori* probability sequence detection over ISI channels.

It is shown in Ungerboeck (1974) that the VA is still applicable when the noise samples are correlated (this is the case when the matched filter replaces the whitening matched filter at the front end of the receiver). A procedure to find the minimum squared distance which characterizes the the performance of the MLSD was introduced in Anderson and Foschini (1975).

For a long CIR, the VA would not be applicable due to a high complexity. There are suboptimal approaches to reduce the complexity of VA; see Duel and Heegard (1989) and Eyuboglu and Qureshi (1988).

The PSP was proposed in Raheli *et al.* (1995) and analyzed in Chugg and Polydoros (1996a,b). A similar approach can also be found in Seshadri (1994).

4 Estimation and detection for fading multipath channels

In the previous chapters, we assumed that the channel is stationary or time-invariant. This would generally be true for wired communications, but it would not be true for wireless communications. In general, wireless channels, which are time-varying due to motion of the transmitter and/or receiver, suffer from fading. Even if the transmitter and receiver are not in motion, the channel can be time-varying if there are moving objects around the transmitter and receiver. Hence, wireless channels are generally fading channels.

In this chapter, we study the channel estimation and detection for fading multipath channels. We will not pay much attention to fading and techniques to mitigate fading although they are important topics in wireless communications. An in-depth discussion of fading channels can be found in Biglieri, Proakis, and Shamai (1998).

The Kalman filter will also be introduced as it plays a key role in the MLSD for random channels. It will be shown that the PSP with Kalman filtering provides a suboptimal solution to the MLSD for random channels.

4.1 Introduction to fading channel modeling

The CIR can be time-varying, in particular due to the motion of transmitter and/or receiver in wireless communications. In addition, the motion of scatterers makes the CIR time-varying.

For a time-varying dispersive channel (i.e. fading multipath channel), the received signal at time l can be modeled as follows:

$$y_l = \sum_{p=0}^{P-1} b_{l-p} h_{p,l} + n_l$$

$$= \mathbf{b}_l^T \mathbf{h}_l + n_l, \tag{4.1}$$

where

$$\mathbf{h}_l = [h_{0,l} \quad h_{1,l} \quad \cdots \quad h_{P-1,l}]^T,$$
$$\mathbf{b}_l = [b_l \quad b_{l-1} \quad \cdots \quad b_{l-P+1}]^T.$$

Here, $\mathbf{h}_l$ is the CIR and n_l is the background zero-mean white Gaussian noise at time l. In addition, b_l stands for the lth symbol. In general, $\{\mathbf{h}_m\}$ is assumed to be a random vector process. To estimate or track time-varying $\mathbf{h}_l$, statistical properties of $\mathbf{h}_l$ should be exploited. In this section, we will characterize the statistical properties of channel processes in wireless or radio communications.

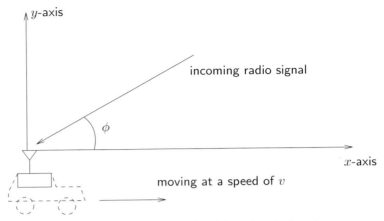

Figure 4.1. Receiver in motion with the received signal through a single path.

4.1.1 Clarke's model

In radio communications, a baseband signal is modulated by a carrier of higher frequency. Let B_W and f_c denote the bandwidth of the baseband signal and the carrier frequency, respectively. The carrier is a sinusoid such that $s_c(t) = \cos(2\pi f_c t)$, where $f_c \gg B_W$. If $x(t)$ denotes a baseband signal, the modulated signal is written as $x(t)s_c(t)$ for a linear modulation. Since $\Re(e^{j2\pi f_c t}) = s_c(t)$, the modulated signal can also be written as $\Re(x(t)e^{j2\pi f_c t})$. Here, $\Re(\cdot)$ stands for the real part. (In addition, $\Im(\cdot)$ will stand for the imaginary part.) A different baseband signal can also be transmitted using the other orthogonal carrier $\sin(2\pi f_c t)$. Let $x_I(t)$ and $x_Q(t)$ denote the baseband signals transmitted by the carriers $\cos(2\pi f_c t)$ and $\sin(2\pi f_c t)$, respectively. Then, the modulated signal is given by

$$x_I(t)\cos(2\pi f_c t) - x_Q(t)\sin(2\pi f_c t) = \Re(x(t)e^{j2\pi f_c t}),$$

where $x(t) = x_I(t) + jx_Q(t)$. It is often convenient to represent the modulated signal by $x(t)e^{j2\pi f_c t}$. Hence, we assume that signals are complex-valued for convenience.

Furthermore, we assume that the noise is complex-valued. Throughout this chapter, it will be assumed that $\Re(n_l)$ and $\Im(n_l)$ are independent white Gaussian random processes with mean zero and variance $N_0/2$. Thus, $\sigma_n^2 = E[|n_l|^2] = N_0$.

To illustrate a fading channel, we consider a receiver in motion as shown in Fig. 4.1. It is assumed that there is a single path.[†] As the receiver travels at a speed of v, the time delay of arrival of a radio signal varies as follows:

$$\tau(t) = \tau_0 - vt\cos\phi, \tag{4.2}$$

where τ_0 represents the time delay at a reference location at $t = 0$ and ϕ is the angle between the direction of the incoming signal and the direction of motion.

[†] Actually, it is assumed that there is a plane wave. If the transmitter is far away from the receiver, the incoming signal can be considered as a plane wave in a free space.

For convenience, let $\tau_0 = 0$. Taking the time-varying propagation delay in Eq. (4.2) into account, the received signal without the background noise is given by

$$\begin{aligned}
\hat{y}(t) &= Ax(t - \tau(t))e^{j2\pi f_c(t-\tau(t))} \\
&\simeq Ax(t)e^{-j2\pi f_c \tau(t)}e^{j2\pi f_c t} \\
&= Ax(t)e^{j2\pi(f_c + f_{sh})t},
\end{aligned} \tag{4.3}$$

where $A \geq 0$ is the attenuation factor and

$$f_{sh} = \frac{v}{\lambda}\cos\phi.$$

Here, $\lambda = c/f_c$ is the wavelength of the carrier, where c is the speed of propagation of the radio signal. In Eq. (4.3), it is assumed that the baseband signal varies slowly compared with the carrier frequency, because the bandwidth of a baseband signal is much smaller than the carrier frequency. Hence, we can ignore the delay of $x(t)$. However, since the carrier frequency is sufficiently high, the delay of the carrier cannot be ignored. As shown in Eq. (4.3), the motion of the receiver results in the carrier frequency shift. This phenomenon is called the Doppler effect.

The received signal is demodulated as follows:

$$\begin{aligned}
y(t) &= \hat{y}(t)e^{-j2\pi f_c t} \\
&= h(t)x(t),
\end{aligned} \tag{4.4}$$

where $h(t) = Ae^{j2\pi f_{sh}t}$ is a time-varying channel process. If the Doppler shift is known, compensation for the time-varying channel process is possible. We can extract the transmitted signal as follows:

$$\hat{x}(t) = e^{-j2\pi f_{sh}t}y(t) = Ax(t).$$

Consider another example. Suppose that there are two paths to a receiver in motion. The demodulated received signal without the background noise is given by

$$y(t) = \left(A_1 e^{j2\pi \frac{v}{\lambda}\cos\phi_1 t} + A_2 e^{j2\pi \frac{v}{\lambda}\cos\phi_2 t}\right)x(t),$$

where $A_p(\geq 0)$ and ϕ_p are the amplitude and angle of incidence of the pth multipath signals, respectively. For convenience, let $\phi_1 = 0$. Then, it follows that

$$y(t) = \underbrace{(A_1 + A_2 e^{j2\pi f_{sh,2}t})}_{=h(t)}x(t),$$

where $f_{sh,2} = (v/\lambda)\cos\phi_2$. The amplitude of the channel process $h(t)$ is now given by

$$|h(t)| = |A_1 + A_2 e^{j2\pi f_{sh,2}t}|.$$

The amplitude now becomes time-varying; it was a constant for the case of a single path (see Eq. (4.4)). We can show that

$$|A_1 - A_2| \leq |h(t)| \leq |A_1 + A_2|.$$

This shows that the signal strength can be very weak if $A_1 \simeq A_2$. In particular, if $A_1 = A_2$, $|h(t)|$ becomes zero periodically, since the two multipath signals can destroy each other.

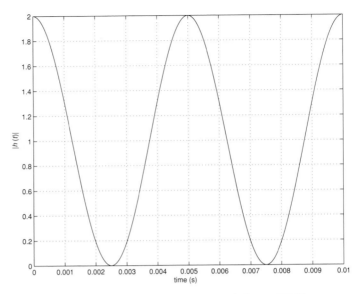

Figure 4.2. Amplitude of $h(t)$, $|h(t)|$, when $A_1 = A_2 = 1$ and $f_{\mathrm{sh},2} = 200\,\mathrm{Hz}$.

Figure 4.2 shows an example of time-varying amplitude of $h(t)$. When $|h(t)|$ is small, the signal experiences a deep fading and the receiver cannot detect any useful signal. In order to overcome deep fading, diversity techniques become very important; see Proakis (1995) for details.

So far, one-path or two-path signals have been considered in order to illustrate fading channels. In general, there can be a number of multipaths. Clarke's two-dimensional isotropic scattering model was proposed to address this case based on a statistical approach. Suppose that a receiver in motion is surrounded by a number of local scatterers which will cause multipaths to the receiver. Each multipath signal can have a different angle of incidence and a different attenuation factor. Assume that the difference among the delays of multipath signals is negligible and that the received signal is a sum of the multipath signals. Then, the fading channel process, $h(t)$, is statistically modeled as follows:

$$h(t) = \alpha e^{\mathrm{j}2\pi \frac{v}{\lambda} \cos \phi t}, \qquad (4.5)$$

where α is a random channel attenuation factor and ϕ is a uniformly distributed random angle of incidence (in a two-dimensional space). If α and ϕ are uncorrelated, the autocorrelation of $h(t)$ can be written as follows:

$$
\begin{aligned}
E[h(t)h^*(t - \tau)] &= E\left[|\alpha|^2 e^{\mathrm{j}2\pi \frac{v}{\lambda} \cos \phi t} e^{-\mathrm{j}2\pi \frac{v}{\lambda} \cos \phi(t-\tau)}\right] \\
&= E[|\alpha|^2] E\left[e^{\mathrm{j}2\pi \frac{v}{\lambda} \cos \phi \tau}\right].
\end{aligned}
\qquad (4.6)
$$

Since

$$
\begin{aligned}
E\left[e^{\mathrm{j}2\pi \frac{v}{\lambda} \tau \cos \phi}\right] &= \frac{1}{2\pi} \int_{-\pi}^{\pi} e^{\mathrm{j}2\pi \frac{v}{\lambda} \tau \cos \phi}\, \mathrm{d}\phi \\
&= J_0(2\pi f_{\mathrm{d}} \tau),
\end{aligned}
$$

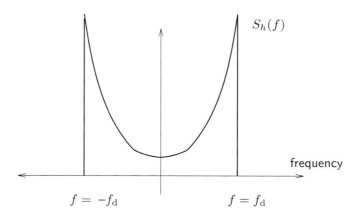

Figure 4.3. Clarke's power spectrum of $h(t)$ with maximum Doppler frequency f_{d}.

where $J_0(x)$ is the Bessel function of the first kind of order 0 and $f_{\mathrm{d}} = v/\lambda$, we can show that

$$\frac{1}{2}E[h(t)h^*(t-\tau)] = \frac{E[|\alpha|^2]}{2}J_0(2\pi f_{\mathrm{d}}\tau). \tag{4.7}$$

In Eq. (4.7), f_{d} is the maximum Doppler shift and is also called the maximum Doppler frequency. From Eq. (4.7), the power spectrum of $h(t)$, called Clarke's power spectrum, can be found as follows:

$$
\begin{aligned}
S_h(f) &= \mathcal{F}\left(\frac{1}{2}E[h(t)h^*(t-\tau)]\right) \\
&= \begin{cases} \dfrac{E[|\alpha|^2]}{2}\dfrac{1}{2\pi f_{\mathrm{d}}}\dfrac{1}{\sqrt{1-(f/f_{\mathrm{d}})^2}}, & |f| \le f_{\mathrm{d}}; \\[2mm] 0, & \text{otherwise,} \end{cases}
\end{aligned} \tag{4.8}
$$

where $\mathcal{F}(\cdot)$ stands for the Fourier transform. Clarke's spectrum, which has a U-shape, is illustrated in Fig. 4.3.

Some properties of $J_n(x)$, the Bessel function of the first kind of order n, are listed in the following.

- For small x (i.e. $x \to 0$) and fixed n:

$$J_n(x) \sim \frac{1}{2^n n!}x^n.$$

- For large x and fixed n:

$$J_n(x) \sim \sqrt{\frac{2}{\pi x}}\cos\left(x - (2n+1)\frac{\pi}{4}\right).$$

- $\frac{\mathrm{d}}{\mathrm{d}x}J_n(x) = \frac{1}{2}(J_{n-1}(x) - J_{n+1}(x))$.
- $J_{-n}(x) = (-1)^n J_n(x)$.
- $J_n(0) = 1$ if $n = 0$. If $n \neq 0$ is an integer, $J_n(0) = 0$.

Figure 4.4 shows a realization of Clarke's fading process when $f_{\mathrm{d}} = 100$ Hz. We can see that a time-varying process, $h(t)$, has deep fading.

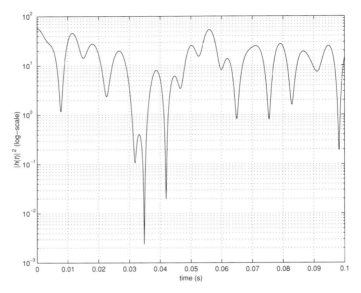

Figure 4.4. Fading process with $f_\mathrm{d} = 100\,\mathrm{Hz}$.

4.1.2 Doppler spread and coherence time

The maximum Doppler frequency is an indicator for the degree of time variation of fading channels. We can readily observe that f_d is proportional to the carrier frequency, f_c, and mobile speed, v. Depending on f_d, the fading channel can be classified into two groups: the fast fading channel and the slow fading channel. In fact, the degree of channel variation is relative to the symbol duration. In order to decide whether the channel is fast or slow, we define the coherence time, denoted by T_c, from the maximum Doppler frequency. The coherence time means the time duration in which the channel variation can be negligible. There are several definitions for the coherence time, two of which we introduce here:

- $T_\mathrm{c} \approx 1/f_\mathrm{d}$;
- $T_\mathrm{c} \approx 9/16\pi f_\mathrm{d}$.

The first definition is simply the inverse of the maximum Doppler frequency, and the second is the time duration in which the normalized correlation is above 0.5. Another definition from the geometric mean of above two coherence times is available:

- $T_\mathrm{c} \approx \sqrt{(1/f_\mathrm{d}) \times (9/16\pi f_\mathrm{d})} = 3/4\sqrt{\pi}\, f_\mathrm{d}$.

With an appropriate definition of the coherence time (depending on the application), we can classify fading channels as follows:

- a fading channel is fast if $1/T < \kappa f_\mathrm{d}$;
- a fading channel is slow if $1/T \geq \kappa f_\mathrm{d}$,

where κ is a constant depending on the definition of the coherence time and the system variables of interest and T is the time interval of interest. For example, T can be the symbol interval.

4.1.3 Frequency-selective fading and coherence bandwidth

If the data rate is sufficiently high or the symbol interval is sufficiently short, the multipath signals can be resolved into multiple symbol intervals. This results in a frequency-selective fading as the channel attenuation varies in the frequency domain.

The impulse response of a multipath channel at time t can be approximated as follows:

$$h(\tau;t) \approx \sum_{k=1}^{K} \alpha_k(t)\delta(\tau - \tau_k(t)), \qquad (4.9)$$

where $\alpha_k(t)$ and $\tau_k(t)$ are the channel attenuation and the (group) time delay of the kth cluster of multipaths at time t, respectively. Suppose that there exist $P_k = E[|\alpha_k(t)|^2]$ and $\tau_k = E[\tau_k(t)]$. Then, there are a few parameters that characterize the frequency-selective fading channel in Eq. (4.9) as follows.

- The mean excess delay:

$$\bar{\tau} = \frac{\sum_k P_k \tau_k}{\sum_k P_k}.$$

- The root mean square (RMS) delay spread:

$$\sigma_\tau = \sqrt{\overline{\tau^2} - (\bar{\tau})^2},$$

 where

$$\overline{\tau^2} = \frac{\sum_k P_k \tau_k^2}{\sum_k P_k}.$$

- The maximum excess delay (X dB), i.e. the time delay during which the multipath energy falls to X dB below the maximum.

Example 4.1.1 Suppose that there are three resolvable multipaths with

$$\{(P_k, \tau_k)\} = \{(1, 0.2), (4, 1), (0.25, 4)\}.$$

The mean excess delay is given by

$$\bar{\tau} = \frac{1 \times 0.2 + 4 \times 1 + 0.25 \times 4}{1 + 4 + 0.25}$$

$$= \frac{5.2}{5.25}$$

$$= 0.9904.$$

To obtain the RMS delay spread, we need to find $\overline{\tau^2}$:

$$\overline{\tau^2} = \frac{1 \times 0.2^2 + 4 \times 1^2 + 0.25 \times 4^2}{1 + 4 + 0.25}$$

$$= \frac{8.04}{5.25}$$

$$= 1.5314.$$

Then, the RMS delay spread is given by

$$\sigma_\tau = \sqrt{1.5314 - 0.9904^2} = 0.7419.$$

The coherence bandwidth, denoted by B_c, is the bandwidth in which the variation of channel in the frequency domain is negligible. There can be different definitions for the coherence bandwidth depending on the application. The coherence bandwidth can be defined as the bandwidth over which the frequency correlation function is above 0.9 when the frequency correlation function is normalized. Then, it becomes

$$B_c \approx \frac{1}{50\sigma_\tau}.$$

For a frequency correlation greater than 0.5, B_c is given by

$$B_c \approx \frac{1}{5\sigma_\tau}.$$

In general, a fading channel becomes frequency-selective if $B_c < B_W$. If $B_c > B_W$, a fading channel becomes frequency-nonselective or flat-fading. Note that there are no exact definitions of the coherence bandwidth and the coherence time. Depending on design and application, different definitions can be applied.

Example 4.1.2   In order to obtain the generic discrete-time model in Eq. (4.1) from Eq. (4.9), a few steps are required. Recall the analog transmitted signal model in Chapter 2:

$$x(t) = \sum_m b_m v(t - mT),$$

where $v(t)$ represents the impulse response of the transmitter filter. For convenience, assume that $v(t)$ is symmetric, i.e. $v(t) = v(-t)$. In addition, assume $g(t) = v(-t) = v(t)$, where $g(t)$ represents the impulse response of the receiver filter. From Eq. (4.9), the received signal is given by

$$r(t) = h(\tau; t) * x(\tau) + n(t)$$
$$= \sum_{k=1}^{K} \alpha_k(t) x(t - \tau_k(t)) + n(t). \tag{4.10}$$

The sampled signal after receiver filtering is given by

$$y_l = y(lT)$$
$$= \int r(\tau) g(lT - \tau) \, d\tau + n_l$$
$$\simeq \sum_{k=1}^{K} \alpha_k(lT) \int x(\tau - \tau_k(lT)) g(lT - \tau) \, d\tau + n_l, \tag{4.11}$$

where $n_l = \int_\tau n(\tau) g(lT - \tau) \, d\tau$. In Eq. (4.11), it is assumed that the channel attenuations and delays at time $t = lT$ approximate those during the time for the integral as the variation is sufficiently slow. Since $x(t) = \sum_m b_m g(t - mT)$, it can be shown that

$$y_l = \sum_m b_m \sum_{k=1}^K \alpha_k(lT) \int g(\tau - mT - \tau_k(lT))g(lT - \tau)\,d\tau + n_l. \qquad (4.12)$$

Letting $lT - \tau = \tau'$, we can show that

$$\int g(\tau - mT - \tau_k(lT))g(lT - \tau)\,d\tau = \int g((l - m)T - \tau_k(lT) - \tau')g(\tau')\,d\tau'.$$

Using this, we can define the time-varying CIR as follows:

$$h_{l-m;l} = \sum_{k=1}^K \alpha_k(lT) \int g((l - m)T - \tau_k(lT) - \tau')g(\tau')\,d\tau'$$

and the sampled received signal is given by

$$y_l = \sum_m b_m h_{l-m;l} + n_l$$

$$= \sum_p b_{l-p} h_{p,l} + n_l. \qquad (4.13)$$

Finally, the discrete-time model in Eq. (4.1) can be found.

4.1.4 Statistical characterization of fading multipath channels

So far, we have considered various cases of fading multipath channels. It is possible to unify these cases into a single model. In the limit of a continuum of multipath components, the received signal can be written as follows:

$$y(t) = \int_{-\infty}^{\infty} h(\tau; t)x(t - \tau)\,d\tau + n(t),$$

where $h(\tau; t)$ is the impulse response of a random multipath channel at time t, which is a doubly spread channel in time and frequency. In general, we assume that $E[h(\tau; t)] = 0$. A detailed description of this fading multipath channel model can be found in Bello (1963).

As $h(\tau; t)$ is a zero-mean random function, a statistical approach is employed to characterize $h(\tau; t)$. If

$$\frac{1}{2}E[h^*(\tau_1; t_1)h(\tau_2; t_2)] = \phi_h(\tau_1; t_1, t_2)\delta(\tau_1 - \tau_2),$$

the channel is said to exhibit delay uncorrelated scattering (US). This means that each multipath component is uncorrelated with the others. In addition, if $\phi_h(\tau; t_1, t_2) = \phi_h(\tau; t_2 - t_1)$, the channel exhibits wide sense stationary uncorrelated scattering (WSSUS). In this case, the second-order statistics can be represented by the delay cross-power density $\phi_h(\tau, \Delta t)$. The scattering function is defined as follows:

$$S_h(\tau; \lambda) = \int \phi_h(\tau; \Delta t)e^{-j2\pi\lambda\Delta t}\,d\Delta t. \qquad (4.14)$$

There are a few more definitions for WSSUS channels as follows.

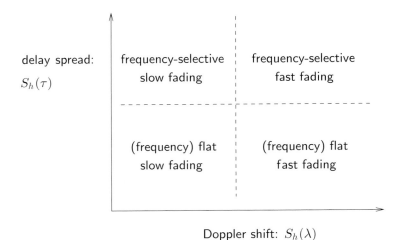

Figure 4.5. Classification of fading channels.

- Delay power spectrum or multipath intensity profile:

$$S_h(\tau) = \int S(\tau; \lambda)\, d\lambda = \phi_h(\tau; 0).$$

- Doppler power spectrum:

$$S_h(\lambda) = \int S(\tau; \lambda)\, d\tau.$$

- Time-correlation function:

$$\bar{\phi}_h(\Delta t) = \int \phi_h(\tau; \Delta t)\, d\tau.$$

The Doppler power spectrum is the Fourier transform of the time correlation function. It can be seen that Clarke's power spectrum in Eq. (4.8) is a special case. Fading channels can be categorized in terms of frequency selectivity and time selectivity as follows.

- Frequency selectivity. If the delay spread of the multipath intensity profile, $S_h(\tau)$, is longer than the symbol interval, the channel exhibits frequency selectivity. Otherwise, it can be considered as frequency-nonselective.
- Time selectivity. If the bandwidth of the Doppler power spectrum is larger than the symbol rate (or the packet rate, depending on applications), the channel exhibits fast fading. Otherwise, the channel becomes slow fading.

A classification of fading channels is illustrated in Fig. 4.5. If the Doppler power spectrum, $S_h(\lambda)$, becomes wider (or the maximum Doppler frequency becomes higher), a fading channel can be classified as a fast fading channel. As the delay spread becomes large (or $S_h(\tau)$ becomes wider), a fading channel can be classified as a frequency-selective fading channel.

4.2 MMSE approach to estimate time-varying channels

For fading multipath channels, the channel process can be considered as random. To estimate a random process, the MMSE or Wiener approach based on a linear estimator can be employed (Papoulis, 1984). In this section, we discuss the estimation of fading channel processes using this MMSE approach.

4.2.1 Channel estimation for flat fading

In this subsection, we introduce the channel estimation for flat fading channels. In Section 4.2.2 we consider the case for which the fading channels are frequency-selective.

Let h_l denote the time-varying channel coefficient at time l. Then, the received signal is given by

$$y_l = h_l b_l + n_l, \quad l = 0, 1, \ldots, L - 1,$$

where L is the length of a packet. In a packet or block of symbols, some symbols are known by the receiver (i.e. they are pre-determined) for the channel estimation. They are called pilot symbols. The resulting transmission scheme that inserts pilot symbols into a packet is called the pilot-symbol-aided (PSA) modulation. In general, pilot symbols are uniformly distributed over a packet to be able to estimate time-varying channel coefficients.

Let I_{ps} denote the index set for pilot symbols within a packet. For convenience, assume $b_l = 1$ if $l \in I_{ps}$. Then, we have

$$y_l = h_l + n_l, \quad l \in I_{ps}. \tag{4.15}$$

Let $\mathbf{h} = [h_0 \; h_1 \; \cdots \; h_{L-1}]^T$. If the variation of the fading channel is not significant, an interpolation approach can be used to estimate $\mathbf{h}$ from y_l for $l \in I_{ps}$. An MMSE approach to estimate and interpolate $\mathbf{h}$ can be formulated as follows:

$$\mathbf{V}_{lm} = \arg \min_{\mathbf{V}} E[\|\mathbf{h} - \mathbf{V}\mathbf{y}_{ps}\|^2], \tag{4.16}$$

where $\mathbf{V}_{lm}$ denotes the linear MMSE (LMMSE) estimator and

$$\mathbf{y}_{ps} = \begin{bmatrix} y_{p(1)} \; y_{p(2)} \; \cdots \; y_{p(L_p)} \end{bmatrix}^T.$$

Here, $p(l)$ is the index for the lth pilot symbol and L_p is the number of pilot symbols in a packet. We can readily show that

$$\mathbf{V}_{lm} = E[\mathbf{h}\mathbf{y}_{ps}^H] (E[\mathbf{y}_{ps}\mathbf{y}_{ps}^H])^{-1}, \tag{4.17}$$

where the superscript H stands for the Hermitian transpose.[†] The LMMSE estimate of $\mathbf{h}$ is given by

$$\hat{\mathbf{h}}_{lm} = \mathbf{V}_{lm}\mathbf{y}_{ps}. \tag{4.18}$$

This is an off-line approach, because $\hat{\mathbf{h}}_{lm}$ is available after $\mathbf{y}_{ps}$ is received.

[†] The Hermitian transpose of vector $\mathbf{x}$ of size $N \times 1$ becomes $\mathbf{x}^H = [x_1^* x_2^* \cdots x_N^*]$. Hence, if $\mathbf{x}$ is real-valued, $\mathbf{x}^H = \mathbf{x}^T$.

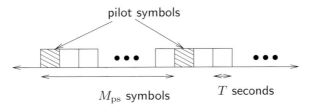

Figure 4.6. Pilot-symbol-aided (PSA) modulation.

If the discrete-time channel process $\{h_m\}$ is sampled from a continuous channel process of Clarke's spectrum, the autocorrelation can be written as follows:

$$R_h(m) = E[h_l h_{l-m}^*]$$
$$= J_0(2\pi f_d T |m|),$$

where T is the symbol duration. We can show that

$$\left[E\left[\mathbf{y}_{ps}\mathbf{y}_{ps}^H\right]\right]_{l,l'} = J_0(2\pi f_d T |p(l) - p(l')|) + N_0\delta_{l,l'}, \quad l,l' = 1, 2, \ldots, L_p$$

and

$$\left[E\left[\mathbf{h}\mathbf{y}_{ps}^H\right]\right]_{l,l'} = J_0(2\pi f_d T |(l-1) - p(l')|), \quad l = 1, 2, \ldots, L; \quad l' = 1, 2, \ldots, L_p.$$

Example 4.2.1   Let $L = 4$ and $I_{ps} = \{0, 2\}$. Then, the zeroth and second symbols are pilot symbols. It can be shown that

$$E\left[\mathbf{y}_{ps}\mathbf{y}_{ps}^H\right] = \begin{bmatrix} 1 + N_0 & J_0(2\pi f_d T) \\ J_0(2\pi f_d T) & 1 + N_0 \end{bmatrix}$$

and

$$E\left[\mathbf{h}\mathbf{y}_{ps}^H\right] = \begin{bmatrix} J_0(2\pi f_d T(0)) & J_0(2\pi f_d T(-2)) \\ J_0(2\pi f_d T(1)) & J_0(2\pi f_d T(-1)) \\ J_0(2\pi f_d T(2)) & J_0(2\pi f_d T(0)) \\ J_0(2\pi f_d T(3)) & J_0(2\pi f_d T(1)) \end{bmatrix}$$

$$= \begin{bmatrix} 1 & J_0(4\pi f_d T) \\ J_0(2\pi f_d T) & J_0(2\pi f_d T) \\ J_0(4\pi f_d T) & 1 \\ J_0(6\pi f_d T) & J_0(2\pi f_d T) \end{bmatrix}.$$

The performance of the LMMSE estimation generally depends on the location of pilot symbols within a data block. The allocation of pilot symbols and its performance analysis can be found in Cavers (1991). It is assumed that pilot symbols are inserted periodically as shown in Fig. 4.6. For example, if M_{ps} (a positive integer) is the pilot symbol spacing, the symbol at time $l = kM_{ps}, k = 0, 1, \ldots, \lfloor L/M_{ps} \rfloor - 1$, becomes a pilot symbol. Since there are less pilot symbols as M_{ps} becomes larger, more data symbols can be transmitted within a packet. Generally, M_{ps} cannot be arbitrarily large when estimating fast fading channel processes. Thus, it is desirable to find the maximum pilot spacing with a reasonably good performance of the channel estimation.

To gain an insight as to how to determine the maximum value of M_{ps}, we can ignore the background noise in Eq. (4.15). Then, we can see that this becomes an interpolation problem. From the following subsampled received signals:

$$y_l = h_l, \ l = kM_{ps}, \ k = 0, 1, \ldots, \left\lfloor \frac{L}{M_{ps}} \right\rfloor - 1, \tag{4.19}$$

the original signals, $h_l, l = 0, 1, \ldots, L - 1$, should be found by an interpolation. The bandwidth of the random process, $\{h_m\}$, can be found from its power spectrum given in Eq. (4.8). As the bandwidth is f_d, the Nyquist rate becomes $2 f_d$ according to the sampling theorem (Oppenheim and Willsky, 1996). This implies that the pilot insertion rate needs to be greater than or equal to the Nyquist rate:

$$\frac{1}{M_{ps} T} \geq 2 f_d$$

or

$$M_{ps} \leq \frac{1}{2 f_d T}, \tag{4.20}$$

where the product term, $f_d T$, is called the fading rate. According to Eq. (4.20), we can see that M_{ps} is closely related to the fading rate. For a detailed account of pilot spacing and performance analysis of PSA modulation, the reader is referred to Cavers (1991). In addition, a tutorial can be found in Tong, Sadler, and Dong (2004).

4.2.2 Channel estimation for frequency-selective fading

To consider the LMMSE estimation for time-varying CIR, assume that the data sequence is known. Stacking all the received signals, from Eq. (4.1), we have

$$\mathbf{y} = [y_0 \ y_1 \ \cdots \ y_{L-1}]^T$$

$$= \underbrace{\begin{bmatrix} \mathbf{b}_0^T & \mathbf{0} & \cdots & \mathbf{0} \\ \mathbf{0} & \mathbf{b}_1^T & \cdots & \mathbf{0} \\ \vdots & \vdots & \ddots & \vdots \\ \mathbf{0} & \mathbf{0} & \cdots & \mathbf{b}_{L-1}^T \end{bmatrix}}_{=\mathbf{X}^T} \underbrace{\begin{bmatrix} \mathbf{h}_0 \\ \mathbf{h}_1 \\ \vdots \\ \mathbf{h}_{L-1} \end{bmatrix}}_{=\mathbf{h}} + \underbrace{\begin{bmatrix} n_0 \\ n_1 \\ \vdots \\ n_{L-1} \end{bmatrix}}_{=\mathbf{w}}$$

$$= \mathbf{X}^T \mathbf{h} + \mathbf{w}. \tag{4.21}$$

The sizes of $\mathbf{X}^T$ and $\mathbf{h}$ are $L \times PL$ and $PL \times 1$, respectively. To estimate $\mathbf{h}$, the LMMSE estimator is given by

$$\mathbf{V}_{lm} = \arg \min_{\mathbf{V}} E[\|\mathbf{h} - \mathbf{V}\mathbf{y}\|^2], \tag{4.22}$$

where $\mathbf{V}$ is a linear estimator of size $PL \times L$ and $\mathbf{V}_{lm}$ is the optimal LMMSE estimator (of the same size). Using the orthogonality principle, we can show that

$$\mathbf{V}_{lm} = E[\mathbf{h}\mathbf{y}^H] \left(E[\mathbf{y}\mathbf{y}^H] \right)^{-1}. \tag{4.23}$$

From Eq. (4.21), we have

$$E[\mathbf{h}\mathbf{y}^{\mathrm{H}}] = \mathbf{R_h}\mathbf{X}^*,$$
$$E[\mathbf{y}\mathbf{y}^{\mathrm{H}}] = \mathbf{X}^{\mathrm{T}}\mathbf{R_h}\mathbf{X}^* + N_0\mathbf{I}, \tag{4.24}$$

where $\mathbf{R_h} = E[\mathbf{h}\mathbf{h}^{\mathrm{H}}]$. Hence, with known $\mathbf{X}^{\mathrm{T}}$, the LMMSE estimator becomes

$$\mathbf{V}_{\mathrm{lm}} = \mathbf{R_h}\mathbf{X}^*\left(\mathbf{X}^{\mathrm{T}}\mathbf{R_h}\mathbf{X}^* + N_0\mathbf{I}\right)^{-1}. \tag{4.25}$$

Consequently, the LMMSE estimate of $\mathbf{h}$ given $\mathbf{y}$ is given by

$$\mathbf{h}_{\mathrm{lm}} = \mathbf{V}_{\mathrm{lm}}\mathbf{y}$$
$$= \mathbf{R_h}\mathbf{X}^*\left(\mathbf{X}^{\mathrm{T}}\mathbf{R_h}\mathbf{X}^* + N_0\mathbf{I}\right)^{-1}\mathbf{y}. \tag{4.26}$$

The LMMSE estimator is closely related to the MAP estimation of time-varying CIR when the channel vector $\mathbf{h}$ is a Gaussian random vector. The MAP estimate of $\mathbf{h}$ that maximizes the *a posteriori* probability can be found as follows:

$$\mathbf{h}_{\mathrm{map}} = \arg\max_{\mathbf{h}} f(\mathbf{h}|\mathbf{y})$$
$$= \arg\max_{\mathbf{h}} f(\mathbf{y}|\mathbf{h})f(\mathbf{h}), \tag{4.27}$$

where $f(\mathbf{y}|\mathbf{h})$ is the likelihood function and $f(\mathbf{h})$ is the (*a priori*) pdf of $\mathbf{h}$. Assume that $\mathbf{h}$ is a Gaussian random vector with mean zero and covariance $\mathbf{R_h} = E[\mathbf{h}\mathbf{h}^{\mathrm{H}}]$. Then, we have

$$\mathbf{h}_{\mathrm{map}} = \arg\max_{\mathbf{h}} \left\{\exp\left(-\frac{1}{N_0}\|\mathbf{y} - \mathbf{X}^{\mathrm{T}}\mathbf{h}\|^2\right)\exp\left(-\mathbf{h}^{\mathrm{H}}\mathbf{R_h}^{-1}\mathbf{h}\right)\right\}$$
$$= \arg\max_{\mathbf{h}} \left\{-\frac{1}{N_0}\|\mathbf{y} - \mathbf{X}^{\mathrm{T}}\mathbf{h}\|^2 - \mathbf{h}^{\mathrm{H}}\mathbf{R_h}^{-1}\mathbf{h}\right\}. \tag{4.28}$$

Note that taking the logarithm does not affect the maximization. Since the function on the right hand side of Eq. (4.28) is a quadratic function, we can easily find the maximum by taking the derivative with respect to $\mathbf{h}$. Then, we can find the solution as follows:

$$\mathbf{h}_{\mathrm{map}} = \frac{1}{N_0}\left(\mathbf{R_h}^{-1} + \frac{1}{N_0}\mathbf{X}^*\mathbf{X}^{\mathrm{T}}\right)^{-1}\mathbf{X}^*\mathbf{y}. \tag{4.29}$$

This MAP estimate is identical to the LMMSE estimate. We can readily show this by applying the matrix inversion lemma.

A major difficulty of the LMMSE estimation is the computational complexity due to the matrix inversion, as shown in Eq. (4.26). Since the size of the matrix $(\mathbf{X}^{\mathrm{T}}\mathbf{R_h}\mathbf{X}^* + N_0\mathbf{I})$ is $L \times L$, the complexity for the matrix inversion is $O(L^3)$. As L increases, this approach becomes prohibitive. In the case that the computatonal complexity is limited, adaptive techniques (i.e. on-line algorithms) including the LMS and RLS algorithm would be preferable.

Another major difficulty of this method is that all data symbols should be known as pilot symbols. This is simply unrealistic in communications. If the PSA modulation is used as a practical method for the channel estimation, the MMSE approach can be modified with an interpolation approach to estimate the channel vector, $\mathbf{h}$. In this case, knowledge of some $\mathbf{b}_l^{\mathrm{T}}$ in Eq. (4.21) are required. As each $\mathbf{b}_l^{\mathrm{T}}$ consists of P data symbols, it requires periodic insertions of at least P consecutive pilot symbols. Consequently, as P becomes larger, more pilot symbols are required, which results in a lower throughput.

As shown above, the LMMSE estimate of **h** in Eq. (4.26) would not be practical unless all the data symbols are known. In the iterative receiver, however, the LMMSE estimation can be useful as statistical information of data symbols would be available through iterations. This is discussed in Chapter 7.

4.3   Adaptive algorithms to track time-varying channels

Both LMS and RLS algorithms can track time-varying channels. It is known that the tracking performance is different from the convergence performance. The convergence behavior is a transient phenomenon, while the tracking behavior is a steady state phenomenon to a time-varying environment. Although the RLS algorithm has a better convergence performance than the LMS algorithm, it is not necessarily true that the RLS algorithm has a better tracking performance than the LMS algorithm; see Eweda (1994) and Haykin *et al.* (1997) for a detailed account of the comparison of tracking performance. In this section, we focus on the LMS algorithm to track time-varying channels of Clarke's power spectrum. Throughout this section, the autocorrelation function of zero-mean fading channel processes is assumed to be given by

$$
\begin{aligned}
R_{h,p}(m) &= E[h_{p,l}h_{q,l-m}^*] \\
&= \sigma_{h,p}^2 J_0(2\pi f_{\mathrm{d}} T |m|)\delta_{p,q},
\end{aligned}
\tag{4.30}
$$

where $\sigma_{h,p}^2$, $p = 0, 1, \ldots, P-1$, denotes the power delay profile.

4.3.1   LMS filter and its tracking performance

Recall the LMS algorithm in Eq. (3.23). The LMS algorithm is given by

$$
\hat{\mathbf{h}}_{l+1} = \hat{\mathbf{h}}_l + 2\mu\mathbf{b}_l\left(y_l - \mathbf{b}_l^{\mathrm{T}}\hat{\mathbf{h}}_l\right)
\tag{4.31}
$$

Note that we denote by $\hat{\mathbf{h}}_{l-1}$ the LMS estimate of $\mathbf{h}_l$ (it was $\mathbf{h}^{(l)}$ in Eq. (3.23)). As the CIR is complex-valued, we consider a complex-valued version of the LMS algorithm. However, since $\mathbf{b}_l$ is real-valued, there is no difference in the LMS algorithm.

As shown in Eq. (4.31), μ decides the dynamics of the LMS filter. For a larger μ, a better tracking ability can be achieved, while a worse noise sensitivity is expected. Therefore, it is desirable to find an optimal value of μ for a tradeoff between tracking ability and noise sensitivity.

Since $y_l = \mathbf{b}_l^{\mathrm{T}}\mathbf{h}_l + n_l$, the estimation error can be decomposed into two terms:

$$
\begin{aligned}
\Delta\mathbf{h}_l &= \hat{\mathbf{h}}_l - \mathbf{h}_l \\
&= \underbrace{\hat{\mathbf{h}}_l - E[\hat{\mathbf{h}}_l]}_{=\Delta\mathbf{h}_{1,l}} + \underbrace{E[\hat{\mathbf{h}}_l] - \mathbf{h}_l}_{=\Delta\mathbf{h}_{2,l}},
\end{aligned}
\tag{4.32}
$$

where $\Delta\mathbf{h}_{1,l}$ denotes the estimation error due to self-noise and $\Delta\mathbf{h}_{2,l}$ denotes the estimation error due to lag. This decomposition is essential in order to clarify sources of tracking errors (Widrow *et al.*, 1976).

Firstly, we consider the estimation error due to lag using a transfer function based approach. Taking the expectation both sides in Eq. (4.31), we have

$$E[\hat{\mathbf{h}}_{l+1}] = E[\hat{\mathbf{h}}_l] + 2\mu(\mathbf{h}_l - E[\hat{\mathbf{h}}_l]). \tag{4.33}$$

Using the $\mathcal{Z}$-transform, we can show that

$$\mathcal{Z}(E[\hat{\mathbf{h}}_l]) = \frac{2\mu}{z - (1 - 2\mu)}\mathbf{h}(z),$$

where $\mathbf{h}(z) = \mathcal{Z}(\mathbf{h}_l)$. Let $\tilde{\mathbf{h}}(z) = \mathcal{Z}(E[\hat{\mathbf{h}}_l]) - \mathbf{h}(z)$, which is the $\mathcal{Z}$-transform of the estimation error due to lag; see Eq. (4.32). It follows that

$$\tilde{\mathbf{h}}(z) = \left(\frac{2\mu}{z - (1 - 2\mu)} - 1\right)\mathbf{h}(z)$$
$$= \psi_{\text{te}}(z)\mathbf{h}(z), \tag{4.34}$$

where

$$\psi_{\text{te}}(z) = \frac{z^{-1} - 1}{1 - (1 - 2\mu)z^{-1}}. \tag{4.35}$$

As shown in Eq. (4.34), the estimation error due to lag can be seen as a filtered signal. The spectra of $\psi_{\text{te}}(z)$ and $\mathbf{h}(z)$ can decide the amount of the estimation error due to lag. Hence, for a given spectrum of $\mathbf{h}(z)$, it is possible to choose $\psi_{\text{te}}(z)$ (actually μ, as shown in Eq. (4.35)) to minimize the estimation error due to lag. It is shown in Lin *et al.* (1995) that the RLS algorithm also has the same transfer function in Eq. (4.35). Thus, the analysis of the RLS algorithm is identical to that of the LMS algorithm, which will follow.

To derive the variance of the estimation error due to lag, we follow the approach in (Lin *et al.*, 1995). From Eq. (4.34), we can show that

$$\sigma^2_{\text{lag},p} = E[|E[\hat{h}_{p,l}] - h_{p,l}|^2]$$
$$= \frac{1}{2\pi}\int_{-\pi}^{\pi} |\psi_{\text{te}}(\omega)|^2 S_p(\omega)\,\mathrm{d}\omega, \tag{4.36}$$

where $S_p(\omega)$ is the power spectrum of the sequence of $h_{p,l}$ that is given as $S_p(\omega) = S_p(z) = \mathcal{Z}(R_{h,p}(m))$ for $z = \mathrm{e}^{\mathrm{j}\omega}$.

It can be shown that

$$|\psi_{\text{te}}(\omega)|^2 = \frac{(\mathrm{e}^{-\mathrm{j}\omega} - 1)(\mathrm{e}^{\mathrm{j}\omega} - 1)}{(1 - \beta\mathrm{e}^{-\mathrm{j}\omega})(1 - \beta\mathrm{e}^{\mathrm{j}\omega})}$$
$$= \frac{2(1 - \cos\omega)}{1 - 2\beta\cos\omega + \beta^2}, \tag{4.37}$$

where $\beta = 1 - 2\mu$. If $|\omega| \le \pi$, $\cos\omega \simeq 1 - \omega^2/2$ and $|\psi_{\text{te}}(\omega)|^2 \simeq \omega^2/(1 - \beta)^2$. From this, we can approximate

$$\int_{-\pi}^{\pi} |\psi_{\text{te}}(\omega)|^2 S_p(\omega)\,\mathrm{d}\omega \simeq \int_{-\pi}^{\pi} \frac{\omega^2}{(1 - \beta)^2} S_p(\omega)\,\mathrm{d}\omega. \tag{4.38}$$

This turns out to be a good approximation since $S_p(\omega)$, which is a Clarke's spectrum, is bandlimited, as shown in Eq. (4.8).

Using the relation between the power spectrum and the autocorrelation function, from Eq. (4.30), it follows that

$$R_{h,p}(m) = \sigma_{h,p}^2 J_0(\omega_d m)$$

$$= \frac{1}{2\pi} \int_{-\pi}^{\pi} S_p(e^{j\omega}) e^{jm\omega} \, d\omega, \tag{4.39}$$

where $\omega_d = 2\pi f_d T$. Taking derivatives twice with respect to m and setting m equal to zero, it can be shown that

$$\sigma_{h,p}^2 J_0''(\omega_d m) \Big|_{m=0} = \frac{1}{2\pi} \int_{-\pi}^{\pi} -\omega^2 S_p(e^{j\omega}) \, d\omega.$$

Using the properties of $J_0(x)$, we can also show that $J_0''(0) = -1/2$. Thus, it can be shown that

$$\int_{-\pi}^{\pi} \omega^2 S_p(e^{j\omega}) \, d\omega = \pi \sigma_{h,p}^2 \omega_d^2. \tag{4.40}$$

Finally, the variance of the estimation error due to lag becomes

$$\sigma_{\text{lag},p}^2 \simeq \frac{\sigma_{h,p}^2 \omega_d^2}{2(1-\beta)^2}. \tag{4.41}$$

The estimation error due to self-noise in Eq. (4.32) is identical to the excess noise for a time-invariant CIR. From Eq. (3.24), it follows that

$$\lim_{l \to \infty} E[(\Delta \mathbf{h}_{1,l})(\Delta \mathbf{h}_{1,l})^H] = \frac{\mu \sigma_n^2}{1-\mu P} \mathbf{I} \simeq \mu \sigma_n^2 \mathbf{I} \quad \text{(for small } \mu P\text{)}. \tag{4.42}$$

The total excess MSE or MSE of the LMS channel estimate becomes

$$\lim_{l \to \infty} E[\|\Delta \mathbf{h}_l\|^2] \simeq \mu P \sigma_n^2 + \frac{\sigma_h^2 \omega_d^2}{8\mu^2}, \tag{4.43}$$

where $\sigma_h^2 = \sum_{p=0}^{P-1} \sigma_{h,p}^2$. Consequently, the optimal μ that minimizes the MSE of the LMS channel estimate can be found as follows:

$$\mu_{\text{opt}} = \left(\frac{\sigma_h^2 \omega_d^2}{4 P \sigma_n^2} \right)^{1/3}. \tag{4.44}$$

This shows that μ increases with ω_d and σ_h^2/σ_n^2.

Figure 4.7 shows the optimal values of μ for different values of fading rate, $f_d T$, and SNR, σ_h^2/σ_n^2. It is shown that if the channel variation becomes faster, μ becomes larger (a larger μ provides a better tracking performance). In addition, if the SNR becomes higher, μ can also be larger as the received signal is more reliable.

As mentioned earlier, the approach in this subsection is based on Lin et al. (1995). Other approaches for the tracking performance of the LMS filter can be found in Eweda (1994), Galdino, Pinto, and de Alencar (2004), Lindbom, Sternad, and Ahlen (2001), Lindbom et al. (2002), and Yousef and Sayed (2001).

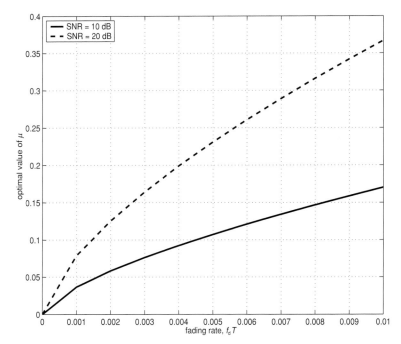

Figure 4.7. Optimal values of μ depending on fading rate and SNR $= \sigma_h^2/\sigma_n^2$ with $P = 4$.

4.3.2 Normalized LMS filter

The optimal value of μ can be determined to minimize the MSE according to Eq. (4.44). However, we need to know the maximum Doppler frequency and SNR to decide μ_{opt}. In practice, these values are not known and must be estimated. To avoid this estimation problem, a robust approach can be used. The normalized LMS (NLMS) filter is a robust approach that can track the variation of channels without knowing parameters of time-varying channels.

The NLMS algorithm can be derived by the following constrained optimization problem (Goodwin and Sin, 1984; Haykin, 1996):

$$\min_{\hat{\mathbf{h}}_{l+1}} \|\hat{\mathbf{h}}_{l+1} - \hat{\mathbf{h}}_l\|^2$$
$$\text{subject to} \quad \mathbf{b}_l^{\mathrm{T}} \hat{\mathbf{h}}_{l+1} = y_l. \tag{4.45}$$

Using the method of Lagrange multipliers, $\hat{\mathbf{h}}_{l+1}$ can be found. For convenience, assume that y_l and $\hat{\mathbf{h}}_l$ are real-valued quantities. The constrained optimization problem becomes an unconstrained optimization problem with Lagrange multiplier λ as follows:

$$\min_{\hat{\mathbf{h}}_{l+1}} \|\hat{\mathbf{h}}_{l+1} - \hat{\mathbf{h}}_l\|^2 + \lambda\left(y_l - \mathbf{b}_l^{\mathrm{T}} \hat{\mathbf{h}}_{l+1}\right). \tag{4.46}$$

The solution becomes

$$\hat{\mathbf{h}}_{l+1}(\lambda) = \hat{\mathbf{h}}_l + \frac{1}{2}\lambda \mathbf{b}_l. \tag{4.47}$$

The value of λ can be decided from the constraint:

$$y_l = \mathbf{b}_l^\mathrm{T} \hat{\mathbf{h}}_{l+1}(\lambda^*)$$

$$= \mathbf{b}_l^\mathrm{T} \hat{\mathbf{h}}_{l+1} + \frac{\lambda^*}{2} \|\mathbf{b}_l\|^2$$

$$\Rightarrow \lambda^* = \frac{2}{\|\mathbf{b}_l\|^2} \left(y_l - \mathbf{b}_l^\mathrm{T} \hat{\mathbf{h}}_l\right).$$

From Eq. (4.47), it follows that

$$\hat{\mathbf{h}}_{l+1} = \hat{\mathbf{h}}_l + \frac{1}{\|\mathbf{b}_l\|^2} \mathbf{b}_l \left(y_l - \mathbf{b}_l^\mathrm{T} \hat{\mathbf{h}}_l\right). \tag{4.48}$$

This is called the NLMS algorithm. It is possible to generalize Eq. (4.48) as follows:

$$\hat{\mathbf{h}}_{l+1} = \hat{\mathbf{h}}_l + \frac{2\tilde{\mu}}{\|\mathbf{b}_l\|^2} \mathbf{b}_l \left(y_l - \mathbf{b}_l^\mathrm{T} \hat{\mathbf{h}}_l\right), \tag{4.49}$$

where $\tilde{\mu}$ is a positive constant. It is known that if $0 < \tilde{\mu} < 1$, the NLMS algorithm in Eq. (4.49) converges (Haykin, 1996). Another modification of the NLMS algorithm is as follows:

$$\hat{\mathbf{h}}_{l+1} = \hat{\mathbf{h}}_l + \frac{2\tilde{\mu}}{\alpha + \|\mathbf{b}_l\|^2} \mathbf{b}_l \left(y_l - \mathbf{b}_l^\mathrm{T} \hat{\mathbf{h}}_l\right), \tag{4.50}$$

where $\alpha > 0$ is introduced to avoid the difficulty caused by a small $\|\mathbf{b}_l\|^2$. In the channel estimation, $\|\mathbf{b}_l\|^2$ will not be zero. However, in some other applications, the term equivalent to $\|\mathbf{b}_l\|^2$ can be small, and it can result in gradient noise amplification.

4.4 Kalman filter for channel tracking

The Kalman filter is a general approach for the recursive estimation; see Anderson and Foschini (1975) for a detailed account of the Kalman filter. However, we confine ourselves to the channel estimation (or tracking). It is essential to build a state-space model for fading channel processes to derive the Kalman filter.

4.4.1 State-space modeling

To model time-varying CIR in Eq. (4.1), an autoregressive (AR) model can be used. For example, the following vector AR(1) model can be considered:

$$\mathbf{h}_{l+1} = \mathbf{A}\mathbf{h}_l + \mathbf{u}_l, \tag{4.51}$$

where $\mathbf{A}$ is a $P \times P$ AR coefficient matrix (in principle, $\mathbf{A}$ can be time-varying) and $\mathbf{u}_l$ is a white noise that drives the AR process, $\{\mathbf{h}_m\}$, assumed to be given by

$$E[\mathbf{u}_l] = \mathbf{0} \quad \text{and} \quad E\left[\mathbf{u}_l \mathbf{u}_l^\mathrm{H}\right] = \mathbf{Q}_l.$$

A pair of state-space and output equations is written as follows:

$$\mathbf{h}_{l+1} = \mathbf{A}\mathbf{h}_l + \mathbf{u}_l,$$
$$y_l = \mathbf{b}_l^\mathrm{T}\mathbf{h}_l + n_l, \tag{4.52}$$

where $\mathbf{h}_l$ is called the state vector and $\{y_m\}$ is called the measurement process in the context of state-space modeling.

A classical problem with Eqs (4.52) is to estimate the (random) state vector, $\mathbf{h}_l$, from observations $\{y_0, y_1, \ldots, y_l\}$ with known parameter vectors, $\{\mathbf{b}_0, \mathbf{b}_1, \ldots, \mathbf{b}_l\}$. The Kalman filter is a *linear* MMSE estimator that estimates $\mathbf{h}_l$ recursively. It is generally assumed that statistical properties of the noise process and parameters of the state-space and output equations (e.g. $\mathbf{A}$) are known.

In above, we adopt an AR(1) (vector) process to model fading channel processes. It can be extended to a general model with an AR(Q) vector process, $Q \geq 1$, such as

$$\mathbf{h}_{l+1} = \sum_{q=0}^{Q-1} \mathbf{A}_q \mathbf{h}_{l-q} + \mathbf{u}_l, \tag{4.53}$$

where $\mathbf{A}_q$ is the qth AR coefficient matrix. Let

$$\mathbf{c}_l = \begin{bmatrix} \mathbf{h}_l^\mathrm{T} & \mathbf{h}_{l-1}^\mathrm{T} & \cdots & \mathbf{h}_{l-Q+1}^\mathrm{T} \end{bmatrix}^\mathrm{T}. \tag{4.54}$$

Then, we can show that

$$\mathbf{c}_{l+1} = \underbrace{\begin{bmatrix} \mathbf{A}_0 & \mathbf{A}_1 & \cdots & \mathbf{A}_{Q-2} & \mathbf{A}_{Q-1} \\ \mathbf{I} & \mathbf{0} & \cdots & \mathbf{0} & \mathbf{0} \\ \mathbf{0} & \mathbf{I} & \cdots & \mathbf{0} & \mathbf{0} \\ \vdots & \vdots & \ddots & \vdots & \vdots \\ \mathbf{0} & \mathbf{0} & \cdots & \mathbf{I} & \mathbf{0} \end{bmatrix}}_{=\bar{\mathbf{A}}} \mathbf{c}_l + \underbrace{\begin{bmatrix} \mathbf{u}_l \\ \mathbf{0} \\ \vdots \\ \mathbf{0} \end{bmatrix}}_{=\bar{\mathbf{u}}_l} \tag{4.55a}$$

$$y_l = \mathbf{d}_l^\mathrm{T}\mathbf{c}_l + n_l, \tag{4.55b}$$

where

$$\mathbf{d}_l = \begin{bmatrix} \mathbf{b}_l^\mathrm{T} & \underbrace{\mathbf{0}^\mathrm{T} \cdots \mathbf{0}^\mathrm{T}}_{Q-1 \text{ times}} \end{bmatrix}^\mathrm{T}$$

and

$$E\left[\bar{\mathbf{u}}_l \bar{\mathbf{u}}_l^\mathrm{T}\right] = \begin{bmatrix} \mathbf{Q}_l & \mathbf{0} & \cdots & \mathbf{0} \\ \mathbf{0} & \mathbf{0} & \cdots & \mathbf{0} \\ \vdots & \vdots & \ddots & \vdots \\ \mathbf{0} & \mathbf{0} & \cdots & \mathbf{0} \end{bmatrix}. \tag{4.56}$$

The AR coefficient matrices can be determined according to spectral properties of fading channel processes. In the following subsections, we derive the Kalman filter by using Eq. (4.52). However, it is also straightforward to extend the Kalman filter using Eq. (4.55b).

4.4.2 Kalman filter

We will develop some key properties of the LMMSE estimation to derive the Kalman filter.

LMMSE estimation and its properties

Suppose that there are two random vectors $\mathbf{x}$ and $\mathbf{z}$ with *known* mean vectors and cross-correlation and covariance matrices as follows:

$$E\begin{bmatrix} \mathbf{x} \\ \mathbf{z} \end{bmatrix} = \begin{bmatrix} \bar{\mathbf{x}} \\ \bar{\mathbf{z}} \end{bmatrix},$$

$$E\left[\begin{bmatrix} \mathbf{x} - \bar{\mathbf{x}} \\ \mathbf{z} - \bar{\mathbf{z}} \end{bmatrix} [(\mathbf{x} - \bar{\mathbf{x}})^H (\mathbf{z} - \bar{\mathbf{z}})^H]\right] = \begin{bmatrix} \mathbf{R}_{xx} & \mathbf{R}_{xz} \\ \mathbf{R}_{zx} & \mathbf{R}_{zz} \end{bmatrix}, \tag{4.57}$$

where

$$\mathbf{R}_{xx} = E[(\mathbf{x} - \bar{\mathbf{x}})(\mathbf{x} - \bar{\mathbf{x}})^H],$$
$$\mathbf{R}_{zz} = E[(\mathbf{z} - \bar{\mathbf{z}})(\mathbf{z} - \bar{\mathbf{z}})^H],$$
$$\mathbf{R}_{xz} = E[(\mathbf{x} - \bar{\mathbf{x}})(\mathbf{z} - \bar{\mathbf{z}})^H],$$
$$\mathbf{R}_{zx} = E[(\mathbf{z} - \bar{\mathbf{z}})(\mathbf{x} - \bar{\mathbf{x}})^H].$$

For convenience, assume that $\bar{\mathbf{z}} = 0$. We consider the LMMSE estimate of $\mathbf{x} - \bar{\mathbf{x}}$ from $\mathbf{z}$. Since the mean vector of $\mathbf{x}$ is known, we do not need to estimate it. Using the orthogonality principle, the LMMSE estimator required to estimate the difference vector $\mathbf{x} - \bar{\mathbf{x}}$ from $\mathbf{z}$ is given by

$$\mathbf{L}_{lm} = \arg\min_{\mathbf{L}} E[\|(\mathbf{x} - \bar{\mathbf{x}}) - \mathbf{L}\mathbf{z}\|^2]$$
$$= \mathbf{R}_{xz}\mathbf{R}_{zz}^{-1}. \tag{4.58}$$

If the inverse of $\mathbf{R}_{zz}$ does not exist, its pseudo-inverse can be used. Then, the LMMSE estimate of $\mathbf{x}$ becomes

$$\hat{\mathbf{x}}_{lm} = \bar{\mathbf{x}} + \mathbf{L}_{lm}\mathbf{z}$$
$$= \bar{\mathbf{x}} + \mathbf{R}_{xz}\mathbf{R}_{zz}^{-1}\mathbf{z}. \tag{4.59}$$

The LMMSE estimate of $\mathbf{x}$ is unbiased and the error covariance matrix of $\hat{\mathbf{x}}_{lm}$ is readily found as

$$Cov(\mathbf{x} - \hat{\mathbf{x}}_{lm}) = E[(\mathbf{x} - \hat{\mathbf{x}}_{lm})(\mathbf{x} - \hat{\mathbf{x}}_{lm})^H]$$
$$= \mathbf{R}_{xx} - \mathbf{R}_{xz}\mathbf{R}_{zz}^{-1}\mathbf{R}_{zx}. \tag{4.60}$$

From Eq. (4.59), we can define an operator for the LMMSE estimation as follows:

$$E^*[\mathbf{x}|\mathbf{z}] = \mathbf{R}_{xz}\mathbf{R}_{zz}^{-1}\mathbf{z}. \tag{4.61}$$

Then, the LMMSE estimate of $\mathbf{x}$ is given by

$$\hat{\mathbf{x}}_{lm} = \bar{\mathbf{x}} + E^*[\mathbf{x}|\mathbf{z}]. \tag{4.62}$$

When multiple observations of $\mathbf{z}$ are available, the LMMSE estimation can be generalized. Suppose that $\mathbf{z}_0^l = \{\mathbf{z}_0, \mathbf{z}_1, \ldots, \mathbf{z}_l\}$ is given. Furthermore, assume that $E[\mathbf{z}_q] = \mathbf{0}, q = 0, 1, \ldots, l$. The LMMSE estimate of $\mathbf{x} - \bar{\mathbf{x}}$ becomes a linear combination of $\mathbf{z}_0, \mathbf{z}_1, \ldots, \mathbf{z}_l$, and $\hat{\mathbf{x}}_{\mathrm{lm}}$ is given by

$$\hat{\mathbf{x}}_{\mathrm{lm}} = \bar{\mathbf{x}} + E^*[\mathbf{x}|\mathbf{z}_0^l]$$
$$= \bar{\mathbf{x}} + \sum_{q=0}^{l} \mathbf{L}_q \mathbf{z}_q, \qquad (4.63)$$

where the $\mathbf{L}_q$'s are the LMMSE coefficient matrices.

Each observation can be decomposed as follows:

$$\mathbf{z}_l = E^*[\mathbf{z}_l|\mathbf{z}_0^{l-1}] + \tilde{\mathbf{z}}_l, \quad \forall l > 0, \qquad (4.64)$$

where $E^*[\mathbf{z}_l|\mathbf{z}_0^{l-1}]$ is the LMMSE estimate of $\mathbf{z}_l$ given $\mathbf{z}_0^{l-1}$ and $\tilde{\mathbf{z}}_l$ is the error vector which is uncorrelated with $\mathbf{z}_0^{l-1}$ (recalling the orthogonality principle, we can easily see that $\tilde{\mathbf{z}}_l$ and $\mathbf{z}_0^{l-1}$ are uncorrelated). Since $\tilde{\mathbf{z}}_l$ and $\mathbf{z}_0^{l-1}$ are uncorrelated (i.e. $E[\tilde{\mathbf{z}}_l \mathbf{z}_q^{\mathrm{H}}] = 0$ for $q < l$), we can show that $\tilde{\mathbf{z}}_l$ and $\tilde{\mathbf{z}}_0^{l-1}$ are also uncorrelated.

Since $E^*[\mathbf{z}_l|\mathbf{z}_0^{l-1}]$ is a linear combination of $\mathbf{z}_0, \mathbf{z}_1, \ldots, \mathbf{z}_{l-1}$, we have

$$\mathbf{z}_l = E^*[\mathbf{z}_l|\tilde{\mathbf{z}}_0^{l-1}] + \tilde{\mathbf{z}}_l \qquad (4.65)$$

by induction. Using Eq. (4.65), it follows that

$$E^*[\mathbf{x}|\mathbf{z}_0^l] = E^*[\mathbf{x}|\tilde{\mathbf{z}}_0^l]. \qquad (4.66)$$

This means that $\mathbf{z}_0^l$ and $\tilde{\mathbf{z}}_0^l$ contain the same information to estimate $\mathbf{x}$ in terms of the LMMSE estimation. (Note that the LMMSE coefficient matrices of the LMMSE operators on each side are different.) Since $\tilde{\mathbf{z}}_q$ is uncorrelated, we can derive the following important observation:

$$\hat{\mathbf{x}}_{\mathrm{lm},l} = \bar{\mathbf{x}} + E^*[\mathbf{x}|\mathbf{z}_0^l]$$
$$= \bar{\mathbf{x}} + E^*[\mathbf{x}|\tilde{\mathbf{z}}_0^l]$$
$$= \bar{\mathbf{x}} + \sum_{q=0}^{l} E^*[\mathbf{x}|\tilde{\mathbf{z}}_q]$$
$$= \hat{\mathbf{x}}_{\mathrm{lm},l-1} + E^*[\mathbf{x}|\tilde{\mathbf{z}}_l], \qquad (4.67)$$

where $\hat{\mathbf{x}}_{\mathrm{lm},l}$ denotes the LMMSE estimate of $\mathbf{x}$ from $\tilde{\mathbf{z}}_0^l$ or $\mathbf{z}_0^l$. Note that the third equality in Eq. (4.67) is valid because the $\tilde{\mathbf{z}}_q$'s are uncorrelated with each other. Equation (4.67) plays a key role in deriving the Kalman filter.

Example 4.4.1 Let x be a Gaussian random variable with $E[x] = 0$ and $E[x^2] = 1$. A random vector $\mathbf{z}$ is given by

$$\mathbf{z} = \begin{bmatrix} \frac{1}{2} \\ -1 \end{bmatrix} x + \mathbf{n},$$

where $\mathbf{n}$ is a 2×1 Gaussian random vector with $E[\mathbf{n}] = \mathbf{0}$ and $E[\mathbf{n}\mathbf{n}^T] = \mathbf{I}$. Assume that $\mathbf{n}$ and x are independent. With $\mathbf{z}$ as an observation, the LMMSE estimate of x is given by $\hat{x}_{lm} = \mathbf{L}_{lm}\mathbf{z}$, where $\mathbf{L}_{lm}$ is found as follows.

Firstly, we need to find $\mathbf{R}_{xz}$ and $\mathbf{R}_{zz}$:

$$\mathbf{R}_{xz} = E[x\mathbf{z}^T] = \begin{bmatrix} \frac{1}{2} & -1 \end{bmatrix}$$

$$\mathbf{R}_{zz} = E[\mathbf{z}\mathbf{z}^T] = \begin{bmatrix} \frac{5}{4} & -\frac{1}{2} \\ -\frac{1}{2} & 2 \end{bmatrix}.$$

Then, it follows that

$$\mathbf{L}_{lm} = \mathbf{R}_{xz}\mathbf{R}_{zz}^{-1} = \begin{bmatrix} \frac{2}{9} & -\frac{4}{9} \end{bmatrix}.$$

Example 4.4.2   In this example, we show that $\tilde{\mathbf{z}}_l$ and $\tilde{\mathbf{z}}_0^{l-1}$ are uncorrelated. Note that to do this, it is sufficient to show that $\tilde{\mathbf{z}}_0^{l-1}$ is a linear combination of $\mathbf{z}_0^{l-1}$.

For convenience, let

$$\mathbf{z} = \begin{bmatrix} \mathbf{z}_0^T & \mathbf{z}_1^T & \cdots & \mathbf{z}_{l-1}^T \end{bmatrix}^T,$$

$$\tilde{\mathbf{z}} = \begin{bmatrix} \tilde{\mathbf{z}}_0^T & \tilde{\mathbf{z}}_1^T & \cdots & \tilde{\mathbf{z}}_{l-1}^T \end{bmatrix}^T.$$

From Eq. (4.64), we can show that there exists an $l \times l$ matrix $\mathbf{D}$ such that

$$\tilde{\mathbf{z}} = \mathbf{D}\mathbf{z}. \tag{4.68}$$

For example, if $l = 2$, it can be shown that

$$\tilde{\mathbf{z}}_0 = \mathbf{z}_0,$$

$$\tilde{\mathbf{z}}_1 = \mathbf{z}_1 - E[\mathbf{z}_1\mathbf{z}_0^H](E[\mathbf{z}_0\mathbf{z}_0^H])^{-1}\mathbf{z}_0,$$

$$\Rightarrow \begin{bmatrix} \tilde{\mathbf{z}}_0 \\ \tilde{\mathbf{z}}_1 \end{bmatrix} = \underbrace{\begin{bmatrix} \mathbf{I} & \mathbf{0} \\ -E[\mathbf{z}_1\mathbf{z}_0^H](E[\mathbf{z}_0\mathbf{z}_0^H])^{-1} & \mathbf{I} \end{bmatrix}}_{=\mathbf{D}} \begin{bmatrix} \mathbf{z}_0 \\ \mathbf{z}_1 \end{bmatrix}.$$

Hence, we can see that $\tilde{\mathbf{z}}$ is a linear combination of $\mathbf{z}$. From this, we can show that

$$E[\tilde{\mathbf{z}}_l\tilde{\mathbf{z}}^H] = \underbrace{E[\tilde{\mathbf{z}}_l\mathbf{z}^H]}_{=\mathbf{0}}\mathbf{D}^H = \mathbf{0}. \tag{4.69}$$

Due to the property in Eq. (4.69), $\{\tilde{\mathbf{z}}_q\}$ is called a pseudo-innovation process.

Example 4.4.3   We can show that Eq. (4.66) is true using some matrix manipulations.

The matrix $\mathbf{D}$ in Eq. (4.68) has the following properties: (1) the diagonal elements of $\mathbf{D}$ are all ones; (2) $\mathbf{D}$ is lower triangle. In this case, $\mathbf{D}^{-1}$ exists. From Eq. (4.62), we have

$$E^*\left[\mathbf{x}|\tilde{\mathbf{z}}_0^{l-1}\right] = E[(\mathbf{x} - \bar{\mathbf{x}})\tilde{\mathbf{z}}^H](E[\tilde{\mathbf{z}}\tilde{\mathbf{z}}^H])^{-1}\tilde{\mathbf{z}}. \tag{4.70}$$

Since $\tilde{\mathbf{z}} = \mathbf{D}\mathbf{z}$, it follows that

$$E[(\mathbf{x} - \bar{\mathbf{x}})\tilde{\mathbf{z}}^H](E[\tilde{\mathbf{z}}\tilde{\mathbf{z}}^H])^{-1} = E[(\mathbf{x} - \bar{\mathbf{x}})\mathbf{z}^H]\mathbf{D}^H(\mathbf{D}E[\mathbf{z}\mathbf{z}^H]\mathbf{D}^H)^{-1}$$

$$= E[(\mathbf{x} - \bar{\mathbf{x}})\mathbf{z}^H](E[\mathbf{z}\mathbf{z}^H])^{-1}\mathbf{D}^{-1}. \tag{4.71}$$

Substituting Eq. (4.71) into Eq. (4.70), we can show that

$$
\begin{aligned}
E^*\left[\mathbf{x}\big|\tilde{\mathbf{z}}_0^{l-1}\right] &= \mathbf{R}_{\mathrm{xz}}\mathbf{R}_{\mathrm{zz}}^{-1}\mathbf{D}^{-1}\tilde{\mathbf{z}} \\
&= \mathbf{R}_{\mathrm{xz}}\mathbf{R}_{\mathrm{zz}}^{-1}\mathbf{z} \\
&= E^*\left[\mathbf{x}\big|\mathbf{z}_0^{l-1}\right].
\end{aligned}
\tag{4.72}
$$

Since it is valid for all l, Eq. (4.72) becomes Eq. (4.66).

Derivation of the Kalman filter

Now we are ready to formulate the Kalman filter to track time-varying channels. Using Eqs (4.60) and (4.67), we can derive the Kalman filter that recursively finds the LMMSE estimate of $\mathbf{h}_l$. Denote

$$
\begin{aligned}
\hat{\mathbf{h}}_{l|l} &= \text{LMMSE estimate of } \mathbf{h}_l \text{ from } \mathbf{y}_0^l; \\
\hat{\mathbf{h}}_{l|l-1} &= \text{LMMSE estimate of } \mathbf{h}_l \text{ from } \mathbf{y}_0^{l-1}; \\
\mathbf{P}_{l|l} &= Cov(\mathbf{h}_l - \hat{\mathbf{h}}_{l|l}); \\
\mathbf{P}_{l|l-1} &= Cov(\mathbf{h}_l - \hat{\mathbf{h}}_{l|l-1}).
\end{aligned}
\tag{4.73}
$$

We will derive the Kalman filtering after showing the first few steps as follows.

(S1) Consider $\mathbf{h}_0$ and y_0. For convenience, the mean vectors and correlation and covariance matrices of random vectors $\mathbf{x}$ and $\mathbf{z}$ are represented as follows:

$$
\begin{bmatrix} \mathbf{x} \\ \mathbf{z} \end{bmatrix} \Rightarrow \begin{bmatrix} \bar{\mathbf{x}} \\ \bar{\mathbf{z}} \end{bmatrix}, \begin{bmatrix} \mathbf{R}_{\mathrm{xx}} & \mathbf{R}_{\mathrm{xz}} \\ \mathbf{R}_{\mathrm{zx}} & \mathbf{R}_{\mathrm{zz}} \end{bmatrix},
$$

where $\bar{\mathbf{x}}$ and $\bar{\mathbf{z}}$ are the mean vectors of $\mathbf{x}$ and $\mathbf{z}$, respectively. Note that $\mathbf{R}_{\mathrm{xx}}$ and $\mathbf{R}_{\mathrm{zz}}$ are the covariance matrices of $\mathbf{x}$ and $\mathbf{z}$, respectively, and $\mathbf{R}_{\mathrm{xz}} = \mathbf{R}_{\mathrm{zx}}^{\mathrm{H}} = E[(\mathbf{x} - \bar{\mathbf{x}})(\mathbf{z} - \bar{\mathbf{z}})^{\mathrm{H}}]$. Then, with $\mathbf{h}_0$ and y_0, we have

$$
\begin{bmatrix} \mathbf{h}_0 \\ y_0 \end{bmatrix} \Rightarrow \begin{bmatrix} \bar{\mathbf{h}}_0 \\ \mathbf{b}_0^{\mathrm{T}}\bar{\mathbf{h}}_0 \end{bmatrix}, \begin{bmatrix} \mathbf{P}_0 & \mathbf{P}_0\mathbf{b}_0 \\ \mathbf{b}_0^{\mathrm{T}}\mathbf{P}_0 & \mathbf{b}_0^{\mathrm{T}}\mathbf{P}_0\mathbf{b}_0 + \sigma_{\mathrm{n}}^2 \end{bmatrix}.
\tag{4.74}
$$

Here, $\sigma_{\mathrm{n}}^2 = E[|n_l|^2]$ denotes the variance of n_l. Generally, it is assumed that the mean vector and covariance matrix of $\mathbf{h}_0$, denoted by $\bar{\mathbf{h}}_0$ and $P_0 = E[(\mathbf{h}_0 - \bar{\mathbf{h}}_0)(\mathbf{h}_0 - \bar{\mathbf{h}}_0)^{\mathrm{H}}]$, respectively, are known. Using Eq. (4.67), it follows that

$$
\begin{aligned}
\hat{\mathbf{h}}_{0|0} &= \bar{\mathbf{h}}_0 + E^*[\mathbf{h}_0|\, y_0] \\
&= \bar{\mathbf{h}}_0 + \mathbf{P}_0\mathbf{b}_0\left(\mathbf{b}_0^{\mathrm{T}}\mathbf{P}_0\mathbf{b}_0 + \sigma_{\mathrm{n}}^2\right)^{-1}\mathbf{b}_0^{\mathrm{T}}\mathbf{P}_0\left(y_0 - \mathbf{b}_0^{\mathrm{T}}\bar{\mathbf{h}}_0\right).
\end{aligned}
\tag{4.75}
$$

In addition, from Eq. (4.60), the *covariance* matrix of $\mathbf{h}_0 - \hat{\mathbf{h}}_{0|0}$ is given by

$$
\begin{aligned}
\mathbf{P}_{0|0} &= Cov(\mathbf{h}_0 - \hat{\mathbf{h}}_{0|0}) \\
&= \mathbf{P}_0 - \mathbf{P}_0\mathbf{b}_0\left(\mathbf{b}_0^{\mathrm{T}}\mathbf{P}_0\mathbf{b}_0 + \sigma_{\mathrm{n}}^2\right)^{-1}\mathbf{b}_0^{\mathrm{T}}\mathbf{P}_0.
\end{aligned}
\tag{4.76}
$$

(S2) Using $\hat{\mathbf{h}}_{0|0}$, the (LMMSE) prediction of $\mathbf{h}_1$ can be found. The prediction of $\mathbf{h}_1$ and its covariance matrix of error are given by

$$\hat{\mathbf{h}}_{1|0} = \mathbf{A}\hat{\mathbf{h}}_{0|0},$$
$$\mathbf{P}_{1|0} = Cov(\mathbf{h}_1 - \hat{\mathbf{h}}_{1|0}) = \mathbf{A}\mathbf{P}_{0|0}\mathbf{A}^{\mathrm{H}} + \mathbf{Q}_1, \tag{4.77}$$

respectively. In addition, the prediction of y_1 and its cross-correlation and covariance become

$$\hat{y}_{1|0} = \mathbf{b}_1^{\mathrm{T}}\hat{\mathbf{h}}_{1|0};$$
$$E[(\mathbf{h}_1 - \hat{\mathbf{h}}_{1|0})(y_1 - \hat{y}_{1|0})] = E[(\mathbf{h}_1 - \hat{\mathbf{h}}_{1|0})(\mathbf{b}_1^{\mathrm{T}}(\mathbf{h}_1 - \hat{\mathbf{h}}_{1|0}) + n_1)];$$
$$= \mathbf{P}_{1|0}\mathbf{b}_1; \tag{4.78}$$
$$Cov(y_1 - \hat{y}_{1|0}) = \mathbf{b}_1^{\mathrm{T}}\mathbf{P}_{1|0}\mathbf{b}_1 + \sigma_{\mathrm{n}}^2.$$

Given the predicted quantities, $\hat{\mathbf{h}}_{1|0}$ and $\hat{y}_{1|0}$, we have

$$\begin{bmatrix} \mathbf{h}_1 \\ y_1 \end{bmatrix} \Rightarrow \begin{bmatrix} \hat{\mathbf{h}}_{1|0} \\ \mathbf{b}_1^{\mathrm{T}}\hat{\mathbf{h}}_{1|0} \end{bmatrix}, \quad \begin{bmatrix} \mathbf{P}_{1|0} & \mathbf{P}_{1|0}\mathbf{b}_1 \\ \mathbf{b}_1^{\mathrm{T}}\mathbf{P}_{1|0} & \mathbf{b}_1^{\mathrm{T}}\mathbf{P}_{1|0}\mathbf{b}_1 + \sigma_{\mathrm{n}}^2 \end{bmatrix}. \tag{4.79}$$

(S3) From Eq. (4.79), the LMMSE estimate of $\mathbf{h}_1$ and its covariance matrix of error given y_1 and $\{\hat{\mathbf{h}}_{1|0}, \hat{y}_{1|0}\}$ can be found using Eqs (4.67) and (4.60) as follows:

$$\hat{\mathbf{h}}_{1|1} = \hat{\mathbf{h}}_{1|0} + E^*[\mathbf{h}_1|y_1 - \hat{y}_{1|0}]$$
$$= \hat{\mathbf{h}}_{1|0} + E^*[\mathbf{h}_1|y_1 - \mathbf{b}_1^{\mathrm{T}}\hat{\mathbf{h}}_{1|0}]$$
$$= \hat{\mathbf{h}}_{1|0} + \mathbf{P}_{1|0}\mathbf{b}_1(\mathbf{b}_1^{\mathrm{T}}\mathbf{P}_{1|0}\mathbf{b}_1 + \sigma_{\mathrm{n}}^2)^{-1}(y_1 - \mathbf{b}_1^{\mathrm{T}}\hat{\mathbf{h}}_{1|0}), \tag{4.80}$$
$$\mathbf{P}_{1|1} = Cov(\mathbf{h}_1 - \hat{\mathbf{h}}_{1|1})$$
$$= \mathbf{P}_{1|0} - \mathbf{P}_{1|0}\mathbf{b}_1(\mathbf{b}_1^{\mathrm{T}}\mathbf{P}_{1|0}\mathbf{b}_1 + \sigma_{\mathrm{n}}^2)^{-1}\mathbf{b}_1^{\mathrm{T}}\mathbf{P}_{1|0}.$$

Using Steps (S1)–(S3), we can derive the Kalman filter. From Eq. (4.67), it follows that

$$\hat{\mathbf{h}}_{l|l} = \hat{\mathbf{h}}_{l|l-1} + E^*[\mathbf{h}_l|\tilde{\mathbf{y}}_l],$$
$$\hat{\mathbf{h}}_{l+1|l} = \mathbf{A}\hat{\mathbf{h}}_{l|l}. \tag{4.81}$$

Accordingly, $\mathbf{P}_{l|l}$ and $\mathbf{P}_{l+1|l}$ can be found recursively. Finally, from Eqs (4.80) and (4.81), we can summarize the Kalman filter to find the LMMSE estimate of $\mathbf{h}_l$ as follows:

$$\hat{\mathbf{h}}_{l|l} = \hat{\mathbf{h}}_{l|l-1} + \mathbf{P}_{l|l-1}\mathbf{b}_l(\mathbf{b}_l^{\mathrm{T}}\mathbf{P}_{l|l-1}\mathbf{b}_l + \sigma_{\mathrm{n}}^2)^{-1}(y_l - \mathbf{b}_l^{\mathrm{T}}\hat{\mathbf{h}}_{l|l-1});$$
$$\mathbf{P}_{l|l} = \mathbf{P}_{l|l-1} - \mathbf{P}_{l|l-1}\mathbf{b}_l(\mathbf{b}_l^{\mathrm{T}}\mathbf{P}_{l|l-1}\mathbf{b}_l + \sigma_{\mathrm{n}}^2)^{-1}\mathbf{b}_l^{\mathrm{T}}\mathbf{P}_{l|l-1};$$
$$\hat{\mathbf{h}}_{l+1|l} = \mathbf{A}\hat{\mathbf{h}}_{l|l}; \tag{4.82}$$
$$\mathbf{P}_{l+1|l} = \mathbf{A}\mathbf{P}_{l|l}\mathbf{A}^{\mathrm{H}} + \mathbf{Q}_l.$$

As shown above, in order to perform the Kalman filter, a state-space model for fading channel processes is required. When the state-space model is not available, the LMS or NLMS filter can be used as they do not require any model for fading channel processes.

4.5   MLSD and PSP with channel tracking for fading channels

Firstly, we discuss the MLSD for fading channels and then we apply the PSP to perform the MLSD.

4.5.1   MLSD for random channels

Recall the MLSD problem in Chapter 3. From Eq. (4.1), the MLSD finds the data sequence that maximizes the likelihood function as

$$\{b_l\}_{\text{mlsd}} = \arg\max_{\{b_l\}} f\left(\mathbf{y}_0^{L-1} \big| \mathbf{b}_0^{L-1}\right), \tag{4.83}$$

where the likelihood function is given by

$$f\left(\mathbf{y}_0^{L-1} \big| \mathbf{b}_0^{L-1}\right) \propto \prod_{l=0}^{L-1} f\left(y_l \big| \mathbf{y}_0^{l-1}, \mathbf{b}_0^{l}\right). \tag{4.84}$$

Here, $\mathbf{y}_0^l$ and $\mathbf{b}_0^l$ are given by

$$\mathbf{y}_0^l = \{y_0, \quad y_1, \quad \ldots, \quad y_l\},$$
$$\mathbf{b}_0^l = \{b_0, \quad b_1, \quad \ldots, \quad b_l\}.$$

Clearly, it is necessary to find the individual likelihood functions:

$$f\left(y_l \big| \mathbf{y}_0^{l-1}, \mathbf{b}_0^{l}\right), \quad l = 0, 1, \ldots, L - 1.$$

Once the likelihood functions are known, the ML sequence can be found by an exhaustive search. However, this becomes impractical because there are 2^L candidate data sequences.

Example 4.5.1   In this example, we will show that even if there is no ISI, the MLSD is necessary to find the ML sequence for unknown random channels, while the symbol-by-symbol ML detection fails to obtain the ML symbols.

Suppose that the received signal is given by

$$y_l = h_l b_l + n_l,$$

where h_l is the random time-varying channel coefficient at time l. In this case, there is no ISI as $P = 1$. Assume that h_l and n_l are independent zero-mean circular complex Gaussian random variables with $E[|h_l|^2] = \sigma_h^2$ and $E[|n_l|^2] = N_0$. Note that if X is a circular complex Gaussian random variable with mean $E[X] = \mu_X$, the real and imaginary components of X are independent and

$$\frac{1}{2} Var(X) = Var(\Re(X)) = Var(\Im(X)) \text{ and } E[(X - \mu_X)^2] = 0.$$

The pdf of X is given by

$$f(x) = \frac{1}{\pi \sigma_X^2} \exp\left(-\frac{1}{\sigma_X^2}|x - \mu_X|^2\right).$$

For the symbol-by-symbol ML detection, the likelihood function of b_l given y_l can be considered. From above, we can show that y_l is a circular complex Gaussian random variable conditioned on b_l. Since

$$E[y_l|b_l] = 0 \text{ and } Var(y_l|b_l) = \sigma_h^2 + N_0,$$

we can show that

$$f(y_l|b_l) = \frac{1}{\pi\left(\sigma_h^2 + N_0\right)} \exp\left(-\frac{|y_l|^2}{\sigma_h^2 + N_0}\right). \tag{4.85}$$

Clearly, this is not useful for detecting b_l, because the likelihood function is not a function of b_l. Consequently, even if $P = 1$, the MLSD with the likelihood function $f(y_l|\mathbf{y}_0^{l-1}, \mathbf{b}_0^l)$ should be considered instead of $f(y_l|b_l)$.

Note that if the h_l's are independent, the MLSD is also unable to detect b_l. Since y_l becomes independent of $\mathbf{y}_0^{l-1}$, $f(y_l|\mathbf{y}_0^{l-1}, \mathbf{b}_0^l) = f(y_l|b_l)$ and the MLSD fails to detect b_l.

4.5.2 Derivation of the likelihood function

To find $f(y_l|\mathbf{y}_0^{l-1}, \mathbf{b}_0^l)$, the distribution of $\mathbf{h}_l$ should be known. Assume that $\{\mathbf{h}_m\}$ is a Gaussian random vector process. In this case, we can show that y_l is a Gaussian random variable conditioned on $\mathbf{y}_0^{l-1}$ and $\mathbf{b}_0^l$. This implies that we just need to find the conditional mean and variance of y_l for the MLSD.

In deriving the Kalman filter, we note that the distribution of the white noise $\mathbf{u}_l$ in Eq. (4.51) is not assumed. If $\mathbf{u}_l$ is Gaussian, $\mathbf{h}_l$ is also Gaussian. In this case, the Kalman filter actually finds the conditional mean and variance of $\mathbf{h}_l$ given $\mathbf{y}_0^{l-1}$. If $\mathbf{h}_l$ and $\{y_0, y_1, \ldots, y_l\}$ are jointly Gaussian, we can show that the conditional means $E[\mathbf{h}_l|\mathbf{y}_0^{l-1}]$ and $E[\mathbf{h}_l|\mathbf{y}_0^l]$ are the LMMSE estimates of $\mathbf{h}_l$, $\hat{\mathbf{h}}_{l|l-1}$, and $\hat{\mathbf{h}}_{l|l}$, from $\mathbf{y}_0^{l-1}$ and $\mathbf{y}_0^l$, respectively. From this, the conditional mean and variance of y_l given $\mathbf{y}_0^{l-1}$, which are required to perform the MLSD as shown in Eq. (4.84), can be found by the Kalman filter. Given $\mathbf{y}_0^{l-1}$ (and $\mathbf{b}_0^l$), we have

$$\begin{aligned}
\mu_{l|l-1}\left(\mathbf{y}_0^{l-1}, \mathbf{b}_0^l\right) &= E\left[y_l|\mathbf{y}_0^{l-1}\right] \\
&= E\left[\mathbf{b}_l^{\mathsf{T}}\mathbf{h}_l + n_l|\mathbf{y}_0^{l-1}\right] \\
&= \mathbf{b}_l^{\mathsf{T}}\hat{\mathbf{h}}_{l|l-1}; \\
\sigma_{l|l-1}^2\left(\mathbf{y}_0^{l-1}, \mathbf{b}_0^l\right) &= Cov\left(\mathbf{b}_l^{\mathsf{T}}\mathbf{h}_l + n_l|\mathbf{y}_0^{l-1}\right) \\
&= \mathbf{b}_l^{\mathsf{T}}\mathbf{P}_{l|l-1}\mathbf{b}_l + \sigma_{\mathrm{n}}^2.
\end{aligned} \tag{4.86}$$

Finally, the conditional pdf of y_l is given by

$$\begin{aligned}
f\left(y_l|\mathbf{y}_0^{l-1}, \mathbf{b}_0^l\right) &= \mathcal{N}\left(\mu_{l|l-1}\left(\mathbf{y}_0^{l-1}, \mathbf{b}_0^l\right), \sigma_{l|l-1}^2\left(\mathbf{y}_0^{l-1}, \mathbf{b}_0^l\right)\right) \\
&= \mathcal{N}\left(\mathbf{b}_l^{\mathsf{T}}\hat{\mathbf{h}}_{l|l-1}, \mathbf{b}_l^{\mathsf{T}}\mathbf{P}_{l|l-1}\mathbf{b}_l + \sigma_{\mathrm{n}}^2\right).
\end{aligned} \tag{4.87}$$

According to Eqs (4.84) and (4.83), the MLSD can be carried out with Kalman filtering to find the individual likelihood function in Eq. (4.87) given $\{y_0, y_1, \ldots, y_l\}$.

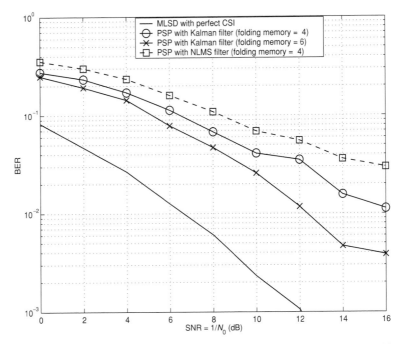

Figure 4.8. Bit error rate performance of the PSP with Kalman filter or NLMS filter (with $\mu = 0.5$) to track time-varying channels.

4.5.3   PSP with Kalman filter and LMS filter

From Eqs (4.83) and (4.87), the log likelihood function (ignoring the constant) can be written as follows:

$$\log f\left(\mathbf{y}_0^{L-1}\big|\mathbf{b}_0^{L-1}\right) = \sum_{l=0}^{L-1} \log f\left(y_l\big|\mathbf{y}_0^{l-1}, \mathbf{b}_0^l\right)$$

$$= -\sum_{l=0}^{L-1} \left(\log\left(\mathbf{b}_l^{\mathrm{T}}\mathbf{P}_{l|l-1}\mathbf{b}_l + \sigma_{\mathrm{n}}^2\right) + \frac{\left|y_l - \mathbf{b}_l^{\mathrm{T}}\hat{\mathbf{h}}_{l|l-1}\right|^2}{\mathbf{b}_l^{\mathrm{T}}\mathbf{P}_{l|l-1}\mathbf{b}_l + \sigma_{\mathrm{n}}^2} \right). \tag{4.88}$$

Since the number of states increases exponentially with time, the VA is not applicable without a forced folding. For a forced folding, the conditional mean and variance of y_l are approximated by functions of $\mathbf{b}_{l-\bar{P}+1}^l$ (not $\mathbf{b}_0^l$) for a finite positive integer $\bar{P}$. Then, we can apply the VA to the MLSD and it results in PSP with Kalman filtering.

Other adaptive filters that have a good tracking performance can replace the Kalman filter; for example, the LMS filter. In this case, a state-space model for fading channel preocesses is not necessary.

Figure 4.8 shows simulation results with two different values of folding memory, $\bar{P}$, in Eq. (4.88) when $P = 4$. A state-space model for fading channel processes is considered for simulations. Based on the following AR(Q) model with $Q = 4$:

$$h_{p,l+1} = 3.8h_{p,l} - 5.415h_{p,l-1} + 3.4295h_{p,l-2} - 0.8145h_{p,l-3} + u_{p,l}, \tag{4.89}$$

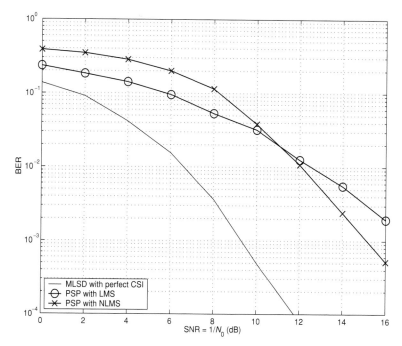

Figure 4.9. Bit error rate performance of the PSP with LMS algorithm: Clarke's fading channel with $f_d T = 0.001$; μ_{opt} in Eq. (4.44) for the LMS algorithm and $\mu = 0.5$ for the NLMS algorithm.

fading channel processes are generated. The variance of $h_{p,l}$ is assumed to be $1/P$ for all p and the variance of $u_{p,l}$ is determined accordingly. As shown in Fig. 4.8, the performance can be improved if a larger folding memory is used at the expense of increasing complexity. When $\bar{P} = P = 4$, there are $2^4 = 16$ states in the PSP, and there are $2^6 = 64$ states when $\bar{P} = 6$.

For comparison purposes, the simulation results of the PSP with the NLMS filter are also shown in Fig. 4.8. The PSP with the NLMS filter does not outperform the PSP with the Kalman filter. However, since the complexity is lower, the PSP with the NLMS filter would be preferable in some practical situations in which the complexity becomes an issue.

To observe the performance difference between the NLMS filter and the LMS filter with optimal μ in the PSP, simulations are carried out, and the results are shown in Fig. 4.9. Fading channel processes of Clarke's spectrum are considered in the simulations; see Eq. (4.30). It is assumed that $f_d T = 0.001$ and $\sigma_{h,p}^2 = 1/P$, where $P = 4$. Note that this fading channel process is different from that generated by Eq. (4.89).

In the PSP, the folding memory is set to $P = 4$ for both LMS and NLMS filters. The optimal value of μ is decided according to Eq. (4.44). It is shown that the PSP with the LMS filter outperforms the PSP with the NLMS filter when the SNR is lower than 11 dB. When the SNR is greater than 11 dB, the PSP with the NLMS filter exhibits the better performance. Note that the LMS filter with optimal μ does not necessarily outperform the NLMS filter since the tracking behavior of the NLMS filter is different from that of the LMS filter.

4.6   Summary and notes

In this chapter we have studied fading channels and adaptive filters that can track time-varying channels. In addition, we introduced the MLSD for fading channels and the PSP with the Kalman filter and other adaptive filters.

Fading channels are very important in wireless communications. A tutorial for fading channels can be found in Biglieri *et al.* (1998). Fading channels and diversity techniques are discussed in Proakis (1995). For the detection over fading channels, the channel tracking becomes crucial. Various adaptive filters can be used; see Anderson and Moore (1979) and Maybeck (1979) for detailed accounts of Kalman filtering and Haykin (1996) for a general theory of adaptive filters.

The PSP is a practical approach of dealing with the MLSD for random channels and has been extensively investigated; see Borah and Hart (1999), Chugg and Polydoros (1996b), and Rollins and Simmons (1997). The optimality issue is addressed in Chugg (1998).

II Iterative signal processing for ISI channels

5 MAP detection and iterative channel estimation

In this chapter, we introduce the maximum *a posteriori* probability (MAP) symbol detection over *static* ISI channels. The MAP symbol detector finds the symbols that maximize the *a posteriori* probabilities. A computationally efficient algorithm, called the BCJR (acronym from the authors' names: Bahl, Cocke, Jelinek, and Raviv) algorithm (Bahl *et al.*, 1974), is introduced for the MAP detection.

In the MAP detection, the CIR should be known. There are various approaches suitable for finding the channel estimation. An approach based on the ML criterion is introduced in this chapter with the expectation and maximization (EM) algorithm (Dempster, Laird, and Rubin, 1977). The EM algorithm is an iterative numerical method employed to solve the ML problem. The EM-based channel estimation is closely related to the MAP symbol detection. The resulting receiver employing the EM-based channel estimation becomes an iterative receiver in which the channel estimation and detection are carried out iteratively. This iterative receiver is also studied in this chapter.

5.1 MAP symbol detection over ISI channels

In this section, we study MAP symbol detection and the BCJR algorithm used to perform the MAP symbol detection.

5.1.1 MAP symbol detection

Recall the received signal model over a static ISI channel. The received signal at time l is given by

$$y_l = \sum_{p=0}^{P-1} h_p\, b_{l-p} + n_l, \quad l = 0, 1, \ldots, L-1, \tag{5.1}$$

where $\{h_p\}$ is the CIR of length P, $\{b_m\}$ is the transmitted data symbol sequence, L is the length of the data symbol sequence, and $\{n_m\}$ is the additive white Gaussian noise sequence with $E[n_l] = 0$ and $E[n_l^2] = \sigma_n^2 = N_0/2$. While the MLSD finds the sequence that maximizes the likelihood function, the MAP symbol detection finds the symbols that maximize the *a posteriori* probabilities individually.

Consider the *a posteriori* probability of b_l given sequence $\{y_m\}$:[†]

$$\Pr(b_l|\{y_m\}).$$

For the BPSK signaling, $b_l \in \{-1, +1\}$, according to the MAP detection principle, the symbol can be detected as follows:

$$\hat{b}_l = \begin{cases} +1, & \text{if } \Pr(b_l = +1|\{y_m\}) \geq \Pr(b_l = -1|\{y_m\}); \\ -1, & \text{if } \Pr(b_l = +1|\{y_m\}) < \Pr(b_l = -1|\{y_m\}), \end{cases} \quad (5.2)$$

or

$$\hat{b}_l = \arg \max_{b_l \in \{-1,+1\}} \Pr(b_l|\{y_m\}).$$

The MAP symbol detection is a symbol-wise optimal detection. The following log-ratio of *a posteriori* probabilities (LAPP) can also be used to detect b_l:

$$L(b_l) = \log \frac{\Pr(b_l = +1|\{y_m\})}{\Pr(b_l = -1|\{y_m\})}. \quad (5.3)$$

The sign of the LAPP gives the hard-decision which is the same as in Eq. (5.2). The LAPP $L(b_l)$ itself can be used as a soft-decision. The absolute value of $L(b_l)$ can indicate the reliability of the decision.

From the likelihood function, the *a posteriori* probability of b_l can be found as follows:

$$\Pr(b_l = +1|\{y_m\}) = C \sum_{\{b_m\},b_l=+1} f(\{y_m\}|\{b_m\}) \Pr(\{b_m\}),$$

$$\Pr(b_l = -1|\{y_m\}) = C \sum_{\{b_m\},b_l=-1} f(\{y_m\}|\{b_m\}) \Pr(\{b_m\}), \quad (5.4)$$

where C is a normalizing constant, $\Pr(\{b_m\})$ denotes the *a priori* probability of the symbol sequence $\{b_m\}$, and $\sum_{\{b_m\},b_l=+1}$ stands for the summation with respect to all the possible symbol sequences with $b_l = +1$. For example, suppose that the length of the bit sequence is 3, i.e. $\{b_0, b_1, b_2\}$, and $l = 1$. Then, the summation $\sum_{\{b_l\},b_l=+1}$ is carried out over the following four sequences:

$$\{+1, \underline{+1}, +1\}, \{+1, \underline{+1}, -1\}, \{-1, \underline{+1}, +1\}, \{-1, \underline{+1}, -1\}.$$

We can define the summation $\sum_{\{b_m\},b_l=-1}$ in a similar way.

Substituting Eq. (5.4) into Eq. (5.3), we obtain the LAPP of b_l as follows:

$$L(b_l) = \log \frac{\sum_{\{b_m\},b_l=+1} f(\{y_m\}|\{b_m\}) \Pr(\{b_m\})}{\sum_{\{b_m\},b_l=-1} f(\{y_m\}|\{b_m\}) \Pr(\{b_m\})}. \quad (5.5)$$

It is clear that the complexity to find the LAPP is prohibitively high, because all the possible combinations should be taken into account in the summations. If the length of the signal sequence is L, there are 2^L signal sequences in computing Eq. (5.5). Hence, the complexity grows exponentially with L. Fortunately, there exist computationally efficient methods, such as the VA. In this section, we introduce one of them, called the BCJR algorithm (Bahl *et al.*, 1974).

[†]   We assume that $\{y_m\}$ for a sequence and **y** for a vector are exchangeable throughout this chapter.

5.1.2 The BCJR algorithm

As in the VA, we can see that the received signal at time l,

$$y_l = \sum_{p=0}^{P-1} h_p b_{l-p} + n_l = h_0 b_l + \sum_{p=1}^{P-1} h_p b_{l-p} + n_l,$$

is decided by the current data symbol, b_l, and the state S_l (of the channel) of the $(P-1)$ past data symbols,

$$\{b_{l-1}, b_{l-2}, \ldots, b_{l-P+1}\}.$$

For convenience, let $\bar{\mathbf{S}}$ denote the set of states ($S_l \in \bar{\mathbf{S}}$). Given b_l and S_l, the likelihood function is readily found as follows:

$$f(y_l | S_l, b_l) = C \exp \left(-\frac{1}{N_0} \left| y_l - \sum_{p=0}^{P-1} h_p b_{l-p} \right|^2 \right), \tag{5.6}$$

where C is a normalizing constant.

To derive the BCJR algorithm for the MAP symbol detection, let us consider the conditional probability of the state transition:

$$\Pr(S_l \to S_{l+1} | \{y_m\}) = \frac{f(S_l \to S_{l+1}, \{y_m\})}{f(\{y_m\})}. \tag{5.7}$$

Then, we can show that

$$\Pr(b_l = +1 | \{y_m\}) = \sum_{ST_{+1}} \Pr(S_l \to S_{l+1} | \{y_m\}),$$

$$\Pr(b_l = -1 | \{y_m\}) = \sum_{ST_{-1}} \Pr(S_l \to S_{l+1} | \{y_m\}),$$

where ST_{+1} and ST_{-1} represent the sets of the state transitions, $S_l \to S_{l+1}$, that are caused by the inputs $b_l = +1$ and $b_l = -1$, respectively, and $\sum_{ST_{+1}}$ and $\sum_{ST_{-1}}$ stand for the summations over the sets ST_{+1} and ST_{-1}, respectively. This is an important observation for reducing the complexity to obtain the LAPP.

Example 5.1.1 Consider an example to find the sets of state transitions. Suppose that $P = 3$. Then, there are $2^{P-1} = 4$ states as follows:

$$\bar{\mathbf{S}} = \{S_l = [b_{l-1}, b_{l-2}] \mid b_{l-1}, b_{l-2} \in \{-1, +1\}\}$$
$$= \{[+1, +1], [+1, -1], [-1, +1], [-1, -1]\}.$$

The set of state transitions caused by $b_l = +1$ input is given by

$$ST_{+1} = \{(S_l \to S_{l+1}) | b_l = +1\}$$
$$= \{([+1, +1] \to [+1, +1]), ([+1, -1] \to [+1, +1]),$$
$$([-1, +1] \to [+1, -1]), ([-1, -1] \to [+1, -1])\}.$$

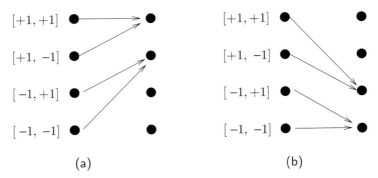

Figure 5.1. Sets of state transitions on the trellis diagram: (a) ST_{+1}, (b) ST_{-1}.

Similarly, ST_{-1} can be found as follows:

$$
\begin{aligned}
ST_{-1} &= \{(\mathcal{S}_l \to \mathcal{S}_{l+1})|b_l = -1\} \\
&= \{([+1, +1] \to [-1, +1]), \; ([+1, -1] \to [-1, +1]), \\
&\quad ([-1, +1] \to [-1, -1]), \; ([-1, -1] \to [-1, -1])\}.
\end{aligned}
$$

The sets of state transitions are shown on the trellis diagram in Fig. 5.1. In general, the number of elements of ST_{+1} or ST_{-1} is 2^{P-1} and the total number of transitions becomes 2^P.

Using Bayes' rule, we can show that the LAPP $L(b_l)$ becomes

$$
\begin{aligned}
L(b_l) &= \log \frac{\sum_{ST_{+1}} \Pr(\mathcal{S}_l \to \mathcal{S}_{l+1}|\{y_m\})}{\sum_{ST_{-1}} \Pr(\mathcal{S}_l \to \mathcal{S}_{l+1}|\{y_m\})} \\
&= \log \frac{\sum_{ST_{+1}} f(\mathcal{S}_l \to \mathcal{S}_{l+1}, \{y_m\})}{\sum_{ST_{-1}} f(\mathcal{S}_l \to \mathcal{S}_{l+1}, \{y_m\})}.
\end{aligned} \tag{5.8}
$$

Comparing this with Eq. (5.5), we can see that Eq. (5.8) can simplify the summation to find the LAPP $L(b_l)$, because the number of terms to be summed in Eq. (5.8) is 2^P (there are 2^P transitions in ST_{+1} and ST_{-1}), while in Eq. (5.5) there are 2^L. Since $P \ll L$ in general, the complexity to find the LAPP in Eq. (5.8) is much lower than that in Eq. (5.5). It is now necessary to find $f(\mathcal{S}_l \to \mathcal{S}_{l+1}, \{y_m\})$ in Eq. (5.8) with low complexity.

It is possible to find $f(\mathcal{S}_l \to \mathcal{S}_{l+1}, \{y_m\})$ with low complexity using recursions. Let

$$
\begin{aligned}
\mathbf{y}_{l,-} &= \{y_0, \quad y_1, \quad \dots, \quad y_{l-1}\}, \\
\mathbf{y}_{l,+} &= \{y_{l+1}, \; y_{l+2}, \quad \dots, \quad y_{L-1}\}.
\end{aligned}
$$

Then, applying the Bayes' rule, we can show that

$$
\begin{aligned}
f(\mathcal{S}_l \to \mathcal{S}_{l+1}, \{y_m\}) &= f(\mathcal{S}_l, \mathcal{S}_{l+1}, \mathbf{y}_{l,+}, \mathbf{y}_{l,-}, y_l) \\
&= f(\mathbf{y}_{l,+}|\mathcal{S}_l, \mathcal{S}_{l+1}, \mathbf{y}_{l,-}, y_l) f(\mathcal{S}_l, \mathcal{S}_{l+1}, \mathbf{y}_{l,-}, y_l) \\
&= f(\mathbf{y}_{l,+}|\mathcal{S}_{l+1}) f(\mathcal{S}_l, \mathcal{S}_{l+1}, \mathbf{y}_{l,-}, y_l) \\
&= f(\mathbf{y}_{l,+}|\mathcal{S}_{l+1}) f(\mathcal{S}_{l+1}, y_l|\mathcal{S}_l, \mathbf{y}_{l,-}) f(\mathcal{S}_l, \mathbf{y}_{l,-}) \\
&= f(\mathbf{y}_{l,+}|\mathcal{S}_{l+1}) f(\mathcal{S}_{l+1}, y_l|\mathcal{S}_l) f(\mathcal{S}_l, \mathbf{y}_{l,-}).
\end{aligned} \tag{5.9}
$$

In Eq. (5.9), we use

$$f(\mathbf{y}_{l,+}|\mathcal{S}_l, \mathcal{S}_{l+1}, \mathbf{y}_{l,-}, y_l) = f(\mathbf{y}_{l,+}|\mathcal{S}_{l+1})$$

(from the second equality to the third equality) since $\mathbf{y}_{l,+}$ depends only on $\mathcal{S}_{l+1}$. In addition, we use

$$f(\mathcal{S}_{l+1}, y_l|\mathcal{S}_l, \mathbf{y}_{l,-}) = f(\mathcal{S}_{l+1}, y_l|\mathcal{S}_l)$$

to obtain the last equality.

Define

$$\gamma(\mathcal{S}_l \rightarrow \mathcal{S}_{l+1}) = f(\mathcal{S}_{l+1}, y_l|\mathcal{S}_l),$$
$$\alpha(\mathcal{S}_l) = f(\mathcal{S}_l, \mathbf{y}_{l,-}), \tag{5.10}$$
$$\beta(\mathcal{S}_{l+1}) = f(\mathbf{y}_{l,+}|\mathcal{S}_{l+1}).$$

We can show that the following recursions exist:

$$\alpha(\mathcal{S}_{l+1}) = \sum_{s_k \in \bar{\mathbf{S}}} \gamma(\mathcal{S}_l = s_k \rightarrow \mathcal{S}_{l+1})\alpha(\mathcal{S}_l = s_k) \tag{5.11}$$

and

$$\beta(\mathcal{S}_l) = \sum_{s_k \in \bar{\mathbf{S}}} \gamma(\mathcal{S}_l \rightarrow \mathcal{S}_{l+1} = s_k)\beta(\mathcal{S}_{l+1} = s_k), \tag{5.12}$$

where the summation takes place over all possible transitions. The recursion in Eq. (5.11) is called the forward recursion, while the recursion in Eq. (5.12) is called the backward recursion due to the updating time direction for each recursion.

To derive Eq. (5.11), using the definitions and Bayes' rule from Eq. (5.10), $\alpha(\mathcal{S}_{l+1})$ is rewritten as follows:

$$\begin{aligned}
\alpha(\mathcal{S}_{l+1}) &= f(\mathcal{S}_{l+1}, \mathbf{y}_{l+1,-}) \\
&= f(\mathcal{S}_{l+1}, \mathbf{y}_{l,-}, y_l) \\
&= \sum_{s_k \in \bar{\mathbf{S}}} f(\mathcal{S}_{l+1}, \mathbf{y}_{l,-}, \mathcal{S}_l = s_k, y_l) \quad \text{(marginalization)} \\
&= \sum_{s_k \in \bar{\mathbf{S}}} f(\mathcal{S}_{l+1}, y_l|\mathcal{S}_l = s_k, \mathbf{y}_{l,-})f(\mathcal{S}_l = s_k, \mathbf{y}_{l,-}).
\end{aligned}$$

Since $\{\mathcal{S}_{l+1}, y_l\}$ depends on the previous state $\mathcal{S}_l$, not $\mathbf{y}_{l,-}$, i.e.

$$f(\mathcal{S}_{l+1}, y_l|\mathcal{S}_l = s_k, \mathbf{y}_{l,-}) = f(\mathcal{S}_{l+1}, y_l|\mathcal{S}_l = s_k),$$

we can show that

$$\alpha(\mathcal{S}_{l+1}) = \sum_{s_k \in \bar{\mathbf{S}}} \underbrace{f(\mathcal{S}_{l+1}, y_l|\mathcal{S}_l = s_k)}_{=\gamma(\mathcal{S}_l = s_k \rightarrow \mathcal{S}_{l+1})} \underbrace{f(\mathcal{S}_l = s_k, \mathbf{y}_{l,-})}_{=\alpha(\mathcal{S}_l = s_k)}.$$

This becomes Eq. (5.11).

The backward recursion in Eq. (5.12) can be derived by a similar approach. By definition, from Eq. (5.10), $\beta(\mathcal{S}_l)$ is rewritten as follows:

$$
\begin{aligned}
\beta(\mathcal{S}_l) &= f(\mathbf{y}_{l-1,+}|\mathcal{S}_l) \\
&= f(y_l, \mathbf{y}_{l,+}|\mathcal{S}_l) \\
&= \sum_{s_k \in \tilde{\mathbf{S}}} f(y_l, \mathbf{y}_{l,+}, \mathcal{S}_{l+1} = s_k|\mathcal{S}_l) \quad \text{(marginalization)} \\
&= \sum_{s_k \in \tilde{\mathbf{S}}} \frac{f(y_l, \mathbf{y}_{l,+}, \mathcal{S}_{l+1} = s_k, \mathcal{S}_l)}{f(\mathcal{S}_l)} \quad \text{(Bayes' rule)}.
\end{aligned}
\tag{5.13}
$$

Applying Bayes' rule repeatedly, the numerator can be rewritten as follows:

$$
\begin{aligned}
f(y_l, \mathbf{y}_{l,+}, \mathcal{S}_{l+1} = s_k, \mathcal{S}_l) &= f(\mathbf{y}_{l,+}, \mathcal{S}_{l+1} = s_k) f(y_l, \mathcal{S}_l|\mathbf{y}_{l,+}, \mathcal{S}_{l+1} = s_k) \\
&= f(\mathbf{y}_{l,+}|\mathcal{S}_{l+1} = s_k) f(\mathcal{S}_{l+1} = s_k) f(y_l, \mathcal{S}_l|\mathbf{y}_{l,+}, \mathcal{S}_{l+1} = s_k).
\end{aligned}
\tag{5.14}
$$

Since

$$
f(y_l, \mathcal{S}_l|\mathbf{y}_{l,+}, \mathcal{S}_{l+1} = s_k) = f(y_l, \mathcal{S}_l|\mathcal{S}_{l+1} = s_k),
$$

it follows that

$$
\begin{aligned}
f(\mathcal{S}_{l+1} = s_k) f(y_l, \mathcal{S}_l|\mathbf{y}_{l,+}, \mathcal{S}_{l+1} = s_k) &= f(\mathcal{S}_{l+1} = s_k) f(y_l, \mathcal{S}_l|\mathcal{S}_{l+1} = s_k) \\
&= f(y_l, \mathcal{S}_l, \mathcal{S}_{l+1} = s_k).
\end{aligned}
\tag{5.15}
$$

Substituting Eq. (5.15) into Eq. (5.14), the numerator becomes

$$
f(y_l, \mathbf{y}_{l,+}, \mathcal{S}_{l+1} = s_k, \mathcal{S}_l) = f(\mathbf{y}_{l,+}|\mathcal{S}_{l+1} = s_k) f(y_l, \mathcal{S}_l, \mathcal{S}_{l+1} = s_k).
$$

Substituting this into Eq. (5.13), we can show that

$$
\begin{aligned}
\beta(\mathcal{S}_l) &= \sum_{s_k \in \tilde{\mathbf{S}}} \frac{f(\mathbf{y}_{l,+}|\mathcal{S}_{l+1} = s_k) f(y_l, \mathcal{S}_l, \mathcal{S}_{l+1} = s_k)}{f(\mathcal{S}_l)} \\
&= \sum_{s_k \in \tilde{\mathbf{S}}} \underbrace{f(\mathbf{y}_{l,+}|\mathcal{S}_{l+1} = s_k)}_{=\beta(\mathcal{S}_{l+1}=s_k)} \underbrace{f(y_l, \mathcal{S}_{l+1} = s_k|\mathcal{S}_l)}_{=\gamma(\mathcal{S}_l \to \mathcal{S}_{l+1}=s_k)}.
\end{aligned}
\tag{5.16}
$$

This becomes Eq. (5.12).

The quantity $\gamma(\mathcal{S}_l \to \mathcal{S}_{l+1})$ is given by

$$
\gamma(\mathcal{S}_l \to \mathcal{S}_{l+1}) = \begin{cases} \Pr(b_l) f(y_l|\mathcal{S}_l, b_l), & \text{for valid transition;} \\ 0, & \text{for invalid transition,} \end{cases}
\tag{5.17}
$$

where $f(y_l|\mathcal{S}_l, b_l)$ is given in Eq. (5.6). Substituting Eqs (5.10) and (5.9) into Eq. (5.8), it is finally shown that

$$
L(b_l) = \log \frac{\sum_{ST_{+1}} \gamma(\mathcal{S}_l \to \mathcal{S}_{l+1}) \alpha(\mathcal{S}_l) \beta(\mathcal{S}_{l+1})}{\sum_{ST_{-1}} \gamma(\mathcal{S}_l \to \mathcal{S}_{l+1}) \alpha(\mathcal{S}_l) \beta(\mathcal{S}_{l+1})}.
\tag{5.18}
$$

In summary, the MAP symbol detection can be carried out as follows.

(i) Compute the quantities $\gamma(S_l \rightarrow S_{l+1})$.
(ii) Compute the α's and β's from the γ's using the forward and backward recursions in Eqs (5.11) and (5.12), respectively.
(iii) Compute the LAPP in Eq. (5.18).

The values for $\alpha(S_0)$ and $\beta(S_L)$ should be determined according to the initial conditions.

Example 5.1.2 Now we show how the BCJR algorithm works with a numerical example.
Suppose that $h_0 = h_1 = 1$, where $P = 2$. In addition, assume that $b_{-1} = 1$ and it is known by the receiver. State S_l of this ISI channel becomes b_{l-1} and has one of two values, -1 or 1. It is assumed that the transmitted signal sequence is $\{b_0, b_1, b_2\} = \{1, -1, 1\}$ and that the noise sequence is $\{n_0, n_1, n_2\} = \{-0.1, 0.1, -0.1\}$. Then, the received signal vector is given by

$$\begin{bmatrix} y_0 \\ y_1 \\ y_2 \end{bmatrix} = \begin{bmatrix} h_0 & 0 & 0 \\ h_1 & h_0 & 0 \\ 0 & h_1 & h_0 \end{bmatrix} \begin{bmatrix} b_0 \\ b_1 \\ b_2 \end{bmatrix} + \begin{bmatrix} n_0 \\ n_1 \\ n_2 \end{bmatrix} + \begin{bmatrix} h_1 b_{-1} \\ 0 \\ 0 \end{bmatrix} = \begin{bmatrix} 1.9 \\ 0.1 \\ -0.1 \end{bmatrix}.$$

In addition, assume that $N_0 = 1$ and $\Pr(b_l = \pm 1) = 1/2$ for all l. The LAPP can be found as follows.

Step 1. To obtain $\gamma(S_l \rightarrow S_{l+1})$, we need to find the likelihood functions as follows:

$$f(y_0|S_0 = \{1\}, b_0 = 1) = C \exp\left(-\frac{1}{N_0}|y_0 - 2|^2\right),$$

$$f(y_0|S_0 = \{1\}, b_0 = -1) = C \exp\left(-\frac{1}{N_0}|y_0 - 0|^2\right),$$

$$f(y_l|S_l = \{1\}, b_l = 1) = C \exp\left(-\frac{1}{N_0}|y_l - 2|^2\right),$$

$$f(y_l|S_l = \{1\}, b_l = -1) = C \exp\left(-\frac{1}{N_0}|y_l - 0|^2\right),$$

$$f(y_l|S_l = \{-1\}, b_l = 1) = C \exp\left(-\frac{1}{N_0}|y_l - 0|^2\right),$$

$$f(y_l|S_l = \{-1\}, b_l = -1) = C \exp\left(-\frac{1}{N_0}|y_l + 2|^2\right), \quad l = 1, 2.$$

For convenience, let $C = 1$. (In fact, the value of C is not 1. However, it does not affect the LAPP because C appears in both the numerator and the denominator in Eq. (5.18).) From $f(y_l|S_l, b_l)$ in the above, we find

$$\gamma(S_0 = \{1\} \rightarrow S_1 = \{1\}) = \tfrac{1}{2}\exp(-|1.9 - 2|^2) = 0.4950,$$
$$\gamma(S_0 = \{1\} \rightarrow S_1 = \{-1\}) = \tfrac{1}{2}\exp(-|1.9 - 0|^2) = 0.0135,$$
$$\gamma(S_1 = \{1\} \rightarrow S_2 = \{1\}) = \tfrac{1}{2}\exp(-|0.1 - 2|^2) = 0.0135,$$
$$\gamma(S_1 = \{1\} \rightarrow S_2 = \{-1\}) = \tfrac{1}{2}\exp(-|0.1 - 0|^2) = 0.4950,$$
$$\gamma(S_1 = \{-1\} \rightarrow S_2 = \{1\}) = \tfrac{1}{2}\exp(-|0.1 - 0|^2) = 0.4950,$$

$$\gamma(\mathcal{S}_1 = \{-1\} \rightarrow \mathcal{S}_2 = \{-1\}) = \tfrac{1}{2}\exp(-|0.1+2|^2) \quad = 0.0061,$$
$$\gamma(\mathcal{S}_2 = \{1\} \rightarrow \mathcal{S}_3 = \{1\}) = \tfrac{1}{2}\exp(-|-0.1-2|^2) = 0.0061,$$
$$\gamma(\mathcal{S}_2 = \{1\} \rightarrow \mathcal{S}_3 = \{-1\}) = \tfrac{1}{2}\exp(-|-0.1-0|^2) = 0.4950,$$
$$\gamma(\mathcal{S}_2 = \{-1\} \rightarrow \mathcal{S}_3 = \{1\}) = \tfrac{1}{2}\exp(-|-0.1-0|^2) = 0.4950,$$
$$\gamma(\mathcal{S}_2 = \{-1\} \rightarrow \mathcal{S}_3 = \{-1\}) = \tfrac{1}{2}\exp(-|-0.1+2|^2) = 0.0135.$$

The forward recursion is carried out as follows:

$$\alpha(\mathcal{S}_0 = \{b_{-1}\} = \{1\}) = 1,$$
$$\alpha(\mathcal{S}_0 = \{b_{-1}\} = \{-1\}) = 0,$$
$$\alpha(\mathcal{S}_1 = \{1\}) = \gamma(\mathcal{S}_0 = \{1\} \rightarrow \mathcal{S}_1 = \{1\}) \times \alpha(\mathcal{S}_0 = \{1\})$$
$$= 0.4950,$$
$$\alpha(\mathcal{S}_1 = \{-1\}) = \gamma(\mathcal{S}_0 = \{1\} \rightarrow \mathcal{S}_1 = \{-1\}) \times \alpha(\mathcal{S}_0 = \{1\})$$
$$= 0.0135,$$
$$\alpha(\mathcal{S}_2 = \{1\}) = \gamma(\mathcal{S}_1 = \{1\} \rightarrow \mathcal{S}_2 = \{1\}) \times \alpha(\mathcal{S}_1 = \{1\})$$
$$+ \gamma(\mathcal{S}_1 = \{-1\} \rightarrow \mathcal{S}_2 = \{1\}) \times \alpha(\mathcal{S}_1 = \{-1\})$$
$$= 0.0134,$$
$$\alpha(\mathcal{S}_2 = \{-1\}) = \gamma(\mathcal{S}_1 = \{1\} \rightarrow \mathcal{S}_2 = \{-1\}) \times \alpha(\mathcal{S}_1 = \{1\})$$
$$+ \gamma(\mathcal{S}_1 = \{-1\} \rightarrow \mathcal{S}_2 = \{-1\}) \times \alpha(\mathcal{S}_1 = \{-1\})$$
$$= 0.2451,$$
$$\alpha(\mathcal{S}_3 = \{1\}) = 0.1214,$$
$$\alpha(\mathcal{S}_3 = \{-1\}) = 0.0099.$$

The results of the backward recursion are obtained as follows:

$$\beta(\mathcal{S}_3 = \{1\}) = \beta(\mathcal{S}_3 = \{-1\}) = 1 \;\text{(initialization)},$$
$$\beta(\mathcal{S}_2 = \{1\}) = \gamma(\mathcal{S}_2 = \{1\} \rightarrow \mathcal{S}_3 = \{1\}) \times \beta(\mathcal{S}_3 = \{1\})$$
$$+ \gamma(\mathcal{S}_2 = \{1\} \rightarrow \mathcal{S}_3 = \{-1\}) \times \beta(\mathcal{S}_3 = \{-1\})$$
$$= 0.5011,$$
$$\beta(\mathcal{S}_2 = \{-1\}) = \gamma(\mathcal{S}_2 = \{-1\} \rightarrow \mathcal{S}_3 = \{1\}) \times \beta(\mathcal{S}_3 = \{1\})$$
$$+ \gamma(\mathcal{S}_2 = \{-1\} \rightarrow \mathcal{S}_3 = \{-1\}) \times \beta(\mathcal{S}_3 = \{-1\})$$
$$= 0.5085,$$
$$\beta(\mathcal{S}_1 = \{1\}) = 0.2585,$$
$$\beta(\mathcal{S}_1 = \{-1\}) = 0.2511,$$
$$\beta(\mathcal{S}_0 = \{1\}) = 0.1313,$$
$$\beta(\mathcal{S}_0 = \{-1\}) = 0 \quad \text{(this probability is zero as } \mathcal{S}_0 = \{-1\} \text{ is not true).}$$

Now, the LAPP of b_l can be found as follows:

$$L(b_0) = \log \frac{\gamma(\mathcal{S}_0 = \{1\} \to \mathcal{S}_1 = \{1\})\alpha(\mathcal{S}_0 = \{1\})\beta(\mathcal{S}_1 = \{1\})}{\gamma(\mathcal{S}_0 = \{1\} \to \mathcal{S}_1 = \{-1\})\alpha(\mathcal{S}_0 = \{1\})\beta(\mathcal{S}_1 = \{-1\})}$$

$$= 3.6309,$$

$$L(b_1) = \log \frac{\left(\begin{array}{c} \gamma(\mathcal{S}_1 = \{1\} \to \mathcal{S}_2 = \{1\})\alpha(\mathcal{S}_1 = \{1\})\beta(\mathcal{S}_2 = \{1\}) \\ +\gamma(\mathcal{S}_1 = \{-1\} \to \mathcal{S}_2 = \{1\})\alpha(\mathcal{S}_1 = \{-1\})\beta(\mathcal{S}_2 = \{1\}) \end{array} \right)}{\left(\begin{array}{c} \gamma(\mathcal{S}_1 = \{1\} \to \mathcal{S}_2 = \{-1\})\alpha(\mathcal{S}_1 = \{1\})\beta(\mathcal{S}_2 = \{-1\}) \\ +\gamma(\mathcal{S}_1 = \{-1\} \to \mathcal{S}_2 = \{-1\})\alpha(\mathcal{S}_1 = \{-1\})\beta(\mathcal{S}_2 = \{-1\}) \end{array} \right)}$$

$$= -2.8807,$$

$$L(b_2) = 2.5024.$$

After taking the signs of $L(b_l)$, we can see that the hard-decisions are correct.

5.1.3 Comparison with MLSD

For an ISI channel with the following CIR:

$$\mathbf{h} = [0.227\ \ 0.460\ \ 0.688\ \ 0.460\ \ 0.227]^{\mathrm{T}}, \tag{5.19}$$

simulations are carried out. The BER results are compared with those of the MLSD for various values of SNR in Fig. 5.2. The SNR is given by

$$\mathrm{SNR} = \frac{\|\mathbf{h}\|^2}{N_0} = \frac{1}{N_0}.$$

It is shown that the MLSD and the MAP symbol detection provide almost the same BER performance. Therefore, as long as the hard-decision is considered, the MLSD would be preferable to the MAP symbol detection, because the VA has less complexity than the BCJR algorithm.

One of the differences from the MLSD is that the MAP symbol detection exploits not only observations, but also prior information. However, even though the standard MLSD relies on observations, the MLSD can be generalized to accommodate prior information about the symbols. In this case, the MLSD becomes the MAP *sequence* detection.

It is straightforward to generalize the MLSD for the MAP sequence detection. Suppose that the *a priori* probability of symbol sequence $\mathbf{b}$, denoted by $\mathrm{Pr}(\mathbf{b})$, is available. Then, the MAP sequence detection can be formulated as follows:

$$\mathbf{b}_{\mathrm{map}} = \arg \max_{\mathbf{b}} f(\mathbf{b}|\mathbf{y})$$

$$= \arg \max_{\mathbf{b}} f(\mathbf{y}|\mathbf{b}) \mathrm{Pr}(\mathbf{b}). \tag{5.20}$$

If the symbols are independent, we can show that

$$\log (f(\mathbf{y}|\mathbf{b}) \mathrm{Pr}(\mathbf{b})) = \log \left(\prod_{l=0}^{L-1} f(y_l|\mathbf{b}_l) \mathrm{Pr}(b_l) \right)$$

$$= \sum_{l=0}^{L-1} \log f(y_l|\mathbf{b}_l) + \log \mathrm{Pr}(b_l), \tag{5.21}$$

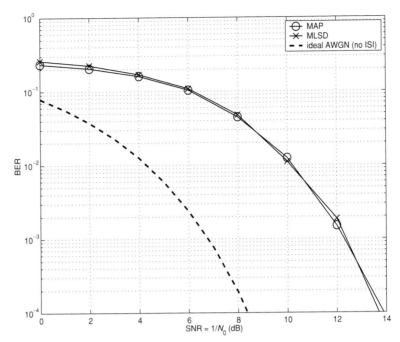

Figure 5.2. Bit error rate results of the MAP symbol detection and MLSD with the ideal BER which can be obtained without ISI.

where

$$\mathbf{b}_l = [b_l \ b_{l-1} \ \dots \ b_{l-P+1}]^{\mathrm{T}}.$$

With the branch metric defined as follows:

$$M_l = \log f(y_l|\mathbf{b}_l) + \log \Pr(b_l), \tag{5.22}$$

the VA can be applied to find the MAP sequence. Comparing with the MLSD detection, the branch metric M_l has an extra term, $\log \Pr(b_l)$, as the *a priori* information about the symbols. Note that this extension of the VA for the MAP sequence detection was discussed in Forney (1973). We have some remarks as follows.

- If the symbols are not independent, Eq. (5.21) is not valid as $\Pr(\mathbf{b}) \neq \prod_{l=0}^{L-1} \Pr(b_l)$ in general. In this case, the VA is not applicable to the MAP sequence detection.
- The BCJR algorithm solves the MAP symbol detection, while the VA can solve the MAP sequence detection. Hence, the results can be different.
- The VA can be generalized to provide soft-outputs (Hagenauer, Offer, and Papke, 1996). This modified approach is called the soft-output Viterbi algorithm (SOVA).
- To perform the MAP symbol or sequence detection, the noise variance is necessary, while the MLSD does not require the noise variance.

The performance of the MAP symbol detection is degraded if the noise variance is different from the true noise variance. In addition, when the CIR is estimated, the performance

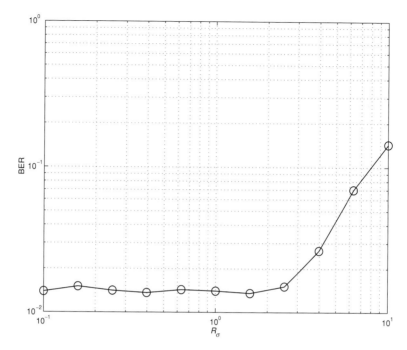

Figure 5.3. Bit error rate versus R_σ.

can be degraded. Firstly, consider the case that the noise variance is different from the true one. It is assumed that the CIR is known exactly and that the symbols are equally likely. Define the ratio as follows:

$$R_\sigma = \frac{\hat{\sigma}_n^2}{\sigma_n^2},$$

where $\hat{\sigma}_n^2$ denotes the estimated noise variance and $\sigma_n^2 = N_0/2$ is the true noise variance. The BER results from simulations are shown in Fig. 5.3 when the SNR is 10 dB with the CIR in Eq. (5.19). If $R_\sigma < 1$ (i.e., the noise variance is underestimated), there is no significant performance degradation. However, when $R_\sigma > 1$ (i.e., the noise variance is overestimated), the performance is degraded. If the noise variance is overestimated, the observation is less weighted and the *a priori* information is overly weighted. Since the data symbols are equally likely, the observation is more important for correct decisions. Thus, we can see that an overestimated noise variance can result in a much greater performance degradation than an underestimated noise variance.

To see the impact of the CIR estimation error, the estimated CIR is modeled as

$$\hat{\mathbf{h}} = \mathbf{h} + \alpha\mathbf{e_h},$$

where $\mathbf{e_h}$ is a zero-mean Gaussian random vector with $E[\mathbf{e_h}\mathbf{e_h^T}] = \mathbf{I}$ and α is a positive constant. The CIR estimation error increases with α and the amount of error in dB is

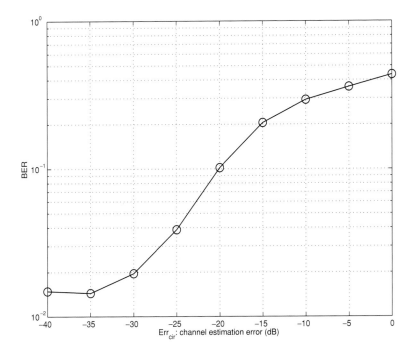

Figure 5.4. Bit error rate versus Err_{cir}.

represented by

$$\text{Err}_{\text{cir}} = 20 \log_{10} \alpha.$$

In Fig. 5.4, the BER increases with Err_{cir}. This shows that the channel estimation is important to maintain good performance.

5.2   ML channel estimation and the EM algorithm

To perform the MAP symbol detection or MLSD, it is necessary to estimate the CIR. In addition, a good channel estimate is desirable so as not to degrade the performance, as shown in Fig. 5.4. To achieve a good channel estimate, the ML channel estimation as an optimal channel estimation can be considered. In general, it is not easy to find a closed form for the optimal solution to the ML channel estimation problem. However, the EM algorithm can be applied to the ML channel estimation problem to find a solution numerically through iterations. In this section, the EM algorithm will be introduced and applied to the ML channel estimation problem.

We also study how the MAP symbol detection in Section 5.1 can be used in the EM-based iterative channel estimation. Since the MAP symbol detection becomes a part of the EM-based iterative channel estimation, we can also obtain the MAP estimates of data symbols.

5.2.1 EM algorithm

The EM algorithm (Dempster *et al.*, 1977) is an iterative numerical method used to solve an ML problem. Suppose that the pdf of a random vector $\mathbf{x}$ is given by

$$f(\mathbf{x}|\theta),$$

where θ is the vector of parameters of interest. The pdf of $\mathbf{x}$ becomes the likelihood function of θ when $\mathbf{x}$ is given. The ML estimate of θ is the solution that maximizes the likelihood function given $\mathbf{x}$ such as

$$
\begin{aligned}
\hat{\theta}_{\mathrm{ml}} &= \arg\max_{\theta} f(\mathbf{x}|\theta) \\
&= \arg\max_{\theta} \log f(\mathbf{x}|\theta),
\end{aligned}
\tag{5.23}
$$

where $\hat{\theta}_{\mathrm{ml}}$ is the ML estimate of θ and is generally a function of $\mathbf{x}$.

The ML estimation problem in Eq. (5.23) can be considered as an optimization problem, and it may not be easily solved if $f(\mathbf{x}|\theta)$ is highly nonlinear with respect to θ. Consider another random vector $\mathbf{z}$ which is not observable ($\mathbf{z}$ is often called the missing data), but related to $\mathbf{x}$. The joint pdf of $\{\mathbf{x}, \mathbf{z}\}$ is given by

$$f(\mathbf{x}, \mathbf{z}|\theta),$$

where $\{\mathbf{x}\}$ is called the incomplete data and $\{\mathbf{x}, \mathbf{z}\}$ is called the complete data. Clearly, using the marginalization, we have

$$f(\mathbf{x}|\theta) = \int f(\mathbf{x}, \mathbf{z}|\theta)\,d\mathbf{z}.$$

In general, the random vector $\mathbf{z}$ is chosen carefully so that the ML estimate of θ can easily be found with $f(\mathbf{x}, \mathbf{z}|\theta)$. The EM algorithm uses the missing data, $\mathbf{z}$, to simplify the optimization process.

To proceed, we need to use the following relation:

$$f(\mathbf{x}, \mathbf{z}|\theta) = f(\mathbf{z}|\mathbf{x}, \theta) f(\mathbf{x}|\theta) \quad \text{(from Bayes' rule)}$$

or

$$\log f(\mathbf{x}|\theta) = \log f(\mathbf{x}, \mathbf{z}|\theta) - \log f(\mathbf{z}|\mathbf{x}, \theta). \tag{5.24}$$

Suppose that we have the conditional pdf of $\mathbf{z}$ given $\mathbf{x}$ and $\theta^{(q)}$, where $\theta^{(q)}$ denotes the estimate of θ at the qth iteration. Note that the EM algorithm is an iterative algorithm. With this conditional pdf, which is denoted by $f(\mathbf{z}|\mathbf{x}, \theta^{(q)})$, from Eq. (5.24), the expectation of $\log f(\mathbf{x}|\theta)$ is given by

$$
\begin{aligned}
E[\log f(\mathbf{x}|\theta)] &= \int \log f(\mathbf{x}|\theta) f(\mathbf{z}|\mathbf{x}, \theta^{(q)})\,d\mathbf{z} \\
&= \int \log f(\mathbf{x}, \mathbf{z}|\theta) f(\mathbf{z}|\mathbf{x}, \theta^{(q)})\,d\mathbf{z} - \int \log f(\mathbf{z}|\mathbf{x}, \theta) f(\mathbf{z}|\mathbf{x}, \theta^{(q)})\,d\mathbf{z}.
\end{aligned}
$$

Since

$$\int \log f(\mathbf{x}|\theta) f(\mathbf{z}|\mathbf{x}, \theta^{(q)})\,d\mathbf{z} = \log f(\mathbf{x}|\theta),$$

i.e. $E[\log f(\mathbf{x}|\theta)] = \log f(\mathbf{x}|\theta)$, it follows that

$$\log f(\mathbf{x}|\theta) = \int \log f(\mathbf{x}, \mathbf{z}|\theta) f(\mathbf{z}|\mathbf{x}, \theta^{(q)}) \, d\mathbf{z} - \int \log f(\mathbf{z}|\mathbf{x}, \theta) f(\mathbf{z}|\mathbf{x}, \theta^{(q)}) \, d\mathbf{z}. \quad (5.25)$$

From this, we can see that the maximization of the right hand side of Eq. (5.25) provides the ML estimate in Eq. (5.23). The relation in Eq. (5.25) plays a key role in deriving the EM algorithm.

For convenience, define

$$Q(\theta|\theta^{(q)}) = \int \log f(\mathbf{x}, \mathbf{z}|\theta) f(\mathbf{z}|\mathbf{x}, \theta^{(q)}) \, d\mathbf{z}$$

and

$$H(\theta|\theta^{(q)}) = \int \log f(\mathbf{z}|\mathbf{x}, \theta) f(\mathbf{z}|\mathbf{x}, \theta^{(q)}) \, d\mathbf{z}.$$

For a given estimate, $\theta^{(q)}$, a better estimate $\theta^{(q+1)}$ can be found if

$$f(\mathbf{x}|\theta^{(q+1)}) \geq f(\mathbf{x}|\theta^{(q)}). \quad (5.26)$$

According to Eq. (5.25), this implies that

$$Q(\theta^{(q+1)}|\theta^{(q)}) - H(\theta^{(q+1)}|\theta^{(q)}) \geq Q(\theta^{(q)}|\theta^{(q)}) - H(\theta^{(q)}|\theta^{(q)}).$$

Note that

$$H(\theta^{(q)}|\theta^{(q)}) - H(\theta^{(q+1)}|\theta^{(q)}) \geq 0 \quad (5.27)$$

for *any* $\theta^{(q+1)}$ by using an inequality of the information divergence (Cover and Thomas, 1991). From this, we can show that

$$Q(\theta^{(q+1)}|\theta^{(q)}) \geq Q(\theta^{(q)}|\theta^{(q)}) \quad (5.28)$$

guarantees the inequality in Eq. (5.26). Hence, if we find any $\theta^{(q+1)}$ which satisfies Eq. (5.28), the next estimate $\theta^{(q+1)}$ is better than the current estimate $\theta^{(q)}$ according to Eq. (5.26) in terms of the likelihood. Through iterations using Eq. (5.28), we can find a sequence of estimates that increases the likelihood. After convergence, a local maximum of $\log f(\mathbf{x}|\theta)$ can be obtained. This iterative method is called the EM algorithm. We can summarize the EM algorithm as follows.

(i) Find an initial estimate of θ, say $\theta^{(0)}$.

(ii) Let $q = 0$.

(iii) (E-step) Compute $Q(\theta|\theta^{(q)})$. That is,

$$\begin{aligned} Q(\theta|\theta^{(q)}) &= E\big[\log f(\mathbf{x}, \mathbf{z}|\theta) | \mathbf{x}, \theta^{(q)} \big] \\ &= \int \log f(\mathbf{x}, \mathbf{z}|\theta) f(\mathbf{z}|\mathbf{x}, \theta^{(q)}) \, d\mathbf{z}, \end{aligned} \quad (5.29)$$

where the expectation is carried out over the missing data.

(iv) (M-step) Find $\theta^{(q+1)}$ that maximizes $Q(\theta|\theta^{(q)})$. That is,

$$\theta^{(q+1)} = \arg\max_{\theta} Q(\theta|\theta^{(q)}). \quad (5.30)$$

(v) If $\|\theta^{(q+1)} - \theta^{(q)}\| < \epsilon$ (or $|Q(\theta^{(q+1)}|\theta^{(q)}) - Q(\theta^{(q)}|\theta^{(q-1)})| < \bar{\epsilon}$), stop. Here, ϵ and $\bar{\epsilon}$ are small positive numbers. Otherwise, let $q = q + 1$ and go to (iii) (the E-step).

The rate of convergence is an important issue for iterative methods. The rate of convergence of the EM algorithm depends on the missing data and the ML problem itself. It is generally known that the EM algorithm has a slow rate of convergence if the dimension of the parameter vector is large (i.e., if there are a large number of parameters to be estimated). Thus, to improve the rate of convergence, a number of variations of the EM algorithm, including the space-alternating generalized EM (SAGE) algorithm (Fessler and Hero, 1994), are proposed; see McLachlan and Krishnan (1997) for details.

Example 5.2.1   To show the inequality in Eq. (5.27), it is sufficient to show that

$$\int \log f(\mathbf{x}) f(\mathbf{x}) \, d\mathbf{x} \geq \int \log g(\mathbf{x}) f(\mathbf{x}) \, d\mathbf{x}$$

or

$$\int f(\mathbf{x}) \log \frac{g(\mathbf{x})}{f(\mathbf{x})} \, d\mathbf{x} \leq 0,$$

where $f(\mathbf{x})$ and $g(\mathbf{x})$ are two different pdfs. Since $\log x \leq x - 1$ for $x \geq 0$, we can show that

$$\int f(\mathbf{x}) \log \frac{g(\mathbf{x})}{f(\mathbf{x})} \, d\mathbf{x} \leq \int f(\mathbf{x}) \left(\frac{g(\mathbf{x})}{f(\mathbf{x})} - 1 \right) \, d\mathbf{x} = 0.$$

Hence, it is verified that Eq. (5.27) is true.

It is noteworthy that the EM algorithm can find a local maximum of $\log f(\mathbf{x}|\theta)$ or $f(\mathbf{x}|\theta)$. If $\log f(\mathbf{x}|\theta)$ is concave, a local maximum becomes the global maximum that corresponds to the ML estimate in Eq. (5.23). If $\log f(\mathbf{x}|\theta)$ has multiple local maxima, the initial estimate $\theta^{(0)}$ is important and needs to be sufficiently close to the global maximum to converge to the global maximum.

We can consider a simple example to see how the EM algorithm works. Consider a received signal sequence over the AWGN channel:

$$y_l = \mu b_l + n_l, \quad l = 0, 1, \ldots, L - 1, \tag{5.31}$$

where μ is the signal amplitude and n_l is the additive white Gaussian noise with zero mean and variance σ^2. Assume that $b_l \in \{-1, +1\}$ is equally likely and independent. Suppose that the two parameters, μ and σ^2, are to be estimated by only the received signal sequence $\{y_0, y_1, \ldots, y_{L-1}\}$. The ML estimation of μ and σ^2 is given by

$$\{\hat{\mu}_{\mathrm{ml}}, \hat{\sigma}^2_{\mathrm{ml}}\} = \arg\max_{\mu, \sigma^2} f(\mathbf{y}|\mu, \sigma^2), \tag{5.32}$$

where $\mathbf{y} = [y_0 \ y_1 \ \cdots \ y_{L-1}]^{\mathrm{T}}$ and $f(\mathbf{y}|\mu, \sigma^2)$ is the likelihood function. It is not easy to find values of μ and σ^2 that maximize the likelihood function because $f(\mathbf{y}|\mu, \sigma^2)$ is complicated. Note that if $\mathbf{b} = [b_0 \ b_1 \ \cdots \ b_{L-1}]$ is known or conditioned, the likelihood function can be easily found as follows:

$$f(\mathbf{y}|\mathbf{b}, \mu, \sigma^2) = \prod_{l=0}^{L-1} \frac{1}{\sqrt{2\pi\sigma^2}} e^{-\frac{1}{2\sigma^2}(y_l - \mu b_l)^2}. \tag{5.33}$$

Then, the likelihood function, $f(\mathbf{y}|\mu, \sigma^2)$, can be obtained from $f(\mathbf{y}|\mathbf{b}, \mu, \sigma^2)$ as follows:

$$
\begin{aligned}
f(\mathbf{y}|\mu, \sigma^2) &= \sum_{\mathbf{b}} f(\mathbf{y}, \mathbf{b}|\mu, \sigma^2) \\
&= \sum_{\mathbf{b}} f(\mathbf{y}|\mathbf{b}, \mu, \sigma^2) \Pr(\mathbf{b}|\mu, \sigma^2) \\
&= \sum_{\mathbf{b}} f(\mathbf{y}|\mathbf{b}, \mu, \sigma^2) \Pr(\mathbf{b}),
\end{aligned}
\tag{5.34}
$$

where the summation is carried out over all the $\mathbf{b}$'s (there are 2^L binary vectors for $\mathbf{b}$). Since b_l is equally likely and independent, $\Pr(\mathbf{b})$ is the same constant for all the possible $\mathbf{b}$'s. As shown in Eq. (5.34), the likelihood function itself is hard to obtain as 2^L functions are to be summed. On the other hand, the conditional likelihood function in Eq. (5.33) is easy to deal with.

With $\mathbf{b}$ as the missing data, the complete data is given by $\{\mathbf{y}, \mathbf{b}\}$. Let $\theta = \{\mu, \sigma^2\}$ and denote the qth estimate by $\theta^{(q)}$. For notational convenience, let $v = \sigma^2$ and let $\hat{\mu}^{(q)}$ and $\hat{v}^{(q)}$ be the qth estimates of μ and $v = \sigma^2$, respectively. In the E-step, we find

$$Q(\theta|\theta^{(q)}) = E[\log f(\mathbf{y}, \mathbf{b}|\theta)|\mathbf{y}, \theta^{(q)}]. \tag{5.35}$$

Since

$$f(\mathbf{y}, \mathbf{b}|\theta) = f(\mathbf{y}|\mathbf{b}, \theta) \Pr(\mathbf{b}|\theta) = f(\mathbf{y}|\mathbf{b}, \theta) \Pr(\mathbf{b})$$

and $\Pr(\mathbf{b}) = 1/2^L$, we have

$$Q(\theta|\theta^{(q)}) = \frac{1}{2^L} \sum_{\mathbf{b}} \log f(\mathbf{y}|\mathbf{b}, \theta) \Pr(\mathbf{b}|\mathbf{y}, \theta^{(q)}). \tag{5.36}$$

The major difficulty in terms of complexity is to find the *a posteriori* probabilities, $\Pr(\mathbf{b}|\mathbf{y}, \theta^{(q)})$, for all the 2^L $\mathbf{b}$'s. To reduce the computational complexity, we can *assume* the following independency:

$$\Pr(\mathbf{b}|\mathbf{y}, \theta^{(q)}) = \prod_{l=0}^{L-1} P_l^{(q)}(b_l), \tag{5.37}$$

where $P_l^{(q)}(b_l) = \Pr(b_l|y_l, \hat{\mu}^{(q)}, \hat{v}^{(q)})$.

With given $\{\mu, \sigma^2\}$, we can obtain the *a posteriori* probability of individual b_l's as follows:

$$\Pr(b_l = 1|y_l, \mu, \sigma^2) = \frac{\exp\left(-\frac{1}{2\sigma^2}(y_l - \mu)^2\right)}{\exp\left(-\frac{1}{2\sigma^2}(y_l - \mu)^2\right) + \exp\left(-\frac{1}{2\sigma^2}(y_l + \mu)^2\right)},$$

$$\Pr(b_l = -1|y_l, \mu, \sigma^2) = 1 - \Pr(b_l = 1|y_l, \mu, \sigma^2), \quad l = 0, 1, \dots, L-1.$$

For convenience, let $P_l^{(q)}(1) = \Pr(b_l = 1|y_l, \hat{\mu}^{(q)}, \hat{v}^{(q)})$ and $P_l^{(q)}(-1) = \Pr(b_l = -1|y_l, \hat{\mu}^{(q)}, \hat{v}^{(q)})$. From Eq. (5.36), using the assumption in Eq. (5.37), it follows that

$$
\begin{aligned}
Q\big(\theta|\theta^{(q)}\big) &= \frac{1}{2^L}\sum_{\mathbf{b}}\left\{\log\left(\prod_{l=0}^{L-1}\frac{1}{\sqrt{2\pi\sigma^2}}\exp\left(-\frac{1}{2\sigma^2}(y_l-\mu b_l)^2\right)\right)\prod_{l=0}^{L-1}P_l^{(q)}(b_l)\right\} \\
&= \frac{1}{2^L}\sum_{\mathbf{b}}\left\{\left(-\sum_{l=0}^{L-1}\left[\frac{1}{2\sigma^2}(y_l-\mu b_l)^2-\frac{1}{2}\log 2\pi\sigma^2\right]\right)\prod_{l=0}^{L-1}P_l^{(q)}(b_l)\right\} \\
&= \frac{1}{2^L}\left\{-\frac{1}{2\sigma^2}\sum_{l=0}^{L-1}\underbrace{\sum_{\mathbf{b}}(y_l-\mu b_l)^2\prod_{l=0}^{L-1}P_l^{(q)}(b_l)}_{=E_{\mathbf{b}}[(y_l-\mu b_l)^2|\mathbf{y},\theta^{(q)}]=E_{b_l}[(y_l-\mu b_l)^2|y_l,\theta^{(q)}]}-\frac{L}{2}\log 2\pi\sigma^2\right\} \\
&= \frac{1}{2^L}\left\{-\frac{1}{2\sigma^2}\sum_{l=0}^{L-1}\left((y_l-\mu)^2P_l^{(q)}(1)+(y_l+\mu)^2P_l^{(q)}(-1)\right)-\frac{L}{2}\log 2\pi\sigma^2\right\}.
\end{aligned}
$$

$$(5.38)$$

In the M-step, the estimates of μ and σ^2 that maximize $Q(\theta|\theta^{(q)})$ must be found. To this end, we need to solve

$$
0=\frac{\mathrm{d}Q\big(\theta|\theta^{(q)}\big)}{\mathrm{d}\mu}=\frac{\mathrm{d}Q\big(\theta|\theta^{(q)}\big)}{\mathrm{d}v}.
$$

Note that $Q(\theta|\theta^{(q)})$ is concave with respect to μ. Hence, the maximum is obtained by taking the derivative and setting it equal to zero. Taking the derivate with respect to μ, we have

$$
\frac{\mathrm{d}Q\big(\theta|\theta^{(q)}\big)}{\mathrm{d}\mu}=-\frac{1}{2^L\sigma^2}\left(-\sum_{l=0}^{L-1}y_l\big(P_l^{(q)}(1)-P_l^{(q)}(-1)\big)+\mu\sum_{l=0}^{L-1}\big(P_l^{(q)}(1)+P_l^{(q)}(-1)\big)\right).
$$

$$(5.39)$$

Since $P_l^{(q)}(1)+P_l^{(q)}(-1)=1$, the next estimate of μ, $\hat{\mu}^{(q+1)}$, is given by

$$
\hat{\mu}^{(q+1)}=\frac{1}{L}\sum_{l=0}^{L-1}y_l\big(P_l^{(q)}(1)-P_l^{(q)}(-1)\big).
$$

$$(5.40)$$

To find $\hat{v}^{(q+1)}$, let

$$
S^{(q+1)}=\sum_{l=0}^{L-1}\left((y_l-\hat{\mu}^{(q+1)})^2P_l^{(q)}(1)+(y_l+\hat{\mu}^{(q+1)})^2P_l^{(q)}(-1)\right).
$$

$$(5.41)$$

From Eq. (5.39), it follows that

$$
\hat{v}^{(q+1)}=\frac{1}{L}S^{(q+1)}.
$$

$$(5.42)$$

The EM algorithm is a powerful means of dealing with the ML estimation problem. As shown above, the missing data plays a crucial role in simplifying the optimization process in the ML estimation. Since the missing data itself is not available, the conditional pdf of the missing data is utilized (see the E-step). In practice, however, it is often difficult to find the conditional pdf of the missing data. In addition, taking the expectation in the E-step with the conditional pdf of the missing data can also be difficult. Thus, a number of variations of the EM algorithm are proposed to overcome these difficulties; see McLachlan and Krishnan (1997) for details.

5.2.2   ML channel estimation using the EM algorithm

The EM algorithm can be applied to the ML channel estimation. Recall the received signal in Eq. (5.1). For convenience, let $\mathbf{y} = [y_0 \ y_1 \ \cdots \ y_{L-1}]^T$, $\mathbf{b} = [b_0 \ b_1 \ \cdots \ b_{L-1}]^T$, and $\mathbf{h} = [h_0 \ h_1 \ \cdots \ h_{P-1}]^T$. Then, the likelihood function is given by

$$f(\mathbf{y}|\mathbf{b}, \mathbf{h}) \propto \exp\left(-\frac{1}{2\sigma_n^2}\sum_{l=0}^{L-1}\left|y_l - \sum_{p=0}^{P-1}h_p b_{l-p}\right|^2\right),\tag{5.43}$$

where $\sigma_n^2 = E[|n_l|^2] = N_0/2$. An ML approach taken to estimate the channel vector $\mathbf{h}$ with the received signal vector $\mathbf{y}$ is given by

$$\hat{\mathbf{h}}_{ml} = \arg\max_{\mathbf{h}} f(\mathbf{y}|\mathbf{h}),\tag{5.44}$$

where the likelihood function of $\mathbf{h}$, $f(\mathbf{y}|\mathbf{h})$, is given by

$$f(\mathbf{y}|\mathbf{h}) = \sum_{\mathbf{b}} f(\mathbf{y}|\mathbf{b}, \mathbf{h}) \Pr(\mathbf{b}|\mathbf{h}).$$

Here, the summation takes place over all the possible vectors of $\mathbf{b}$. Since the binary vector $\mathbf{b}$ is independent of $\mathbf{h}$, the likelihood function becomes

$$f(\mathbf{y}|\mathbf{h}) = \sum_{\mathbf{b}} f(\mathbf{y}|\mathbf{b}, \mathbf{h}) \Pr(\mathbf{b}).$$

The channel estimation problem in Eq. (5.44) can be considered as a blind channel estimation problem as there is no training sequence; see Ding and Li (2001) for detailed a discussion on blind channel estimation. The advantage of blind channel estimation is that the data throughput can increase. However, as will be shown later through an example, the likelihood function can have multiple local maxima and the EM algorithm may not converge to the global maximum. Hence, the initial estimate needs to be close to the global maximum. A good initial estimate can be found by using a training sequence (as a result, the overall channel estimation problem becomes semi-blind in this case).

To derive the EM algorithm for the channel estimation problem in Eq. (5.44), let $\{\mathbf{y}\}$ be the incomplete data and let $\{\mathbf{y}, \mathbf{b}\}$ be the complete data. Suppose that an initial estimate of $\mathbf{h}$, denoted by $\hat{\mathbf{h}}^{(0)}$, is available. In addition, denote by $\hat{\mathbf{h}}^{(q)}$ the estimate of $\mathbf{h}$ at the qth iteration. Then, the E-step taken to find the expectation of the log-likelihood is as follows:

$$Q(\mathbf{h}|\hat{\mathbf{h}}^{(q)}) = E[\log f(\mathbf{b}, \mathbf{y}|\mathbf{h})|\mathbf{y}, \hat{\mathbf{h}}^{(q)}],\tag{5.45}$$

where the expectation takes place over $\mathbf{b}$. To carry out the E-step, we need to find $\log f(\mathbf{b}, \mathbf{y}|\mathbf{h})$. We can easily show that

$$\log f(\mathbf{b}, \mathbf{y}|\mathbf{h}) = \log(f(\mathbf{y}|\mathbf{b}, \mathbf{h}) \Pr(\mathbf{b}|\mathbf{h}))$$
$$= \log f(\mathbf{y}|\mathbf{b}, \mathbf{h}) + \log \Pr(\mathbf{b}|\mathbf{h}).\tag{5.46}$$

Since $\mathbf{b}$ and $\mathbf{h}$ are independent, $\log \Pr(\mathbf{b}|\mathbf{h}) = \log \Pr(\mathbf{b})$ and it is not a function of $\mathbf{h}$. Hence, we only need to find the first term on the right hand side of Eq. (5.46). It follows that

$$E\big[\log f(\mathbf{y}|\mathbf{b}, \mathbf{h})|\mathbf{y}, \hat{\mathbf{h}}^{(q)}\big] = c - \frac{1}{2\sigma_{\mathrm{n}}^2} E_{\mathbf{b}}\big[\|\mathbf{y} - \mathbf{B}^{\mathrm{T}}\mathbf{h}\|^2|\mathbf{y}, \hat{\mathbf{h}}^{(q)}\big], \qquad (5.47)$$

where c is a constant and

$$\mathbf{B} = [\mathbf{b}_0 \ \ \mathbf{b}_1 \ \ \cdots \ \ \mathbf{b}_{L-1}].$$

Here, $\mathbf{b}_l = [b_l \ \ b_{l-1} \ \ \cdots \ \ b_{l-P+1}]^{\mathrm{T}}$, where $b_l = 0$ for $l < 0$. From this, we have

$$Q\big(\mathbf{h}|\hat{\mathbf{h}}^{(q)}\big) = -\frac{1}{2\sigma_{\mathrm{n}}^2} E_{\mathbf{b}}\big[\|\mathbf{y} - \mathbf{B}^{\mathrm{T}}\mathbf{h}\|^2|\mathbf{y}, \hat{\mathbf{h}}^{(q)}\big] + c',$$

where c' is a constant. Deleting unnecessary terms, we can redefine $Q(\mathbf{h}|\hat{\mathbf{h}}^{(q)})$ as follows:

$$Q\big(\mathbf{h}|\hat{\mathbf{h}}^{(q)}\big) = -E_{\mathbf{b}}\big[\|\mathbf{y} - \mathbf{B}^{\mathrm{T}}\mathbf{h}\|^2|\mathbf{y}, \hat{\mathbf{h}}^{(q)}\big]. \qquad (5.48)$$

It follows that

$$Q\big(\mathbf{h}|\hat{\mathbf{h}}^{(q)}\big) = -\sum_{\mathbf{b}} \|\mathbf{y} - \mathbf{B}^{\mathrm{T}}\mathbf{h}\|^2 \Pr\big(\mathbf{b}|\mathbf{y}, \hat{\mathbf{h}}^{(q)}\big). \qquad (5.49)$$

Note that we need to find the *a posteriori* probability of $\mathbf{b}$ conditioned on $\mathbf{y}$ and $\mathbf{h}^{(q)}$ for the E-step.

In the M-step, the vector $\mathbf{h}$ that maximizes $Q(\mathbf{h}|\hat{\mathbf{h}}^{(q)})$ must be found as follows:

$$\hat{\mathbf{h}}^{(q+1)} = \arg\max_{\mathbf{h}} Q\big(\mathbf{h}|\hat{\mathbf{h}}^{(q)}\big)$$

$$= \arg\min_{\mathbf{h}} E_{\mathbf{b}}\big[\|\mathbf{y} - \mathbf{B}^{\mathrm{T}}\mathbf{h}\|^2|\mathbf{y}, \hat{\mathbf{h}}^{(q)}\big]. \qquad (5.50)$$

We can perform the E-step and M-step iteratively to find the ML estimate of $\mathbf{h}$ in Eq. (5.44).

The *a posteriori* probability of $\mathbf{b}$ is required in the E-step. We can explicitly show that

$$\Pr\big(\mathbf{b}|\mathbf{y}, \hat{\mathbf{h}}^{(q)}\big) = \frac{f\big(\mathbf{y}|\mathbf{b}, \hat{\mathbf{h}}^{(q)}\big) \Pr\big(\mathbf{b}, \hat{\mathbf{h}}^{(q)}\big)}{\Pr\big(\mathbf{y}, \hat{\mathbf{h}}^{(q)}\big)}$$

$$\propto f\big(\mathbf{y}|\mathbf{b}, \hat{\mathbf{h}}^{(q)}\big) \Pr\big(\mathbf{b}, \hat{\mathbf{h}}^{(q)}\big). \qquad (5.51)$$

In Eq. (5.51), we ignore $\Pr(\mathbf{y}, \hat{\mathbf{h}}^{(q)})$, because it is independent of $\mathbf{b}$. Furthermore, since $\mathbf{b}$ and $\hat{\mathbf{h}}^{(q)}$ are independent, it follows that

$$\Pr\big(\mathbf{b}|\mathbf{y}, \hat{\mathbf{h}}^{(q)}\big) \propto f\big(\mathbf{y}|\mathbf{b}, \hat{\mathbf{h}}^{(q)}\big) \Pr(\mathbf{b}) \Pr\big(\hat{\mathbf{h}}^{(q)}\big)$$

$$\propto f\big(\mathbf{y}|\mathbf{b}, \hat{\mathbf{h}}^{(q)}\big) \Pr(\mathbf{b}). \qquad (5.52)$$

If the binary signal, b_l, is independent and equally likely, we can see that $\Pr(\mathbf{b})$ becomes independent of $\mathbf{b}$, and finally we have

$$\Pr\big(\mathbf{b}|\mathbf{y}, \hat{\mathbf{h}}^{(q)}\big) \propto f\big(\mathbf{y}|\mathbf{b}, \hat{\mathbf{h}}^{(q)}\big) \propto \exp\left(-\frac{1}{2\sigma_{\mathrm{n}}^2}\|\mathbf{y} - \mathbf{B}^{\mathrm{T}}\mathbf{h}^{(q)}\|^2\right). \qquad (5.53)$$

In this EM-based channel estimation, the major complexity burden occurs in computing the *a posteriori* probability of $\mathbf{b}$. Since there are L samples, we have 2^L binary vectors for

Table 5.1. *Numerical results of the EM algorithm with* $\hat{\mathbf{h}}^{(0)} = [0.5\ 0]^T$

q	0th	1st	2nd	3rd	4th	5th
h_0	0.5000	1.0083	0.9851	0.9849	0.9849	0.9849
h_1	0	−0.3560	−0.4977	−0.4982	−0.4982	−0.4982
Q		−3.8064	−0.5419	−0.4752	−0.4752	−0.4752

Table 5.2. *Numerical results of the EM algorithm with* $\hat{\mathbf{h}}^{(0)} = [0\ 1]^T$

q	0th	1st	2nd	3rd	4th	5th
h_0	0	0.0386	0.0669	0.0789	0.0837	0.0855
h_1	1.0000	1.2195	1.2292	1.2310	1.2315	1.2317
Q	−2.5906	−2.3536	−2.3365	−2.3304	−2.3282	−2.3273

b. This shows that even though the ML estimate of **h** in Eq. (5.44) can be obtained by the EM algorithm, the complexity is still prohibitively high, especially when the length of the signal symbol sequence, L, is great.

Example 5.2.2   Suppose that the true CIR is given by $\{h_0,\ h_1\} = \{1,\ -0.5\}$. The transmitted binary sequence is $\{b_0, b_1, b_2, b_3\} = \{1, -1, 1, 1, -1\}$ and we assume $b_{-1} = 0$. There are five received signal samples with the noise samples as follows:

$$\mathbf{y} = \begin{bmatrix} y_0 \\ y_1 \\ y_2 \\ y_3 \\ y_4 \end{bmatrix} = \begin{bmatrix} 1.0571 \\ -0.9653 \\ 1.5297 \\ 0.4521 \\ -1.9172 \end{bmatrix} \text{ and } \mathbf{n} = \begin{bmatrix} n_0 \\ n_1 \\ n_2 \\ n_3 \\ n_4 \end{bmatrix} = \begin{bmatrix} 0.0571 \\ 0.5347 \\ 0.0297 \\ -0.0479 \\ -0.4172 \end{bmatrix}.$$

The noise has been generated by a Gaussian random number generator with zero mean and variance 0.2512 so that the SNR, $E[|b_l|^2]/\sigma_n^2 = 1/\sigma_n^2$, is 6 dB. The simulation results are shown in Table 5.1.

The initial channel vector $\hat{\mathbf{h}}^{(0)} = [0.5\ 0]^T$. After three iterations, the EM algorithm converges. Note that the EM algorithm is guaranteed to converge to a local maximum. Hence, it is important to have an initial channel vector $\hat{\mathbf{h}}^{(0)}$ which is close to the global maximum. If $\mathbf{h}^{(0)}$ is not close to the global maximum, the EM algorithm will converge to an undesirable local maximum. For example, if $\hat{\mathbf{h}}^{(0)} = [0\ 1]^T$, we have an undesirable result, as shown in Table 5.2.

After a few more iterations, the EM algorithm converges and provides $\hat{h}_0 = 0.0867$ and $\hat{h}_1 = 1.2318$. Obviously, this is not close to the true channel vector and is an undesirable result. In communication systems, a pilot sequence can help to achieve a good initial channel estimate for the EM algorithm so that it can converge to the desirable ML estimate.

Previously, it was assumed that the noise variance, σ_n^2, is known in developing the EM algorithm for the ML channel estimation. In general, however, since the noise variance,

σ_n^2, is unknown, it has to be estimated. In this case, the EM algorithm needs to estimate not only $\mathbf{h}$, but also σ_n^2. For notational convenience, let $v = \sigma_n^2$. The E-step is modified as follows:

$$Q(\mathbf{h}, v | \hat{\mathbf{h}}^{(q)}, \hat{v}^{(q)}) = E_\mathbf{b}\big[\log f(\mathbf{b}, \mathbf{y} | \mathbf{h}, v) | \mathbf{y}, \hat{\mathbf{h}}^{(q)}, \hat{v}^{(q)}\big]$$
$$= -\frac{L \log v}{2} - \frac{1}{2v} E_\mathbf{b}\big[\|\mathbf{y} - \mathbf{B}^\mathsf{T}\mathbf{h}\|^2 | \mathbf{y}, \hat{\mathbf{h}}^{(q)}, \hat{v}^{(q)}\big] + C, \qquad (5.54)$$

where $\hat{v}^{(q)}$ represents the qth estimate of $v = \sigma_n^2$ and C is a constant. Since C does not affect the M-step, we ignore it. The M-step becomes

$$\{\hat{\mathbf{h}}^{(q+1)}, \hat{v}^{(q+1)}\} = \arg\max_{\mathbf{h}, v} Q(\mathbf{h}, v | \hat{\mathbf{h}}^{(q)}, \hat{v}^{(q)}), \qquad (5.55)$$

where $\hat{\mathbf{h}}^{(q+1)}$ and $\hat{v}^{(q+1)}$ must be found jointly. Fortunately, the next channel estimate, $\hat{\mathbf{h}}^{(q+1)}$, can be obtained independently from $\hat{v}^{(q+1)}$; then $\hat{v}^{(q+1)}$ can be readily found. Finally, the M-step is as follows:

$$\hat{\mathbf{h}}^{(q+1)} = \arg\max_{\mathbf{h}} Q(\mathbf{h}, v | \hat{\mathbf{h}}^{(q)}, \hat{v}^{(q)})$$
$$= \arg\min_{\mathbf{h}} E_\mathbf{b}[\|\mathbf{y} - \mathbf{B}^\mathsf{T}\mathbf{h}\|^2 | \mathbf{y}, \hat{\mathbf{h}}^{(q)}, \hat{v}^{(q)}], \qquad (5.56a)$$

$$\hat{v}^{(q+1)} = \arg\max_{v} \left\{ -\frac{L \log v}{2} - \frac{1}{2v} E_\mathbf{b}\big[\|\mathbf{y} - \mathbf{B}^\mathsf{T}\hat{\mathbf{h}}^{(q+1)}\|^2 | \mathbf{y}, \hat{\mathbf{h}}^{(q)}, \hat{v}^{(q)}\big] \right\}$$
$$= \frac{1}{L} E_\mathbf{b}\big[\|\mathbf{y} - \mathbf{B}^\mathsf{T}\hat{\mathbf{h}}^{(q+1)}\|^2 | \mathbf{y}, \hat{\mathbf{h}}^{(q)}, \hat{v}^{(q)}\big]. \qquad (5.56b)$$

5.3 Iterative channel estimation with MAP detection

In the EM-based iterative channel estimation, we need to compute the *a posteriori* probability $\Pr(\mathbf{b} | \mathbf{y}, \hat{\mathbf{h}}^{(q)})$. In general, the complexity to find all the *a posteriori* probabilities of a binary sequence grows exponentially with the length of the data symbol sequence. However, if we assume that the data symbols are independent of each other given $\mathbf{y}$ and $\hat{\mathbf{h}}^{(q)}$, the complexity can be reduced. That is, to reduce the complexity, assume that

$$\Pr(\mathbf{b} | \mathbf{y}, \hat{\mathbf{h}}^{(q)}) = \prod_{l=0}^{L-1} \Pr(b_l | \mathbf{y}, \hat{\mathbf{h}}^{(q)}). \qquad (5.57)$$

Note that this may not be true and that the independence of the b_l's (i.e., $\Pr(\mathbf{b}) = \prod_{l=0}^{L-1} \Pr(b_l)$) does not guarantee the independence in Eq. (5.57). Under the assumption in Eq. (5.57), the MAP symbol detector in Section 5.1 can be used to obtain the *a posteriori* probability of each b_l.

A block diagram for the iterative channel estimation based on the EM algorithm with the MAP symbol detector is depicted in Fig. 5.5. Through the EM iteration, we can obtain the ML channel estimate as well as the (approximate) MAP symbol estimate from $\Pr(b_l | \mathbf{y}, \hat{\mathbf{h}}_{\mathrm{ml}})$.

We note that the signal detection with unknown CIR can be carried out following two different approaches. The first approach is that studied in this section. Firstly, the channel

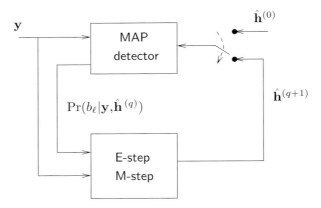

Figure 5.5. Iterative channel estimation with the MAP symbol detector.

estimation is considered. Once an estimated CIR is available, the MAP symbol detection or MLSD can be applied. In fact, as shown above, the MAP symbol detection is involved in the ML channel estimation when the EM algorithm is used. This approach is referred to as the *explicit* channel estimation–*implicit* data detection (ExC–ImD) method.

The second approach is the MLSD with unknown CIR in Chapter 3. The PSP (with Kalman filtering; see Chapter 4 for details) can be used to perform the sequence detection with channel estimation. This approach is referred to as the *explicit* data detection–*implicit* channel estimation (ExD–ImC) method.

In general, it is not easy to analyze the performance of the EM-based channel estimation. Hence, simulation results are presented to observe the performance. With the CIR in Eq. (5.19), the EM algorithm to find the ML estimates of the CIR and noise variance is performed and the simulation results are shown in Figs 5.6 and 5.7. The SNR is defined as follows:

$$\text{SNR} = \frac{\|\mathbf{h}\|^2}{N_0} = \frac{1}{N_0}.$$

It is assumed that a data block consists of 10 pilot symbols and 400 data symbols. With 10 pilot symbols, an initial channel estimate and a noise variance estimate are obtained. To see the difference between the estimated CIR and true CIR, the sum squared error (SSE) defined as

$$\text{SSE}^{(q)} = \left\|\mathbf{h} - \hat{\mathbf{h}}^{(q)}\right\|^2$$

is considered. An averaged SSE is used to estimate the MSE with 1000 realizations. As shown in Fig. 5.6, the MSE decreases with the number of iterations. In particular, there is a large performance difference from the first iteration to the second iteration. In the second iteration, the EM algorithm starts to use data symbols for a better channel estimate. Since there are 400 data symbols, a much lower MSE of the estimated CIR is expected. It is also observed that the performance improvement for each iteration increases with the SNR.

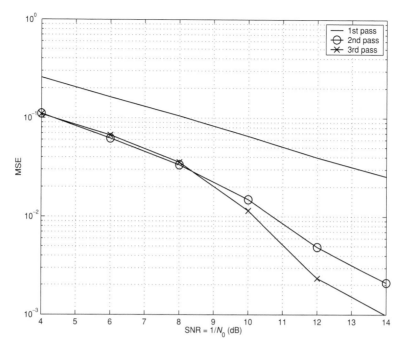

Figure 5.6. (Empirical) MSE versus SNR.

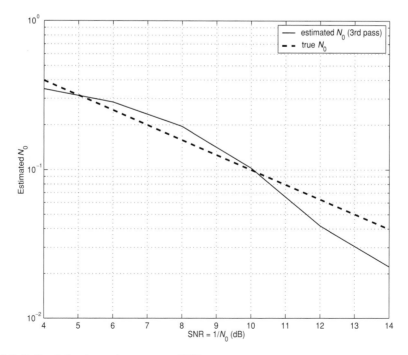

Figure 5.7. Estimated noise variance versus SNR.

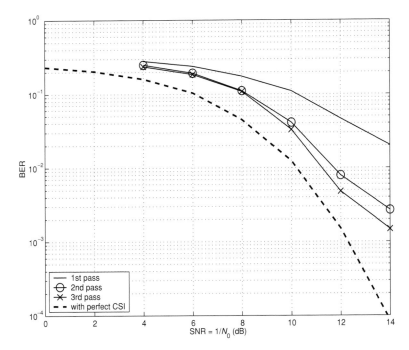

Figure 5.8. Bit error rate versus SNR.

Figure 5.6 shows that the MSE after three iterations is not always smaller than that after two iterations, especially when the SNR is low. This is due to the assumption in Eq. (5.57). In fact, $\prod_{l=0}^{L-1} \Pr(b_l|\mathbf{y}, \hat{\mathbf{h}}^{(q)})$ is an approximation of $\Pr(\mathbf{b}|\mathbf{y}, \hat{\mathbf{h}}^{(q)})$. However, as mentioned earlier, this approximation allows the use of the MAP symbol detection that can be carried out by the BCJR algorithm with low complexity.

Figure 5.7 shows the simulation results for the estimation of noise variance. After three iterations, the estimated noise variance is sufficiently close to the true noise variance.

In the EM algorithm for the ML channel estimation, we do not explicitly detect symbols. However, it is necessary to find the *a posteriori* probability of data symbols with estimated CIR and noise variance for the E-step. From the *a posteriori* probability of symbols obtained by the MAP symbol detection, symbol estimates can be obtained. The BER of the symbol estimate from the *a posteriori* probability is obtained by simulations, and the results are shown in Fig. 5.8. It is shown that the BER decreases with the number of iterations since better estimated CIR and noise variance are available.

The EM algorithm for the ML channel estimation is a fundamental approach taken to deal with the signal detection problem with unknown CIR. This approach is an off-line one as it is necessary to use an entire received signal sequence. On the other hand, the PSP with adaptive channel estimation, which is also applicable to the signal detection with unknown CIR, can be considered as an on-line approach. Since the EM algorithm requires several MAP detection operations, the complexity can be higher than that of the PSP with adaptive channel estimation.

5.4 ML data detection via the EM algorithm

The EM algorithm is also applicable to ML data detection. Consider the ML data detection as follows:

$$\hat{\mathbf{b}}_{ml} = \arg\max_{\mathbf{b}} f(\mathbf{y}|\mathbf{b}). \tag{5.58}$$

Assume that $\mathbf{h}$ is an (unknown) random vector and its pdf, denoted by $f(\mathbf{h})$, is available. Note that $\mathbf{h}$ was considered as an (unknown) deterministic vector in the ML channel estimation in Section 5.2.

Suppose that $\{\mathbf{y}, \mathbf{h}\}$ is the complete data, while $\{\mathbf{y}\}$ is the incomplete data. Let $\mathbf{b}^{(q)}$ denote the estimate of $\mathbf{b}$ at the qth EM iteration. The E-step becomes

$$\begin{aligned}
Q(\mathbf{b}|\mathbf{b}^{(q)}) &= E_{\mathbf{h}}\left[f(\mathbf{y}, \mathbf{h}|\mathbf{b})|\mathbf{y}, \mathbf{b}^{(q)} \right] \\
&= \int \log f(\mathbf{y}, \mathbf{h}|\mathbf{b}) f(\mathbf{h}|\mathbf{y}, \mathbf{b}^{(q)}) \, d\mathbf{h}.
\end{aligned} \tag{5.59}$$

Since

$$f(\mathbf{y}, \mathbf{h}|\mathbf{b}) = f(\mathbf{y}|\mathbf{b}, \mathbf{h}) f(\mathbf{h}|\mathbf{b}) = f(\mathbf{y}|\mathbf{b}, \mathbf{h}) f(\mathbf{h}),$$

we have

$$Q(\mathbf{b}|\mathbf{b}^{(q)}) = -\frac{1}{N_0} E_{\mathbf{h}}\left[\|\mathbf{y} - \mathbf{B}^{\mathsf{T}}\mathbf{h}\|^2 | \mathbf{y}, \mathbf{b}^{(q)} \right] + E_{\mathbf{h}}\left[\log f(\mathbf{h})|\mathbf{y}, \mathbf{b}^{(q)} \right] + C, \tag{5.60}$$

where C is a constant. As $E_{\mathbf{h}}[\log f(\mathbf{h})|\mathbf{y}, \mathbf{b}^{(q)}]$ is not a function of $\mathbf{b}$, we can ignore it, and the resulting E-step is to find

$$\begin{aligned}
Q(\mathbf{b}|\mathbf{b}^{(q)}) &= -\frac{1}{N_0} E_{\mathbf{h}}\left[\|\mathbf{y} - \mathbf{B}^{\mathsf{T}}\mathbf{h}\|^2 | \mathbf{y}, \mathbf{b}^{(q)} \right] \\
&= -\frac{1}{N_0} \sum_{l=0}^{L-1} E_{\mathbf{h}}\left[|y_l - \mathbf{b}_l^{\mathsf{T}}\mathbf{h}|^2 | \mathbf{y}, \mathbf{b}^{(q)} \right] \\
&= -\frac{1}{N_0} \sum_{l=0}^{L-1} |y_l|^2 - 2\mathbf{b}_l^{\mathsf{T}}\bar{\mathbf{h}}^{(q)} y_l + \mathbf{b}_l^{\mathsf{T}}\mathbf{R}_{\mathbf{h}}^{(q)}\mathbf{b}_l,
\end{aligned} \tag{5.61}$$

where

$$\begin{aligned}
\bar{\mathbf{h}}^{(q)} &= E\left[\mathbf{h}|\mathbf{y}, \mathbf{b}^{(q)} \right], \\
\mathbf{R}_{\mathbf{h}}^{(q)} &= E\left[\mathbf{h}\mathbf{h}^{\mathsf{T}}|\mathbf{y}, \mathbf{b}^{(q)} \right].
\end{aligned} \tag{5.62}$$

The M-step is to find $\mathbf{b}$ that maximizes $Q(\mathbf{b}|\mathbf{b}^{(q)})$:

$$\mathbf{b}^{(q+1)} = \arg\max_{\mathbf{b}} Q(\mathbf{b}|\mathbf{b}^{(q)}). \tag{5.63}$$

This can be done using the VA. Let the (total) cost function be given by

$$V(\mathbf{b}) = \sum_{l=0}^{L-1} |y_l|^2 - 2\mathbf{b}_l^{\mathsf{T}}\bar{\mathbf{h}}^{(q)} y_l + \mathbf{b}_l^{\mathsf{T}}\mathbf{R}_{\mathbf{h}}^{(q)}\mathbf{b}_l. \tag{5.64}$$

Then, the branch metric becomes

$$M_l = \mathcal{M}_l(b_l, \mathcal{S}_l; y_l)$$
$$= |y_l|^2 - 2\mathbf{b}_l^{\mathsf{T}}\bar{\mathbf{h}}^{(q)}y_l + \mathbf{b}_l^{\mathsf{T}}\mathbf{R}_{\mathbf{h}}^{(q)}\mathbf{b}_l. \tag{5.65}$$

We note that $\mathbf{b}_l$ consists of the current input b_l and the current state $\mathcal{S}_l$ (which is decided by the $(P-1)$ past inputs, $b_{l-1}, b_{l-2}, \ldots, b_{l-P+1}$). Using Eq. (5.65), we see that the VA is applicable for the M-step.

The EM algorithm for the ML data detection must find $\bar{\mathbf{h}}^{(q)}$ and $\mathbf{R}_{\mathbf{h}}^{(q)}$ in Eqs (5.62). Using Bayes' rule, we can show that

$$f(\mathbf{h}|\mathbf{y}, \mathbf{b}) = \frac{f(\mathbf{y}|\mathbf{h}, \mathbf{b})f(\mathbf{h})}{f(\mathbf{y}|\mathbf{b})}$$
$$\propto f(\mathbf{y}|\mathbf{h}, \mathbf{b})f(\mathbf{h}). \tag{5.66}$$

If $\mathbf{h}$ is Gaussian, we have

$$f(\mathbf{h}) = \frac{1}{\sqrt{2\pi|\mathbf{C}_{\mathbf{h}}|}}\exp\left(\frac{1}{2}(\mathbf{h} - \bar{\mathbf{h}})^{\mathsf{T}}\mathbf{C}_{\mathbf{h}}^{-1}(\mathbf{h} - \bar{\mathbf{h}})\right), \tag{5.67}$$

where $\bar{\mathbf{h}} = E[\mathbf{h}]$ and $\mathbf{C}_{\mathbf{h}} = E[(\mathbf{h} - \bar{\mathbf{h}})(\mathbf{h} - \bar{\mathbf{h}})^{\mathsf{T}}]$. Then, it can be shown that

$$f(\mathbf{y}|\mathbf{h}, \mathbf{b})f(\mathbf{h}) = C'\exp\left(-\frac{1}{N_0}\|\mathbf{y} - \mathbf{B}^{\mathsf{T}}\mathbf{h}\|^2 - \frac{1}{2}(\mathbf{h} - \bar{\mathbf{h}})^{\mathsf{T}}\mathbf{C}_{\mathbf{h}}^{-1}(\mathbf{h} - \bar{\mathbf{h}})\right)$$
$$= C''\exp\left(\frac{1}{2}(\mathbf{h} - \mathbf{m})^{\mathsf{T}}\mathbf{K}^{-1}(\mathbf{h} - \mathbf{m})\right), \tag{5.68}$$

where C' and C'' are normalizing constants[†] and

$$\mathbf{K} = \left(\mathbf{C}_{\mathbf{h}}^{-1} + \frac{2}{N_0}\mathbf{B}\mathbf{B}^{\mathsf{T}}\right)^{-1}$$
$$\mathbf{m} = \left(\mathbf{C}_{\mathbf{h}}^{-1} + \frac{2}{N_0}\mathbf{B}\mathbf{B}^{\mathsf{T}}\right)^{-1}\left(\mathbf{C}_{\mathbf{h}}^{-1}\bar{\mathbf{h}} + \frac{2}{N_0}\mathbf{B}\mathbf{y}\right). \tag{5.69}$$

From Eqs (5.68) and (5.66), it seems that $\mathbf{h}$ conditioned on $\mathbf{y}$ and $\mathbf{b}$ is Gaussian as its conditional pdf is a Gaussian pdf. It is true because $\mathbf{y}$ and $\mathbf{n}$ are Gaussian. Thus, letting $\mathbf{B} = \mathbf{B}^{(q)}$, we have

$$E\left[\mathbf{h}|\mathbf{y}, \mathbf{b}^{(q)}\right] = \mathbf{m}$$
$$= \left(\mathbf{C}_{\mathbf{h}}^{-1} + \frac{2}{N_0}\mathbf{B}^{(q)}(\mathbf{B}^{(q)})^{\mathsf{T}}\right)^{-1}\left(\mathbf{C}_{\mathbf{h}}^{-1}\bar{\mathbf{h}} + \frac{2}{N_0}\mathbf{B}^{(q)}\mathbf{y}\right),$$
$$E\left[\tilde{\mathbf{h}}\tilde{\mathbf{h}}^{\mathsf{T}}|\mathbf{y}, \mathbf{b}^{(q)}\right] = \mathbf{K}$$
$$= \left(\mathbf{C}_{\mathbf{h}}^{-1} + \frac{2}{N_0}\mathbf{B}^{(q)}(\mathbf{B}^{(q)})^{\mathsf{T}}\right)^{-1}, \tag{5.70}$$

where $\tilde{\mathbf{h}} = \mathbf{h} - \mathbf{m}$. We can find $\bar{\mathbf{h}}^{(q)}$ and $\mathbf{R}_{\mathbf{h}}^{(q)}$ from Eqs (5.70).

[†] The normalizing constants C' and C'' can be seen as functions of $\mathbf{y}$ and $\mathbf{b}$ (or $\mathbf{B}$). However, since $\mathbf{y}$ and $\mathbf{b}$ are given, we consider C' and C'' as constants (and not functions of $\mathbf{h}$).

We may not have any prior information of $\mathbf{h}$, i.e. $f(\mathbf{h})$ may not be available. In this case, we can assume that $\mathbf{C_h}^{-1}$ approaches $\mathbf{0}$ in Eqs (5.70). Then, from Eqs (5.70), we can show that

$$E\left[\mathbf{h}|\mathbf{y}, \mathbf{b}^{(q)}\right] = \left(\mathbf{B}^{(q)}\left(\mathbf{B}^{(q)}\right)^{\mathrm{T}}\right)^{-1}\mathbf{B}^{(q)}\mathbf{y},$$

$$E\left[\tilde{\mathbf{h}}\tilde{\mathbf{h}}^{\mathrm{T}}|\mathbf{y}, \mathbf{b}^{(q)}\right] = \frac{N_0}{2}\left(\mathbf{B}^{(q)}\left(\mathbf{B}^{(q)}\right)^{\mathrm{T}}\right)^{-1}. \tag{5.71}$$

5.5 Summary and notes

Two important algorithms were studied in this chapter: the BCJR algorithm for the MAP symbol detection and the EM algorithm for the ML channel estimation. These are essential algorithms for signal detection and channel estimation over an unknown ISI channel.

In Chapters 3 and 4, the MLSD was introduced to detect symbol sequences over an unknown ISI channel, in which the channel estimation is a part of the MLSD. The same problem was discussed in this chapter with a different approach. The ML channel estimation was considered with the EM algorithm, in which the detection becomes a part of the channel estimation. Clearly, the MLSD provides an optimal performance in terms of detection, while the ML channel estimation provides an optimal performance in terms of channel estimation. Since the performance of detection is degraded by the channel estimation error, both data detection and channel estimation are closely related. Hence, a good channel estimate can provide a satisfactory performance of detection.

The implementation of the EM algorithm for data detection and channel estimation can be different depending on how the ML estimation problem is formulated and how the missing data are defined. For example, other approaches are discussed in Bapat (1998), Georghiades and Han (1997), and Zamiri-Jafarian and Pasupathy (1999).

6 Iterative receivers over static ISI channels

In this chapter, we will study the iterative receiver over *static* and *known* ISI channels. When a coded signal is transmitted, the conventional receiver performs the channel equalization (or detection) and the channel decoding separately. Although this approach is not optimal, it has been widely used because of low complexity and simple implementation. On the other hand, a joint processing for the equalization and decoding that can provide an optimal performance cannot be widely used due to the complexity growing exponentially with the length of coded sequences.

Fortunately, using the turbo principle, which was originally proposed for channel decoding (Berrou, Glavieux, and Thitimajshima, 1993; Hagenauer *et al.*, 1996), it becomes possible to build a receiver that can provide a near optimal performance, while maintaining a moderate complexity. It is often called the turbo equalizer (Koetter, Singer, and Tuchler, 2004; Laot, Glavieux, and Labat, 2001; Tuchler, Koetter, and Singer, 2002a) as the turbo principle is adopted. However, we refer to it as the iterative receiver since the turbo equalizer is not only for channel equalization, but also for channel decoding. According to the turbo principle, the channel equalizer and channel decoder exchange the information of bits to achieve a better performance through iterations.

In this chapter, we review information theory and convolutional codes. Both subjects are necessary to understand the iterative receiver. Then, the structure and operation of the iterative receiver are studied. Several different channel equalizers within the iterative receiver and an approach for the performance analysis are also discussed.

6.1 A brief overview of information theory

In this section, we briefly introduce the entropy, channel capacity, and mutual information. The concept of mutual information will be used later to analyze the iterative receiver. The reader is referred to Cover and Thomas (1991) for a detailed account of information theory.

6.1.1 Entropy

Let us consider a probabilistic experiment, the outcome of which is one of the alphabet $\{s_0, s_1, \ldots, s_{K-1}\}$. Let $\Pr(S = s_k) = p_k$ denote the probability that symbol s_k is observed,

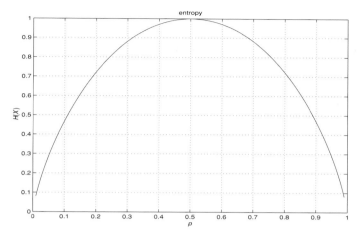

Figure 6.1. The entropy of a binary random variable.

where S is the random variable for the probabilistic experiment. It should be satisfied that $\sum_{k=0}^{K-1} p_k = 1$ and $p_k \geq 0$ for all k. Then, the information of s_k is defined as follows:

$$I(s_k) = \log_2 \left(\frac{1}{\Pr(S = s_k)} \right)$$
$$= -\log_2 \left(\Pr(S = s_k) \right) \tag{6.1}$$

and the *entropy* of random variable S is given by

$$H(S) = E[I(S)]$$
$$= -\sum_{k=0}^{K-1} p_k \log_2 \left(p_k \right). \tag{6.2}$$

It is assumed that $0 \cdot \log(\infty) = 0$. The dimension of the entropy is bits per symbol, where "bit" is an acronym for binary information digit. The entropy can be understood in various ways; it can indicate randomness or uncertainty of random variables. As the entropy increases, a random variable can be more random or uncertain. A random variable of a higher entropy is regarded as a random variable of more information.

Example 6.1.1 Consider a binary random variable $X \in \{0, 1\}$, where $\Pr(X = 0) = p$ and $\Pr(X = 1) = 1 - p$. Then, the entropy of X is given by

$$H(X) = E[-\log_2 \Pr(X)]$$
$$= \Pr(X = 0) \times -\log_2 \Pr(X = 0) + \Pr(X = 1) \times -\log_2 \Pr(X = 1)$$
$$= -p \log_2 p - (1 - p) \log_2 (1 - p).$$

Figure 6.1 shows the entropy curve in terms of the probability p. The entropy becomes the maximum when p is $1/2$, i.e. the uncertainty is maximized when $p = 1/2$. If there is no uncertainty (i.e. $p = 0$ or $p = 1$), the entropy becomes zero.

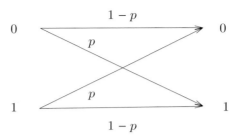

Figure 6.2. Binary symmetric channel (BSC).

6.1.2 Mutual information and channel capacity

A digital communication system is often simplified as a memoryless system in which the current output is dependent on the current input. If the input and output signals are discrete, the memoryless channel is called the discrete memoryless channel (DMC).

The binary symmetric channel (BSC) is a special case of the DMC, where the input and output are bits. A BSC is illustrated in Fig. 6.2. In this case, the BSC is fully characterized by the following conditional probability:

$$\Pr(y|x) = f_{Y|X}(y|x) = \begin{cases} 1 - p, & \text{if } y = x; \\ p, & \text{if } y \neq x. \end{cases}$$

where x and y stand for the input and output, respectively. Note that the input distribution $\Pr(X = x) = f_X(x)$ has not yet been given.

Suppose that we observe $Y = y$ from the BSC. The entropy or uncertainty of X with observation $Y = y$ can be measured as follows:

$$H(X|Y = y) = E[-\log_2 f_{X|Y}(X|y)].$$

Note that since $Y = y$, y is not a random variable and the expectation is carried out over X. That is,

$$H(X|Y = y) = -\log_2 f_{X|Y}(X = 0|y)f_{X|Y}(X = 0|y) \\ - \log_2 f_{X|Y}(X = 1|y)f_{X|Y}(X = 1|y).$$

If $H(X|Y = y)$ is close to zero, the decision based on $Y = y$ will be correct with a high probability. Otherwise (i.e. if close to 1), incorrect decisions will happen with a higher probability.

Example 6.1.2   In the BSC, let $Y = y = 0$. Then, it follows that

$$\begin{aligned} H(X|Y = y) &= E[-\log_2 f_{X|Y}(X|y)] \\ &= -\log_2 f_{X|Y}(X = 0|y = 0)f_{X|Y}(X = 0|y = 0) \\ &\quad - \log_2 f_{X|Y}(X = 1|y = 0)f_{X|Y}(X = 1|y = 0) \\ &= -\log_2(1 - p) \times f_{X|Y}(X = 0|y = 0) \\ &\quad - \log_2 p \times f_{X|Y}(X = 1|y = 0). \end{aligned}$$

In the above, we consider the particular case that $Y = y$. With a random variable Y, the *conditional entropy* is defined as follows:

$$H(X|Y) = E_{X,Y}[-\log_2 f_{X|Y}(X|Y)]$$

$$= \sum_x \sum_y -\log_2 f_{X|Y}(x|y) f_{X,Y}(X = x, Y = y).$$

If X and Y are independent, then

$$H(X|Y) = E[-\log_2 f_{X|Y}(X|Y)]$$

$$= E\left[-\log_2 \frac{f_{X,Y}(X, Y)}{f_Y(Y)}\right]$$

$$= E\left[-\log_2 \frac{f_X(X) f_Y(Y)}{f_Y(Y)}\right]$$

$$= E\left[-\log_2 f_X(X)\right]$$

$$= H(X).$$

This means that there is no entropy or uncertainty reduction by observing Y. On the other hand, if $X = Y$, then

$$H(X|Y) = E[-\log_2 f_{X|Y}(X|Y)]$$

$$= E\left[-\log_2 \frac{f_Y(Y)}{f_Y(Y)}\right]$$

$$= 0.$$

That is, the entropy becomes zero by observing Y. Thus, there is no uncertainty in the information of X when Y is given.

Define the *mutual information* as

$$I(X; Y) = H(X) - H(X|Y),$$

which is the difference between the entropy of X (the original entropy) and the conditional entropy of X given Y. We can show that

$$I(X; Y) = H(X) - H(X|Y)$$

$$= E[-\log_2 f_X(X)] - E\left[-\log_2 \frac{f_{X,Y}(X, Y)}{f_Y(Y)}\right]$$

$$= E\left[-\log_2 \frac{f_X(X) f_Y(Y)}{f_{X,Y}(X, Y)}\right]$$

$$= H(X) + H(Y) - H(X, Y)$$

$$= I(Y; X)$$

$$= H(Y) - H(Y|X).$$

Note that if X and Y are independent, we have

$$
\begin{aligned}
H(X, Y) &= E[-\log_2 f_{X,Y}(X, Y)] \\
&= E[-\log_2(f_X(X) f_Y(Y))] \\
&= E[-\log_2 f_X(X)] + E[-\log_2 f_Y(Y)] \\
&= H(X) + H(Y).
\end{aligned}
$$

Hence, in this case, it follows that

$$
I(X; Y) = 0.
$$

If $X = Y$, then

$$
I(X; Y) = H(X).
$$

Note that the dimension of $I(X; Y)$ is the same as $H(X)$, i.e. bits per symbol.

Suppose that X and Y denote the input and output of a channel, respectively. The mutual information shows the amount of information that can be delivered reliably over the channel. The entropy $H(X)$ is the amount of information of X and the conditional entropy $H(X|Y)$ is the amount of uncertainty of X given Y. Thus, the difference, $I(X; Y)$, becomes the amount of information of X from Y when Y is observed.

We can show that

$$
I(X; Y) \le H(X).
$$

This inequality is a consequence of the fact that the amount of information reliably transmitted over the channel cannot exceed the original amount of information of X.

The mutual information can be maximized with respect to the input distribution, $f_X(x)$. The *channel capacity* is the maximum of mutual information and is defined as

$$
C = \max_{f_X(x)} I(X; Y) \quad \text{(bits per symbol)}.
$$

It is important to note that for a given channel of capacity C, the channel input X cannot be transmitted without any loss if the entropy $H(x)$ is greater than C.

6.1.3 Differential entropy and mutual information for continuous random variables

Previously we only considered discrete random variables. For a continuous random variable, the entropy can be defined in a similar way as follows:

$$
\begin{aligned}
h(X) &= E[-\log_2 f_X(X)] \\
&= -\int_{-\infty}^{\infty} f_X(x) \log_2 f_X(x) \, dx.
\end{aligned}
\tag{6.3}
$$

This is called the *differential entropy* of X. Consider an example. Let $X \sim \mathcal{N}(\mu, \sigma^2)$. The differential entropy of a Gaussian random variable becomes

$$h(X) = \int_{-\infty}^{\infty} \frac{1}{\sqrt{2\pi\sigma^2}} \exp\left(-\frac{(x-\mu)^2}{2\sigma^2}\right) \left[\log_2(\sqrt{2\pi\sigma^2}) + (\log_2 e)\frac{(x-\mu)^2}{2\sigma^2}\right] dx$$

$$= \log_2(\sqrt{2\pi\sigma^2}) + (\log_2 e)\frac{1}{2}$$

$$= \frac{1}{2}\log_2(2\pi e\sigma^2). \tag{6.4}$$

Interestingly, the mean μ does not affect the differential entropy, while the variance σ^2 does. The differential entropy increases with σ^2.

Although $H(X) \geq 0$ for a discrete random variable X, the differential entropy $h(X)$ for a continuous random variable X can have a negative value. For example, let $\sigma^2 = 1/4\pi e$ in Eq. (6.4). Then, $h(X) = -0.5$.

The mutual information between two continuous random variables X and Y can be defined as follows:

$$I(X;Y) = \int_{-\infty}^{\infty} \int_{-\infty}^{\infty} f_{X,Y}(x,y) \log_2 \frac{f_{X|Y}(x|y)}{f_X(x)} dx\, dy. \tag{6.5}$$

There are some important properties of the mutual information as follows.

(i) $I(X;Y) = I(Y;X)$. That is,

$$I(X;Y) = h(X) - h(X|Y)$$
$$= h(Y) - h(Y|X), \tag{6.6}$$

where

$$h(X|Y) = \int_{-\infty}^{\infty} \int_{-\infty}^{\infty} f_{X,Y}(x,y) \log_2 \frac{1}{f_{X|Y}(x|y)} dx\, dy. \tag{6.7}$$

(ii) $I(X;Y) \geq 0$.

(iii) Suppose that X and N are independent continuous random variables. We have

$$h(X+N|X) = h(N).$$

(iv) Consider two pdfs $f_X(x)$ and $f_Y(x)$. Using $\log_2 x \leq x - 1, x \geq 0$, it can be shown that

$$\int_{-\infty}^{\infty} f_Y(x) \log_2 \frac{f_X(x)}{f_Y(x)} dx \leq 0. \tag{6.8}$$

Another important result is that the differential entropy of a Gaussian random variable X is always greater than or equal to that of any other continuous random variable Y which has the same mean and variance as X. That is,

$$h(Y) \leq h(X),$$

where $E[X] = E[Y]$ and $Var(X) = Var(Y)$. To show this, we need to use Eq. (6.8). It can be shown that

$$-\int f_Y(x) \log_2 f_Y(x)\, dx \leq -\int f_Y(x) \log_2 f_X(x)\, dx. \tag{6.9}$$

The term on the left hand side of Eq. (6.9) is $h(Y)$. Let μ and σ^2 be the mean and variance, respectively. Then, the term on the right hand side of Eq. (6.9) becomes

$$
-\int f_Y(x) \log_2 f_X(x)\, dx = \int f_Y(x) \left[\frac{1}{2} \log_2(2\pi\sigma^2) + \log_2 e \frac{(x-\mu)^2}{2\sigma^2} \right] dx
$$

$$
= \frac{1}{2} \log_2(2\pi\sigma^2) + \frac{\log_2 e}{2\sigma^2} \underbrace{\int f_Y(x)(x-\mu)^2\, dx}_{=E[(Y-\mu)^2]}
$$

$$
= \frac{1}{2} \log_2(2\pi\sigma^2) + \frac{\log_2 e}{2}
$$

$$
= \frac{1}{2} \log_2(2\pi e \sigma^2)
$$

$$
= h(X).
$$

Thus, we can confirm that

$$
h(Y) \le \frac{1}{2} \log_2(2\pi e \sigma^2). \tag{6.10}
$$

6.1.4 Information capacity theorem

Consider the AWGN channel. The received signal is given by

$$
Y = X + N,
$$

where X is the transmitted signal which is a continuous random variable and N is the white Gaussian noise with zero mean and variance σ^2. The mutual information becomes

$$
\begin{aligned}
I(X; Y) &= h(Y) - h(Y|X) \\
&= h(Y) - h(N), \tag{6.11}
\end{aligned}
$$

where N and X are independent of each other. Suppose that X has zero mean and variance P. The capacity of the AWGN channel becomes

$$
\begin{aligned}
C_{\text{awgn}} &= \max_{f_X(x)} I(X, Y) \\
&= \max_{f_X(x)} h(Y) - h(N). \tag{6.12}
\end{aligned}
$$

The variance of the random variable Y is $P + \sigma^2$ and the mean is zero. From Eq. (6.10), it can be shown that

$$
h(Y) \le \frac{1}{2} \log_2 \left(2\pi e (P + \sigma^2) \right)
$$

and the upper bound is achieved when Y is a Gaussian random variable. This means that X should be a Gaussian random variable, i.e. $X \sim \mathcal{N}(0, P)$, to achieve the channel capacity.

Consequently, the channel capacity is given by

$$
\begin{aligned}
C_{\text{awgn}} &= \frac{1}{2} \log_2 \left(\frac{P + \sigma^2}{\sigma^2} \right) \\
&= \frac{1}{2} \log_2 \left(1 + \frac{P}{\sigma^2} \right) \quad \text{(bits per symbol)}.
\end{aligned}
\tag{6.13}
$$

This is the maximum rate (of bits per symbol) that can be transmitted over the AWGN channel.

6.1.5 Mutual information of binary signals over the AWGN channel

The mutual information of binary signaling over the AWGN channel is of practical importance for digital communications. Since the channel capacity cannot be achieved with binary signaling, the mutual information becomes a practical indicator for the achievable transmission rate.

The signal X is represented as μb, where $b \in \{-1, +1\}$ is a binary signal and μ is a constant. The received signal Y is given by

$$
\begin{aligned}
Y &= X + N \\
&= \mu b + N,
\end{aligned}
$$

where N is the white Gaussian noise with zero mean and variance σ_{n}^2. Since X is not a Gaussian random variable, the mutual information between X and Y is less than C_{awgn}. Using the definition of the mutual information, we can show that

$$
\begin{aligned}
I(Y; b) &= H(Y) - H(Y|b) \\
&= -\int f_Y(y) \log_2 f_Y(y)\, \mathrm{d}y + \sum_{b \in \{+1, -1\}} \int f_{Y,b}(Y, b) \log_2 f_{Y|b}(y|b)\, \mathrm{d}y.
\end{aligned}
\tag{6.14}
$$

Assume that b is equally likely. Then, it follows that

$$
\begin{aligned}
f_Y(y) &= \frac{1}{2} \left(f_{Y|b}(y|b = +1) + f_{Y|b}(y|b = -1) \right) \\
f_{Y,b}(y, b) &= \frac{1}{2} f_{Y|b}(y|b) \\
&= \frac{1}{2} \frac{1}{\sqrt{2\pi}\,\sigma_n} \exp \left(-\frac{1}{2\sigma_{\text{n}}^2} (y - \mu b)^2 \right).
\end{aligned}
$$

After some manipulation, we can show that

$$
\begin{aligned}
I(Y; b) &= \frac{1}{2} \sum_b \int f_{Y|b}(y|b) \log_2 \frac{2 f_{Y|b}(y|b)}{f_{Y|b}(y|b = +1) + f_{Y|b}(y|b = -1)}\, \mathrm{d}y \\
&= 1 - \frac{1}{2} \sum_b \int f_{Y|b}(y|b) \log_2 \left(1 + \frac{f_{Y|b}(y| - b)}{f_{Y|b}(y|b)} \right) \mathrm{d}y.
\end{aligned}
\tag{6.15}
$$

As $f_{Y|b}(y|b)$ is a Gaussian pdf, it can be shown that

$$\frac{f_{Y|b}(y|-b)}{f_{Y|b}(y|b)} = \frac{e^{-\frac{1}{2\sigma_{\bar{n}}^2}(y+\mu b)^2}}{e^{-\frac{1}{2\sigma_{\bar{n}}^2}(y-\mu b)^2}} = \exp\left(-\frac{2\mu}{\sigma_{\mathrm{n}}^2}yb\right).$$

Then, it follows that

$$I(Y;b) = 1 - \frac{1}{2}\sum_b \int \frac{1}{\sqrt{2\pi}\sigma_{\mathrm{n}}} e^{-\frac{1}{2\sigma_{\bar{n}}^2}(y-\mu b)^2} \log_2\left(1 + e^{-\frac{2\mu}{\sigma_{\bar{n}}^2}yb}\right) dy. \qquad (6.16)$$

Let $y = \bar{y}b$. Then, it can be shown that

$$e^{-\frac{1}{2\sigma_{\bar{n}}^2}(y-\mu b)^2} \log_2\left(1 + e^{-\frac{2\mu}{\sigma_{\bar{n}}^2}yb}\right) = e^{-\frac{1}{2\sigma_{\bar{n}}^2}(\bar{y}-\mu)^2} \log_2\left(1 + e^{-\frac{2\mu}{\sigma_{\bar{n}}^2}\bar{y}}\right)$$

for $b \in \{-1, +1\}$. Using this, the mutual information in Eq. (6.16) is rewritten as follows:

$$I(Y;b) = 1 - \int \frac{1}{\sqrt{2\pi}\sigma_{\mathrm{n}}} e^{-\frac{1}{2\sigma_{\bar{n}}^2}(y-\mu)^2} \log_2\left(1 + e^{-\frac{2\mu}{\sigma_{\bar{n}}^2}y}\right) dy. \qquad (6.17)$$

This mutual information is invariant with respect to a constant scaling. That is,

$$I(Y;b) = I(\bar{Y};b),$$

where $\bar{Y} = aY$ for constant a. Let us choose a as follows:

$$a = \frac{2\mu}{\sigma_{\mathrm{n}}^2}.$$

Then, we can show that

$$\bar{Y} = \bar{\mu}b + \bar{n},$$

where $\bar{\mu} = 2\mu^2/\sigma_{\mathrm{n}}^2$ and $\bar{\sigma}_n^2 = E[|\bar{n}|^2] = 4\mu^2/\sigma_{\mathrm{n}}^2$. In this case, $2\bar{\mu} = \bar{\sigma}_{\mathrm{n}}^2$ and the mutual information can be expressed by a function of a single parameter as follows:

$$\begin{aligned}
I(Y;b) &= I(\bar{Y};b) \\
&= J(\bar{\sigma}_{\mathrm{n}}^2) \\
&\triangleq 1 - \int \frac{1}{\sqrt{2\pi}\bar{\sigma}_{\mathrm{n}}} e^{-\frac{1}{2\bar{\sigma}_{\bar{n}}^2}(y-\bar{\sigma}_{\mathrm{n}}^2/2)^2} \log_2(1 + e^{-y}) dy. \qquad (6.18)
\end{aligned}$$

To compare the channel capacity and the mutual information, consider the following received signal:

$$Y = X + N,$$

where N is a zero-mean Gaussian random variable with variance σ_{n}^2. Let $X = \mu b$, where $b \in \{-1, +1\}$ is equally likely and μ is a constant. The mutual information becomes $J(\bar{\sigma}_{\mathrm{n}}^2)$, where $\bar{\sigma}_{\mathrm{n}}^2 = 4\mu^2/\sigma_{\mathrm{n}}^2$, while the channel capacity becomes $\frac{1}{2}\log_2(1 + \mu^2/\sigma_{\mathrm{n}}^2)$, which can be achieved when X is a Gaussian random variable (not a scaled binary random variable). The channel capacity and mutual information are shown in Fig. 6.3 for different values of the SNR $(= \mu^2/\sigma_{\mathrm{n}}^2)$.

It is interesting to see that there is no big difference between the channel capacity and mutual information when the SNR is low $(\leq 0$ dB$)$, where $I(Y;b) \leq 0.5$ bits per symbol.

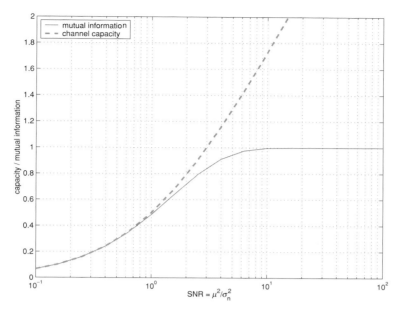

Figure 6.3. Channel capacity and mutual information in terms of the SNR $= \mu^2/\sigma_n^2$.

As the SNR increases, the difference becomes larger. Note that since the maximum data rate is one bit per symbol with binary signaling, it is impossible to have a higher rate than one bit per symbol. Thus, if the SNR is high, the size of the symbol alphabet should be large to increase the data rate.

6.2 Convolutional codes

There are two different types of channel codes: block codes and convolutional codes. Since we do not attempt to introduce a general theory of channel codes, we discuss a special class of convolutional codes in this section to introduce the iterative receiver. The reader is referred to Lin and Costello (1983) and Proakis (1995) for a detailed account of channel codes.

6.2.1 Convolutional encoders

Prior to introducing convolutional codes, it would be useful to clarify some terminology. A bit sequence means a number sequence consisting of "0" and "1," while a data symbol sequence means a number sequence consisting of the symbols in a symbol alphabet. In this book, we confine ourselves to binary convolutional codes whose inputs and outputs are bit sequences. In addition, we only consider binary data symbol sequences with a binary alphabet $\{-1, +1\}$ for transmission.

A convolutional encoder is a finite state machine (FSM) which consists of shift registers and modulo-2 adders. The numbers of inputs and outputs are denoted by k_c and n_c,

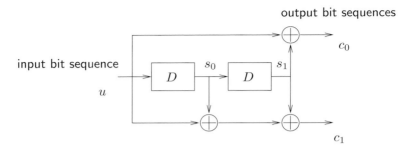

Figure 6.4. A $(2, 1, 3)$ convolutional encoder. In the block diagram, D stands for a register which is a memory element.

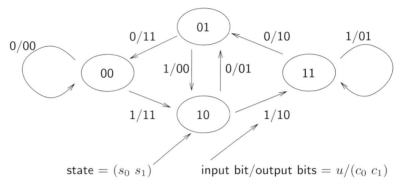

Figure 6.5. The state diagram of a $(2, 1, 3)$ convolutional encoder.

respectively. The *code rate* is defined as $r_c = k_c/n_c$. For example, if $k_c = 1$ (one input) and $n_c = 2$ (two outputs), the code rate becomes $r_c = 1/2$. Since channel codes provide a redundancy to protect data sequences, there are more outputs than inputs. That is, the code rate r_c is usually less than 1 as $k_c < n_c$. We will assume that $k_c = 1$ for the sake of simplicity.

A convolutional encoder is depicted in Fig. 6.4, where u and (c_0, c_1) represent the input bit and the output bits, respectively. There is another important parameter of convolutional encoders: the *constraint length*, denoted by m_c, is defined as "$m_c = 1 +$ number of memory elements" when $k_c = 1$. Thus, the set of parameters required to describe a convolutional encoder is (n_c, k_c, m_c). For the convolutional encoder shown in Fig. 6.4, we have $m_c = 3$ and $(n_c, k_c, m_c) = (2, 1, 3)$. Together with the set of parameters, (n_c, k_c, m_c), the configuration of connections from inputs to outputs through memory elements must be described to describe fully a convolutional encoder. The connections are represented by sets of binary numbers, which are called the generators. From the input to output c_0, the generator is given by $\mathbf{g}_0 = [1\ 0\ 1]$ (from left to right), where 1 and 0 stand for connection and no connection, respectively. The generator to output c_1 is given by $\mathbf{g}_1 = [1\ 1\ 1]$. For convenience, the generators can be represented in octal. For $\mathbf{g}_0 = [1\ 0\ 1]$ and $\mathbf{g}_1 = [1\ 1\ 1]$, we have $\mathbf{g}_0 = 5$ and $\mathbf{g}_1 = 7$ in octal.

Since a convolutional encoder is an FSM, it can be described by a state diagram. Figure 6.5 shows the state diagram for the convolutional encoder in Fig. 6.4, where the numbers in a

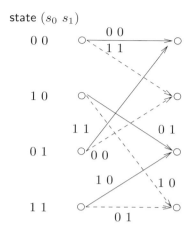

Figure 6.6. The trellis diagram of a $(2, 1, 3)$ convolutional encoder (the solid line represents the branch due to input "0" and the dashed line represents the branch due to input "1").

circle represent a state (i.e. the contents of two registers). The input bit $u \in \{0, 1\}$ changes the state. The outputs, c_0 and c_1, depend on the current state as well as the current input. For example, if the state, $(s_0 \; s_1)$, is "01" and the input is "0," then $c_0 = \mathbf{g}_0 \cdot [u \; s_0 \; s_1]^{\mathrm{T}} = (1 \cdot 0) \oplus (0 \cdot 0) \oplus (1 \cdot 1) = 1$. It can also be shown that $c_1 = \mathbf{g}_1 \cdot [u \; s_0 \; s_1]^{\mathrm{T}} = 1$. Thus, the output bits become $(c_0 \; c_1) = (1 \; 1)$ and the next state becomes $(s_0 \; s_1) = (0 \; 0)$. The state diagram is shown in Fig. 6.5.

The state transition of a convolutional encoder can be represented by a trellis diagram. The possible transitions of states are depicted in Fig. 6.6 for the convolutional encoder in Fig. 6.4. Note that a transition from state $(0 \; 1)$ to state $(0 \; 1)$ or $(1 \; 1)$ is not allowed. The trellis diagram is very useful in the decoding process. It is noteworthy that the state diagram can be obtained from the trellis diagram, and vice versa.

The input to a convolutional encoder is a bit sequence rather than a bit. Hence, we denote by $\{u_l\}$, $l = 0, 1, \ldots, K_c - 1$, the input bit sequence, where K_c is the length of the input bit sequence. The n_c output bit sequences are denoted by $\{c_{l,t}\}$, $t = 0, 1, \ldots, n_c - 1$, and the total length of the coded bit sequences becomes $K_c n_c$.

Example 6.2.1   Consider the convolutional encoder in Fig. 6.4.

(a) Suppose the information bit sequence is as follows:

$$\{u_l\} = \{u_1, u_2, \ldots, u_5\} = \{0, 1, 0, 1, 1\}.$$

When the initial state of the convolutional encoder is $(s_0 \; s_1) = (0 \; 0)$, the coded bit sequence is given by

$$\{(c_{l,0}, c_{l,1})\} = \{(c_{1,0}, c_{1,1}), \; (c_{2,0}, c_{2,1}), \ldots, (c_{5,0}, c_{5,1})\}$$
$$= \{(00), (11), (01), (00), (10)\},$$

where $(c_{l,0}, \; c_{l,1})$ denotes the coded bits. The coded sequence can be obtained using the trellis diagram shown in Fig. 6.7.

Table 6.1. *Coded sequences*

Input sequences	Coded sequences
000	(00), (00), (00)
001	(00), (00), (11)
010	(00), (11), (01)
011	(00), (11), (10)
100	(11), (01), (11)
101	(11), (01), (00)
110	(11), (10), (10)
111	(11), (10), (01)

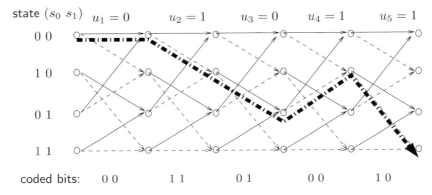

Figure 6.7. Generating a coded sequence using the trellis diagram in Example 6.2.1.

(b) With initial state $(s_0\ s_1) = (0\ 0)$, the list of all possible coded bit sequences for all the input sequences of length 3 is given in Table 6.1.

6.2.2   *Decoding algorithms for convolutional codes*

A coded sequence is transmitted over a channel after a proper modulation. For BPSK signaling, coded bit sequences, $c_{l,t}$, can be converted as $\bar{c}_{l,t} = 1 - 2c_{l,t}$. In this case, coded bit 0 becomes binary data symbol "+1" and coded bit 1 becomes binary data symbol "−1." The resulting coded data symbol sequence can be denoted by $\{\bar{c}_{l,t}\} = \{1 - 2c_{l,t}\}$. A coded sequence can be transmitted in any order as long as the receiver can reorder the sequence.

The Viterbi algorithm for ML decoding

Suppose that the received signal sequence is corrupted by a white Gaussian noise process. The received signal is given by

$$z_{l,t} = \varphi_{l,t}\bar{c}_{l,t} + e_{l,t}, \quad t = 0, 1, \ldots, n_c - 1; \ \ l = 0, 1, \ldots, K_c - 1, \tag{6.19}$$

where $\varphi_{l,t}$ represents the (possibly time-varying) channel fading for coded data symbol $\bar{c}_{l,t}$ and $e_{l,t}$ is a white Gaussian noise with zero mean and variance $N_0/2$. From Eq. (6.19), the likelihood function becomes

$$f(\{z_{l,t}\}|\{\bar{c}_{l,t}\}) = \prod_{l=0}^{K_c-1} \prod_{t=0}^{n_c-1} \frac{1}{\sqrt{\pi N_0}} \exp\left(-\frac{1}{N_0}(z_{l,t} - \varphi_{l,t}\bar{c}_{l,t})^2\right)$$

$$\propto \exp\left(-\frac{1}{N_0} \sum_{l=0}^{K_c-1} \sum_{t=0}^{n_c-1}(z_{l,t} - \varphi_{l,t}\bar{c}_{l,t})^2\right).$$

The ML decoding finds the coded data symbol sequence that minimizes the following cost function:

$$V(\{\bar{c}_{l,t}\}) = \sum_{l=0}^{K_c-1} \sum_{t=0}^{n_c-1}(z_{l,t} - \varphi_{l,t}\bar{c}_{l,t})^2$$

$$= \sum_{l=0}^{K_c-1} \sum_{t=0}^{n_c-1}(z_{l,t} - \varphi_{l,t}(1 - 2c_{l,t}))^2. \tag{6.20}$$

The VA can be applied to the ML decoding as the convolutional code has the trellis structure. Note that any coded sequence can be represented as a valid path along the trellis diagram.

From Eq. (6.20), the branch metric can be defined as follows:

$$M_l = \sum_{t=0}^{n_c-1}(z_{l,t} - \varphi_{l,t}(1 - 2c_{l,t}))^2. \tag{6.21}$$

Denoting by $\mathcal{S}_l$ the current state, the branch metric becomes a function of the input bit u_l and the current state since the coded bits are decided by $\mathcal{S}_l$ and u_l. Then, the branch metric can be rewritten as follows:

$$M_l = \mathcal{M}(u_l, \mathcal{S}_l; \mathbf{z}_l)$$

$$= \sum_{t=0}^{n_c-1}(z_{l,t} - \varphi_{l,t}(1 - 2c_{l,t}))^2. \tag{6.22}$$

The survival path that minimizes the accumulated cost for each state can be found along the trellis diagram. As the VA is discussed for the MLSD in Chapter 3, we provide no further details.

The BCJR algorithm for MAP decoding

The BCJR algorithm can also be used for channel decoding under the MAP criterion. In the iterative receiver, the input signal of the MAP decoder may not be given as in Eq. (6.19). Rather, we may have the LLR of message bit u_l from the LLRs of coded bits, $\{c_{l,0}, c_{l,1}, \ldots, c_{l,n_c-1}\}$ as the input signal of the MAP decoder.

Let $\bar{u}_l = 1 - 2u_l$ and $\mathrm{LLR}(\mathcal{S}_l, \bar{u}_l)$ denote the LLR of $\mathcal{S}_l$ and $\bar{u}_l$, which is given by

$$\mathrm{LLR}(\mathcal{S}_l, \bar{u}_l) = \log \frac{f(\mathbf{z}_l|\mathcal{S}_l, \bar{u}_l = +1)}{f(\mathbf{z}_l|\mathcal{S}_l, \bar{u}_l = -1)}, \tag{6.23}$$

where $\mathbf{z}_l = [z_{l,0} \ \cdots \ z_{l,n_c-1}]^T$ represents the (virtual) received signal. In Eq. (6.23), $f(\mathbf{z}_l|\mathcal{S}_l, \bar{u}_l)$ stands for the (virtual) likelihood of $\mathcal{S}_l$ and $\bar{u}_l$ for a given received signal

z_l. For normalization purposes, assuming that

$$1 = f(\mathbf{z}_l|\mathcal{S}_l, \bar{u}_l = +1) + f(\mathbf{z}_l|\mathcal{S}_l, \bar{u}_l = -1),$$

we can readily show that

$$f(\mathbf{z}_l|\mathcal{S}_l, \bar{u}_l = +1) = \frac{1}{1 + \exp(-\mathrm{LLR}(\mathcal{S}_l, \bar{u}_l))},$$

$$f(\mathbf{z}_l|\mathcal{S}_l, \bar{u}_l = -1) = \frac{1}{1 + \exp(\mathrm{LLR}(\mathcal{S}_l, \bar{u}_l))}. \tag{6.24}$$

Once the likelihoods are obtained, $\gamma(\mathcal{S}_l \to \mathcal{S}_{l+1})$ can be found and the rest of the BCJR algorithm is straightforward; see Section 5.1 for details.

The LLR of $\mathcal{S}_l$ and $\bar{u}_l$ can be obtained from the LLRs of $\mathcal{S}_l$ and $\bar{c}_{l,t}, t = 0, 1, \ldots, n_c - 1$, as follows:

$$\mathrm{LLR}(\mathcal{S}_l, \bar{u}_l = +1) = \sum_{c_{l,t}:u_l=0} \mathrm{LLR}(\mathcal{S}_l, \bar{c}_{l,t}),$$

$$\mathrm{LLR}(\mathcal{S}_l, \bar{u}_l = -1) = \sum_{c_{l,t}:u_l=1} \mathrm{LLR}(\mathcal{S}_l, \bar{c}_{l,t}),$$

where $\sum_{c_{l,t}:u_l}$ denotes the summation over coded bits, $c_{l,t}, t = 0, 1, \ldots, n_c - 1$, which result from the input u_l with the current $\mathcal{S}_l$. In the iterative receiver, the LLR of $\bar{c}_{l,t}$ is usually the input to the channel decoder.

Figure 6.8 shows the simulation results for a coded BER. A rate-half convolutional code with generator polynomial $(5, 7)$ in octal is used. For decoding, both the VA and the BCJR algorithms are used. We can see that these two decoding algorithms provide almost the same performance. In Fig. 6.8, E_b stands for the bit energy, which is the energy for transmitting one message bit. For coded signals, we can also define the coded bit energy. For a rate-k_c/n_c convolutional code, we have

$$k_c E_b = n_c E_{b,\mathrm{coded}},$$

where $E_{b,\mathrm{coded}}$ stands for the coded bit energy. Thus, the bit energy becomes

$$E_b = \frac{n_c}{k_c} E_{b,\mathrm{coded}}.$$

If the code rate is $1/2$, the coded bit energy becomes one-half of the bit energy. In Fig. 6.8, we can see that coded signals require a lower E_b/N_0 than uncoded signals to achieve a practical range of BER (say, 10^{-2} to 10^{-5}).

6.2.3 Error analysis of convolutional codes

The subject of error analysis[†] is important in the understanding and design of convolutional codes. In this subsection, we will study an error analysis for the ML decoding. Any coded

---

[†] It is not necessary to read Section 6.2.3 in order to understand the rest of the chapter since a different approach will be used for performance analysis of the iterative receiver. However, to complete the overview of convolutional codes, an error analysis is included. This subsection will be needed, however, in Chapter 11 when we discuss coded OFDM.

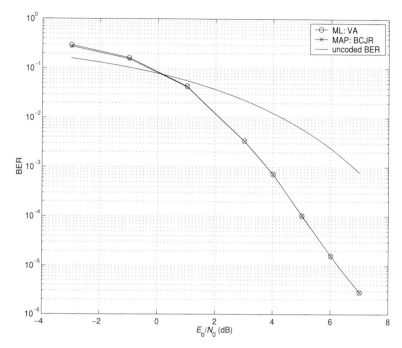

Figure 6.8. Bit error rate performance of the ML and MAP decoding for a rate-half convolutional code with generator polynomial $(7, 5)$ in octal; $L = 1024$.

sequence can be represented in a trellis diagram as a valid path. The ML decoding must choose the valid path that has the smallest cost in Eq. (6.20). An error occurs in the ML decoding when a wrong (but valid) path is chosen along the trellis diagram. We make the following remarks.

- If an incorrect path is chosen, it causes multiple bit errors.
- A convolutional code is linear. This means that the error analysis for an all-zero sequence as the transmitted sequence is valid for all other possible sequences.

Assume that an all-zero sequence is transmitted and that the starting and terminating states are the state of all zeros. For the error analysis, the paths starting and terminating at the state of all zeros with no intermediate returns, which are called the fundamental paths, are important. Since all the paths from the state of all zeros to the state of all zeros can be uniquely decomposed into a sequence of fundamental paths, we only focus on the fundamental paths (which yield incorrect decisions or bit errors) for the error analysis.

For the analysis, we need to develop the transfer function for the fundamental paths. Consider the state diagram of the convolutional encoder in Fig. 6.4 as an example. Any fundamental path must start from the state of zeros and terminate at the state of zeros. Hence, the state of zeros can be split into the starting and terminating states as in Fig. 6.9.

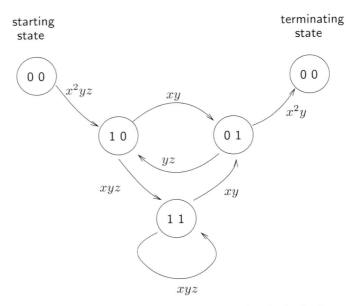

Figure 6.9. The state diagram required to find the transfer function for the fundamental paths of the convolutional encoder in Fig. 6.4.

Define the variable for a transition as

$$x^i y^j z^k,$$

where i denotes the output (coded bit) Hamming distance, j denotes the number of transitions, and k denotes the input (information bit) Hamming distance. Here, the *Hamming distance* between two bit sequences is defined as the number of bits in the difference. For example, the Hamming distance between 0010 and 1100 is 3. Note that for each transition, $j = 1$.

   Consider a state transition from the starting state of zeros to the state of (1 0). The output Hamming distance is 2 (because the coded bits are (1 1) and the Hamming distance from (0 0) is 2) and the input Hamming distance is 1 according to Fig. 6.5. Then, for this transition, we have

$$x^2 y^1 z^1 = x^2 yz.$$

Note that the number of transitions is 1 (i.e. $j = 1$). The variables for the other transitions are shown in Fig. 6.9.

   The state diagram in Fig. 6.9 is helpful if we need to find the information of a fundamental path. For example, consider the fundamental path $(0\ 0) \to (1\ 0) \to (0\ 1) \to (0\ 0)$. The transfer function is the product of all the associated variables, given by

$$(x^2 yz) \cdot (xy) \cdot (x^2 y) = x^5 y^3 z.$$

This shows that this fundamental path has an output of Hamming distance 5, one input Hamming distance of 1, and three transitions. By applying Mason's method to obtain the transfer function from the starting state of zeros to the terminating state of zeros, we have

$$T(x, y, z) = \frac{x^5 y^3 z}{1 - xy(1 + y)z}$$

$$= x^5 y^3 z \left(1 + xy(1 + y)z + (xy(1 + y)z)^2 + \cdots \right)$$

$$= x^5 y^3 z + x^6 y^4 z^2 + x^6 y^5 z^2 + \cdots, \tag{6.25}$$

using $1/(1 - x) = 1 + x + x^2 + x^3 + \cdots$. This yields the fact that all fundamental paths have the minimum (output) Hamming distance that is greater than or equal to 5. This Hamming distance is called the *free distance* and is denoted by d_{free}. The free distance is an important parameter, necessary to understand the performance of the convolutional code. The free distance is the smallest distance between the all-zero sequence (this sequence is assumed to be the transmitted sequence) and any possibly incorrectly decoded sequence. Hence, the error probability of choosing an incorrect path decreases with the free distance since more errors are involved.

Let $\mathbf{e}(d)$ denote an error sequence corresponding to a fundamental path of d output Hamming distance (from the all-zero sequence). In addition, let $\Pr(\mathbf{e}(d))$ denote the error probability that the error sequence $\mathbf{e}(d)$ is chosen in the ML decoding. It is assumed that this error probability is invariant with respect to the starting time of the error sequence, there is no error before the starting time, and it only depends on the number of errors, d. Then, using the union bound, an upper bound of the probability of error can be found as follows:

$$\Pr(\text{Error}) \le \sum_{d=d_{\text{free}}}^{\infty} A_d \Pr(\mathbf{e}(d)), \tag{6.26}$$

where A_d is the number of fundamental paths of d (output) Hamming distance. Note that $\{A_d, d\}_{d=d_{\text{free}}}^{\infty}$ is called the distance spectrum of a convolutional code and plays a key role in analyzing the performance. Since $\Pr(\mathbf{e}(d))$ decreases quickly as d increases in general, a good approximation of the probability of error in Eq. (6.26) is given by

$$\Pr(\text{Error}) \simeq A_{d_{\text{free}}} \Pr(\mathbf{e}(d_{\text{free}})).$$

We can also find the BER. Let $B_{q,d}$ denote the number of bit errors of the qth fundamental path of d Hamming distance. An approximate average number of bit errors (from all the error sequences starting at a given time) is given by

$$E[\text{number of bit errors}] \le \sum_{d=d_{\text{free}}}^{\infty} \sum_{q} B_{q,d} \Pr(\mathbf{e}(d))$$

$$= \sum_{d=d_{\text{free}}}^{\infty} B_d \Pr(\mathbf{e}(d))$$

$$\simeq B_{d_{\text{free}}} \Pr(\mathbf{e}(d_{\text{free}})),$$

where $B_d = \sum_q B_{q,d}$. At each time, since there are k_{c} input bits, we can show that

$$E[\text{number of bit errors}] = k_{\text{c}} P_{\text{b}},$$

Figure 6.10. Block diagram of the transmitter for coded signal transmission.

where P_b denotes the BER. Then, the BER can be approximated as follows:

$$P_b = \frac{1}{k_c} E[\text{number of bit errors}]$$

$$\simeq \frac{1}{k_c} \sum_{d=d_{\text{free}}}^{\infty} B_d \Pr(\mathbf{e}(d))$$

$$\simeq \frac{1}{k_c} B_{d_{\text{free}}} \Pr(\mathbf{e}(d_{\text{free}})). \tag{6.27}$$

According to Eq. (6.25), for example, $A_{d_{\text{free}}} = A_5 = 1$, for the convolutional code in Fig. 6.9. In addition, $B_{d_{\text{free}}} = \sum_q B_{q,d_{\text{free}}} = B_{1,5} = 1$. Hence, the BER is approximated as follows:

$$P_b \simeq \Pr(\mathbf{e}(d_{\text{free}})). \tag{6.28}$$

6.3   Iterative receivers

When a coded sequence is transmitted over an ISI channel, a conventional receiver assumes that the transmitted signal is an uncoded signal for equalization or detection. After equalization or detection of signals, a channel decoder is applied to extract the message bit sequence. Generally, the equalizer has to mitigate the ISI as much as possible to obtain a satisfactory performance, and the overall performance is dependent on the performance of the equalizer since the residual interference will degrade the performance of the following channel decoder.

The iterative receiver can deal with the ISI effectively through feedback and can provide an excellent performance. The transmitter is depicted in Fig. 6.10. A random bit interleaver is an important component for the operation of the iterative receiver. This will be explained later. Note that the resulting transmission scheme is called the bit interleaved coded modulation (BICM), which is studied extensively in Caire, Taricco, and Biglieri (1998).

6.3.1   *Structure and operation of iterative receivers*

Generally, the iterative receiver consists of a soft-input soft-output (SISO) equalizer (or detector), a SISO decoder, an (bit) interleaver, and a deinterleaver, as shown in Fig. 6.11. The SISO equalizer in the iterative receiver should exploit the *a priori* information of bits which will be provided by the SISO decoder. Roughly, the operation of the iterative receiver

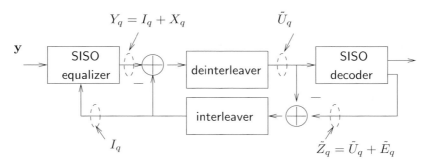

Figure 6.11. Block diagram for the iterative receiver.

can be understood as follows: the (extrinsic) bit information from the channel decoder is fed back to the channel equalizer to suppress the ISI effectively so that the equalizer can provide more reliable information to the channel decoder for the next iteration. Through iterations, more reliable information can be exchanged between the equalizer and decoder. An ideal performance can be achieved when the ISI is completely removed and an optimally combined signal is provided to the channel decoder.

In the iterative receiver, statistics are exchanged through iterations for a better performance. The input and output statistics of a SISO equalizer and a decoder are shown in Fig. 6.11. Suppose that the SISO equalizer provides the following output at the qth iteration:

$$Y_q = I_q + X_q, \tag{6.29}$$

where I_q is the input which denotes the extrinsic information of the data sequence from the channel decoder and X_q denotes the information of the transmitted data sequence from the received signal sequence. Generally, X_q depends on I_q. It is expected that a SISO equalizer provides more reliable information X_q for a data sequence the greater the reliability of I_q. With a more reliable I_q, we assume that the SISO equalizer can mitigate the ISI more effectively and that the resulting output will be close to that without the ISI. Note that Y_q, X_q, and I_q denote statistics; precise definitions will be given later.

The output of the channel decoder can be written as follows:

$$\tilde{Z}_q = \tilde{U}_q + \tilde{E}_q, \tag{6.30}$$

where $\tilde{U}_q$ is the input information to the channel decoder and $\tilde{E}_q$ is the additional information gained by channel decoding. The quantities with $\tilde{\ }$ denote variables after deinterleaving, while the quantities without $\tilde{\ }$ denote variables after interleaving. For example, $\tilde{U}_q$ denotes the output of the deinterleaver, while U_q denotes the input. According to Fig. 6.11, U_q becomes X_q as follows:

$$U_q = Y_q - I_q$$
$$= X_q$$
$$\tilde{U}_q = \tilde{X}_q.$$

In the next iteration, we have

$$I_{q+1} = E_q$$
$$\tilde{Z}_{q+1} = \tilde{X}_{q+1} + \tilde{E}_{q+1}. \tag{6.31}$$

Through iterations, if X_q approaches a limit that can be obtained with no ISI, the output information of the channel decoder, $\tilde{Z}_q$, can also approach an ideal outcome, which would be the decoder's output with no ISI. Thus, it is expected that the iterative receiver provides an ideal performance that can be achieved without the ISI.

Note that the subtractions of the extrinsic information are important to avoid divergence due to accumulation of information. For example, suppose that the extrinsic information is not removed. Then, we have

$$I_{q+1} = X_q + I_q + E_q$$
$$\tilde{Z}_{q+1} = (\tilde{X}_{q+1} + \tilde{I}_{q+1}) + \tilde{E}_{q+1}$$
$$= \tilde{X}_{q+1} + \tilde{X}_q + \tilde{I}_q + \tilde{E}_q + \tilde{E}_{q+1}.$$

Clearly, in $\tilde{Z}_q$, the accumulated terms, $\sum_{k=0}^{q} \tilde{X}_k$ and $\sum_{k=0}^{q} \tilde{E}_k$, which diverge as $q \to \infty$, appear through $\tilde{I}_q$. Thus, the iterative receiver can diverge.

The quantities X_q, I_q, and E_q are not clearly defined yet. For the SISO equalizer, Y_q denotes the LAPP, while X_q and I_q denote the LLR and log-ratio of *a priori* probability, respectively. For example, for a given symbol b and received signal y, we can show that

$$\underbrace{\log \frac{\Pr(b=+1|y)}{\Pr(b=-1|y)}}_{=Y} = \underbrace{\log \frac{\Pr(y|b=+1)}{\Pr(y|b=-1)}}_{=X} + \underbrace{\log \frac{\Pr(b=+1)}{\Pr(b=-1)}}_{I}.$$

Given a symbol sequence and a received signal sequence over an ISI channel, we can define Y_q, X_q, and I_q by the same approach. In this case, the MAP symbol detection introduced in Section 5.1 can be used for the SISO equalization. In general, any SISO equalizer that can exploit the *a priori* information in order to remove the ISI and improve the performance can be used. The MAP channel decoder is appropriate to provide the output in Eq. (6.30).

6.3.2   Ideal performance of the iterative receiver

An ideal performance of the iterative receiver can be achieved when the ISI is eliminated. Recall that the received signal y_l at time l over an ISI channel is as follows:

$$y_l = \sum_{p=0}^{P-1} h_p b_{l-p} + n_l, \quad l = 0, 1, \ldots, L-1, \tag{6.32}$$

where $\{h_p\}$ is a (real-valued) CIR of length P, $\{b_m\}$ is a transmitted (coded and interleaved) symbol sequence, and n_l is a (real-valued) white Gaussian noise with $E[n_l] = 0$ and $E[|n_l|^2] = N_0/2$. Here, L denotes the length of the data sequence. To consider the ideal performance, define $\mathbf{y}_l$ as follows:

$$\mathbf{y}_l = [y_l \ y_{l+1} \ \cdots \ y_{l+P-1}]^\mathrm{T}$$

$$= \underbrace{\begin{bmatrix} h_{P-1} & h_{P-2} & \cdots & h_0 & \cdots & 0 \\ 0 & h_{P-1} & \cdots & & \vdots & \cdots & 0 \\ \vdots & & \ddots & & \vdots & \ddots & \vdots \\ 0 & & \cdots & h_{P-1} & \cdots & h_0 \end{bmatrix}}_{=\mathbf{H}} \underbrace{\begin{bmatrix} b_{l-P+1} \\ b_{l-P+2} \\ \vdots \\ b_{l+P-1} \end{bmatrix}}_{=\mathbf{b}_l} + \underbrace{\begin{bmatrix} n_l \\ n_{l+1} \\ \vdots \\ n_{l+P-1} \end{bmatrix}}_{=\mathbf{n}_l}, \qquad (6.33)$$

where $\mathbf{H}$ is a Toeplitz matrix of size $P \times (2P - 1)$. The matrix $\mathbf{H}$ can be decomposed into three submatrices as follows:

$$\mathbf{H} = \begin{bmatrix} \mathbf{H}_- & \vdots & \mathbf{h} & \vdots & \mathbf{H}_+ \end{bmatrix},$$

where $\mathbf{H}_-$ and $\mathbf{H}_+$ are the submatrices obtained by taking the first $(P - 1)$ column vectors and the last $(P - 1)$ column vectors of $\mathbf{H}$, respectively, and $\mathbf{h}$ is the Pth column vector, which is given by $\mathbf{h} = [h_0 \ h_1 \ \cdots \ h_{P-1}]^\mathrm{T}$. We can decompose the data symbol vector $\mathbf{b}_l$ accordingly and denote by $\mathbf{b}_{l,-}$ and $\mathbf{b}_{l,+}$ for $\mathbf{H}_-$ and $\mathbf{H}_+$, respectively. Then, Eq. (6.33) is rewritten as follows:

$$\mathbf{y}_l = \mathbf{h}b_l + \mathbf{H}_+\mathbf{b}_{l,+} + \mathbf{H}_-\mathbf{b}_{l,-} + \mathbf{n}_l. \qquad (6.34)$$

If the ISI components can be eliminated (using an ideal *a priori* information), we can have

$$\mathbf{r}_l = \mathbf{y}_l - (\mathbf{H}_+\mathbf{b}_{l,+} + \mathbf{H}_-\mathbf{b}_{l,-})$$
$$= \mathbf{h}b_l + \mathbf{n}_l. \qquad (6.35)$$

To the channel decoder, the LLR of b_l from $\mathbf{r}_l$ can be the input as a soft-decision. The LLR is given by

$$\mathrm{LLR}_{\mathrm{ideal}}(b_l; \mathbf{r}_l) = \log \frac{f(\mathbf{r}_l | b_l = +1)}{f(\mathbf{r}_l | b_l = -1)}$$
$$= -\frac{1}{N_0}||\mathbf{r}_l - \mathbf{h}||^2 + \frac{1}{N_0}||\mathbf{r}_l + \mathbf{h}||^2$$
$$= \frac{4}{N_0}\mathbf{h}^\mathrm{T}\mathbf{r}_l$$
$$= \mu_{\mathrm{aw}}b_l + \eta_l, \qquad (6.36)$$

where $\mu_{\mathrm{aw}} = (4/N_0)||\mathbf{h}||^2$ and $\eta_l = (4/N_0)\mathbf{h}^\mathrm{T}\mathbf{n}_l$ is a zero-mean Gaussian noise with variance as follows:

$$\sigma_\eta^2 = E[|\eta_l|^2] = 2\mu_{\mathrm{aw}}.$$

This relationship between the mean and variance is called the consistency condition (Brink, 2001). The SNR of the LLR is given by

$$\mathrm{SNR} = \frac{\mu_{\mathrm{aw}}^2}{\sigma_\eta^2} = \frac{1}{2}\mu_{\mathrm{aw}} = \frac{||\mathbf{h}||^2}{N_0/2}.$$

The coded BER obtained with the LLR in Eq. (6.36) is referred to as the matched filter bound (MFB) as the SNR of the input signal to the decoder is the same as that which can be obtained by matched filtering (after eliminating the ISI components).

6.3.3 Equalizers within the iterative receiver

In general, a coded sequence from a convolutional encoder is correlated. The correlation can be broken by a (random) bit interleaver. Once a bit interleaver is applied, a SISO equalizer can assume that the b_l's are independent. Hence, any SISO equalizer that provides an optimal performance for independent symbol sequences can be used to provide a good performance within the iterative receiver.

The MAP equalizer (or detector) can be used for SISO equalization within the iterative receiver. The output of the MAP equalizer is the LAPP given by

$$L(b_l) = \log \frac{\Pr(b_l = +1|\{y_m\})}{\Pr(b_l = -1|\{y_m\})}. \tag{6.37}$$

Since the LLR becomes the input to the channel decoder, the LLR must be obtained. If the *a priori* probability is denoted by $\Pr(b_l)$, the LLR is given by

$$
\begin{aligned}
\text{LLR}(b_l) &= \log \frac{f(\{y_m\}|b_l = +1)}{f(\{y_m\}|b_l = -1)} \\
&= \log \frac{\Pr(b_l = +1|\{y_m\})/\Pr(b_l = +1)}{\Pr(b_l = -1|\{y_m\})/\Pr(b_l = -1)} \\
&= \log \frac{\Pr(b_l = +1|\{y_m\})}{\Pr(b_l = -1|\{y_m\})} - \log \frac{\Pr(b_l = +1)}{\Pr(b_l = -1)} \\
&= L(b_l) - \log \frac{\Pr(b_l = +1)}{\Pr(b_l = -1)}, \tag{6.38}
\end{aligned}
$$

where $\log(\Pr(b_l = +1)/\Pr(b_l = -1))$ is the log-ratio of the *a priori* probability (LAPIP).

Example 6.3.1 Consider an FIR channel with $P = 2$ and $\sigma_n^2 = N_0/2 = 0.5$. Let $h_0 = 1$ and $h_1 = 0.5$. We are interested in finding the LLR of b_1 when $y_1 = 2$ and $y_2 = -1$. We assume that the b_l's are independent and that $\Pr(b_0 = +1) = 0.7$, $\Pr(b_1 = +1) = 0.4$, and $\Pr(b_2 = +1) = 0.2$.

To find the LLR of b_1, we need to find the *a posteriori* probability as follows:

$$
\begin{aligned}
\Pr(b_1|y_1, y_2) &= \sum_{b_0, b_2} \Pr(b_0, b_1, b_2|y_1, y_2) \\
&\propto \sum_{b_0, b_2} f(y_1, y_2|b_0, b_1, b_2) \Pr(b_0, b_1, b_2) \\
&= \sum_{b_0, b_2} f(y_1|b_0, b_1) f(y_2|b_1, b_2) \Pr(b_0) \Pr(b_1) \Pr(b_2). \tag{6.39}
\end{aligned}
$$

Since $N_0 = 1$, we have

$$
\begin{aligned}
f(y_1|b_0, b_1) &= c \cdot \exp\left(-(2 - h_0 b_1 - h_1 b_0)^2\right) \\
f(y_2|b_1, b_2) &= c \cdot \exp\left(-(-1 - h_0 b_2 - h_1 b_1)^2\right),
\end{aligned}
$$

where c is a normalizing constant. Then, it follows that

$$\Pr(b_1 = +1|y_1, y_2) = 0.9964 \text{ and } \Pr(b_1 = -1|y_1, y_2) = 0.0036,$$

and the LLR becomes

$$\text{LLR}(b_1) = \log \frac{0.9964}{0.0036} - \log \frac{0.4}{0.6} = 6.0227.$$

In general, the complexity of the MAP equalizer grows exponentially with the length of the CIR. Hence, the MAP equalizer would not be practical when P is large. There are other computationally efficient methods to replace the MAP equalizer. An MMSE approach with soft (interference) cancelation (SC) is proposed in Wang and Poor (1999) and is used for the equalizer over ISI channels in Tuchler *et al.* (2002a,b).

To explain the MMSE approach with SC, we need to consider again the signal vector $\mathbf{y}_l$ in Eq. (6.33). If the *a priori* probabilities of the b_l's are available, the ISI components can be mitigated by SC and filtering. The mean value of b_l with respect to the *a priori* probability is given by

$$\bar{b}_l = E[b_l] = \Pr(b_l = +1) - \Pr(b_l = -1), \quad l = 0, 1, \dots, L - 1. \tag{6.40}$$

Let $\bar{\mathbf{b}}_{l,+}$ and $\bar{\mathbf{b}}_{l,-}$ denote the vectors $\mathbf{b}_{l,+}$ and $\mathbf{b}_{l,-}$, respectively, after replacing the b_l's by the $\bar{b}_l$'s. The difference vectors are defined as follows:

$$\begin{aligned} \tilde{\mathbf{b}}_{l,+} &= \mathbf{b}_{l,+} - \bar{\mathbf{b}}_{l,+} \\ \tilde{\mathbf{b}}_{l,-} &= \mathbf{b}_{l,-} - \bar{\mathbf{b}}_{l,-}. \end{aligned} \tag{6.41}$$

The difference vectors are random vectors and their mean vectors are zero vectors. Assuming the data symbols are uncorrelated, the covariance matrices can be written as follows:

$$\begin{aligned} \mathbf{Q}_{l,+} &= E\left[\tilde{\mathbf{b}}_{l,+}\tilde{\mathbf{b}}_{l,+}^T\right] \\ &= \text{Diag}\left(1 - (\bar{b}_{l+1})^2, \ 1 - (\bar{b}_{l+2})^2, \dots, 1 - (\bar{b}_{l+P-1})^2\right) \\ \mathbf{Q}_{l,-} &= E\left[\tilde{\mathbf{b}}_{l,-}\tilde{\mathbf{b}}_{l,-}^T\right] \\ &= \text{Diag}\left(1 - (\bar{b}_{l-P+1})^2, \ 1 - (\bar{b}_{l-P+2})^2, \dots, 1 - (\bar{b}_{l-1})^2\right). \end{aligned} \tag{6.42}$$

In the MMSE approach with SC, firstly the ISI components are suppressed by SC as follows:

$$\begin{aligned} \mathbf{r}_l &= \mathbf{y}_l - \left(\mathbf{H}_+\bar{\mathbf{b}}_{l,+} + \mathbf{H}_-\bar{\mathbf{b}}_{l,-}\right) \\ &= \mathbf{h}b_l + \mathbf{H}_+\tilde{\mathbf{b}}_{l,+} + \mathbf{H}_-\tilde{\mathbf{b}}_{l,-} + \mathbf{n}_l. \end{aligned} \tag{6.43}$$

Secondly, the residual ISI components, $\tilde{\mathbf{b}}_{l,+}$ and $\tilde{\mathbf{b}}_{l,-}$, are further mitigated by the MMSE filtering. The MMSE problem with $\mathbf{r}_l$ is given by

$$\mathbf{w}_{\text{mmse},l} = \arg \min_{\mathbf{w}_l} E\left[\left\|b_l - \mathbf{w}_l^T\mathbf{r}_l\right\|^2\right], \tag{6.44}$$

where $\mathbf{w}_{\text{mmse},l}$ is the MMSE filtering vector readily given by

$$\begin{aligned} \mathbf{w}_{\text{mmse},l} &= \left(E\left[\mathbf{r}_l\mathbf{r}_l^T\right]\right)^{-1} E[\mathbf{r}_l b_l] \\ &= \left(\mathbf{h}\mathbf{h}^T + \mathbf{H}_+\mathbf{Q}_{l,+}\mathbf{H}_+^T + \mathbf{H}_-\mathbf{Q}_{l,-}\mathbf{H}_-^T + \frac{N_0}{2}\mathbf{I}\right)^{-1} \mathbf{h}. \end{aligned} \tag{6.45}$$

Then, the MMSE filtered output is given by

$$\hat{b}_{\text{mmse},l} = \mathbf{w}_{\text{mmse},l}^{\text{T}} \mathbf{r}_l$$
$$= \mathbf{w}_{\text{mmse},l}^{\text{T}} (\mathbf{y}_l - \mathbf{H}_+ \bar{\mathbf{b}}_{l,+} - \mathbf{H}_- \bar{\mathbf{b}}_{l,-}). \tag{6.46}$$

This approach is referred to as the MMSE equalizer with SC or the MMSE-SC equalizer. Note that if the b_l's are equally likely (i.e., $\Pr(b_l) = 1/2$ for $b_l \in \{+1, -1\}$), the MMSE-SC equalizer becomes the conventional MMSE equalizer. This is the case when there is no prior information.

The LLR of $\hat{b}_{\text{mmse},l}$ in Eq. (6.46) becomes the input to the channel decoder. Equation (6.46) can be rewritten as follows:

$$\hat{b}_{\text{mmse},l} = \mathbf{w}_{\text{mmse},l}^{\text{T}} (\mathbf{h} b_l + \mathbf{H}_+ \tilde{\mathbf{b}}_{l,+} + \mathbf{H}_- \tilde{\mathbf{b}}_{l,-} + \mathbf{n}_l). \tag{6.47}$$

To find the LLR of $\hat{b}_{\text{mmse},l}$, we need to derive the pdf of $\hat{b}_{\text{mmse},l}$. As this is not easy, we can consider an approximation. In Wang and Poor (1999), it is assumed that $\hat{b}_{\text{mmse},l}$ can be expressed as follows:

$$\hat{b}_{\text{mmse},l} = \mu_l b_l + \eta_l, \tag{6.48}$$

where μ_l is the conditional mean and η_l is a zero-mean Gaussian random variable. Then, the likelihood function is given by

$$f(\hat{b}_{\text{mmse},l} | b_l) = \frac{1}{\sqrt{2\pi} \sigma_{\eta,l}} \exp \left(-\frac{1}{2\sigma_{\eta,l}^2} (\hat{b}_{\text{mmse},l} - \mu_l b_l)^2 \right),$$

where $\sigma_{\eta,l}^2 = E[|\eta_l|^2]$. The approximation or modeling for $\hat{b}_{\text{mmse},l}$ in Eq. (6.48) is called the Gaussian approximation. Now we need to find μ_l and $\sigma_{\eta,l}^2$. From Eq. (6.47), after some manipulation we can show that

$$\mu_l = E[\hat{b}_{\text{mmse},l} b_l]$$
$$= E\left[\mathbf{w}_{\text{mmse},l}^{\text{T}} (\mathbf{h} + \mathbf{H}_+ \tilde{\mathbf{b}}_{l,+} b_l + \mathbf{H}_- \tilde{\mathbf{b}}_{l,-} b_l + \mathbf{n}_l b_l) \right]$$
$$= \mathbf{w}_{\text{mmse},l}^{\text{T}} \mathbf{h}$$
$$\sigma_{\eta,l}^2 = E[(\hat{b}_{\text{mmse},l} - \mu_l b_l)^2]$$
$$= \mu_l - \mu_l^2. \tag{6.49}$$

Finally, the LLR of b_l from $\hat{b}_{\text{mmse},l}$ is given by

$$\text{LLR}(b_l; \hat{b}_{\text{mmse},l}) = \log \frac{f(\hat{b}_{\text{mmse},l} | b_l = +1)}{f(\hat{b}_{\text{mmse},l} | b_l = -1)}$$
$$= -\frac{1}{2\sigma_{\eta,l}^2} (\hat{b}_{\text{mmse},l} - \mu_l)^2 + \frac{1}{2\sigma_{\eta,l}^2} (\hat{b}_{\text{mmse},l} + \mu_l)^2$$
$$= \frac{2 \hat{b}_{\text{mmse},l}}{1 - \mu_l}. \tag{6.50}$$

Under the Gaussian assumption, the LLR from $\hat{b}_{\mathrm{mmse},l}$ satisfies the consistency condition. If the LLR in Eq. (6.50) is rewritten as

$$\mathrm{LLR}(b_l; \hat{b}_{\mathrm{mmse},l}) = \mu_{\mathrm{mmse},l} b_l + \eta_{\mathrm{mmse},l}, \tag{6.51}$$

we can readily show that

$$E[|\eta_{\mathrm{mmse},l}|^2] = 2\mu_{\mathrm{mmse},l},$$

where $\mu_{\mathrm{mmse},l} = 2\mu_l/(1 - \mu_l)$.

Example 6.3.2 (Example 6.3.1 revisited) The LLR of b_1 from $\hat{b}_{\mathrm{mmse},1}$ can also be found. To find $\hat{b}_{\mathrm{mmse},1}$, the MMSE filtering vector should be found:

$$\mathbf{w}_{\mathrm{mmse},1} = \left(\mathbf{h}\mathbf{h}^{\mathrm{T}} + \mathbf{H}_+ \mathbf{Q}_{1,+} \mathbf{H}_+^{\mathrm{T}} + \mathbf{H}_- \mathbf{Q}_{1,-} \mathbf{H}_-^{\mathrm{T}} + \frac{N_0}{2}\mathbf{I} \right)^{-1} \mathbf{h}$$

$$= \begin{bmatrix} 1.71 & 0.5 \\ 0.5 & 1.39 \end{bmatrix}^{-1} \begin{bmatrix} 1 \\ 0.5 \end{bmatrix} = \begin{bmatrix} 0.5360 \\ 0.1669 \end{bmatrix},$$

where $\mathbf{h} = [1 \ 0.5]^{\mathrm{T}}$, $\mathbf{H}_- = [0.5 \ 0]^{\mathrm{T}}$, and $\mathbf{H}_+ = [0 \ 1]^{\mathrm{T}}$. In addition, $\mathbf{Q}_{1,+} = 1 - (-0.6)^2$ and $\mathbf{Q}_{1,-} = 1 - (0.4)^2$ since $\bar{\mathbf{b}}_{l,+} = \bar{b}_2 = -0.6$ and $\bar{\mathbf{b}}_{l,-} = \bar{b}_0 = 0.4$. In addition, $\mu_1 = \mathbf{w}_{\mathrm{mmse},1}^{\mathrm{T}}\mathbf{h} = 0.6194$. Finally, we find

$$\hat{b}_{\mathrm{mmse},1} = \mathbf{w}_{\mathrm{mmse},1}^{\mathrm{T}}(\mathbf{y}_1 - \mathbf{H}_+ \bar{\mathbf{b}}_{1,+} - \mathbf{H}_- \bar{\mathbf{b}}_{1,-})$$

$$= 0.8980,$$

$$\mathrm{LLR}(b_1; \hat{b}_{\mathrm{mmse},1}) = 4.7195,$$

where $\mathbf{y}_1 = [y_1 \ y_2]^{\mathrm{T}} = [2 \ -1]^{\mathrm{T}}$.

In the iterative receiver, we can use either the MAP equalizer or the MMSE-SC equalizer. The MAP equalizer is able to provide the LLR from $\mathbf{y}_l$, while the MMSE-SC equalizer provides the LLR from $\hat{b}_{\mathrm{mmse},l}$. The MMSE-SC equalizer uses an indirect approach to provide the LLR and therefore its performance would be worse than that produced by the MAP equalizer. To observe the performance difference, simulations are carried out with the following CIR:

$$\mathbf{h} = [0.227 \ 0.46 \ 0.688 \ 0.46 \ 0.227]^{\mathrm{T}}. \tag{6.52}$$

The SNR is defined as follows:

$$\mathrm{SNR} = \frac{\|\mathbf{h}\|^2}{N_0} = \frac{1}{N_0}.$$

For channel coding, a rate-half convolutional code with generator polynomial $(7, 5)$ in octal is employed. The simulation results of the iterative receiver with $L = 2^{12}$ are shown in Fig. 6.12. It is shown that the iterative receiver with the MAP equalizer performs better than that with the MMSE-SC equalizer.

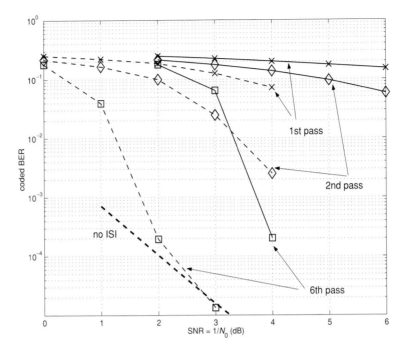

Figure 6.12. Bit error rate performance of the iterative receiver with the MAP and MMSE-SC equal-izers when $L = 2^{12}$. Dashed lines denote the iterative receiver with the MAP equalizer; solid lines denote the iterative receiver with the MMSE-SC equalizer. Note that for the iterative receiver with the MAP equalizer, simulation results are obtained for a range of SNR from 0 dB to 4 dB, while simulation results for the iterative receiver with the MMSE-SC equalizer are obtained for a range of SNR from 2 dB to 6 dB.

6.4   EXIT chart analysis

Various approaches for the analysis of the iterative receiver have been proposed. In this section, we introduce one of them, which is based on the extrinsic information transfer (EXIT) chart. The EXIT chart was originally proposed in Brink (2001) to analyze the turbo code. As shown in Tuchler *et al.* (2002a,b), it is also applicable to the performance analysis of the iterative receiver.

6.4.1   EXIT charts for iterative receivers

The EXIT charts for a SISO equalizer and an MAP decoder are the transfer functions of the mutual information of the input and output extrinsic information. As shown in Fig. 6.11, the extrinsic bit information is exchanged between the equalizer and decoder. Through the mutual information, we can quantify the reliability of the extrinsic bit information. Thus, from the EXIT charts, the convergence behavior of the iterative receiver can be studied through the mutual information of the exchanged extrinsic bit information.

Firstly, we consider the EXIT chart for an MAP decoder. The input extrinsic bit information to the MAP decoder is the LLR, modeled as

$$\text{LLR}_{\text{in},l} = z_l$$
$$= \mu b_l + \eta_l, \tag{6.53}$$

where $2\mu = \sigma_\eta^2 = E[|\eta_l|^2]$. The noise η_l is a Gaussian random variable. With the LLR as the input to the MAP decoder, the BCJR algorithm can be performed to obtain the output extrinsic bit information, which can be modeled as

$$\text{LLR}_{\text{out},l} = \bar{z}_l$$
$$= \varphi b_l + e_l, \tag{6.54}$$

where φ becomes the (nominal) signal gain and e_l is assumed to be a Gaussian random variable. As $\bar{z}_l$ is an LLR, the consistency condition that $2\varphi = E[|e_l|^2]$ should hold. It is generally known that the approximation in Eq. (6.54) is practically reasonable.

For the input LLR in Eq. (6.53), the mutual information becomes $I_{\text{in}} = J(\sigma_\eta^2)$ as shown in Eq. (6.18). Note that $0 \le I_{\text{in}} \le 1$. For a particular value of I_{in}, the mutual information of the output extrinsic bit information can be found. Then, pairs of $(I_{\text{in}}, I_{\text{out}})$ can construct a curve for the EXIT chart (Brink, 2001). Note that as an LLR becomes more reliable, its mutual information becomes higher. In general, the EXIT chart is an increasing function. That is, for an input LLR with a higher mutual information, an output LLR of a higher mutual information is expected.

An EXIT chart can also be found for a SISO equalizer. For the CIR in Eq. (6.52), an EXIT chart for the MAP equalizer is obtained at an SNR of 4 dB and shown in Fig. 6.13 with the EXIT chart for the MAP decoder that is used in Fig. 6.12: the solid line represents the EXIT chart for the MAP decoder and the dashed line represents the EXIT chart for the MAP equalizer. The EXIT chart for the MAP decoder depicts the mutual information of the output LLR along the x-axis and that for the input LLR along the y-axis. The converse is true for the EXIT chart for the MAP equalizer. Note that the output LLR of the MAP equalizer becomes the input LLR of the MAP decoder and the output LLR of the MAP decoder becomes the input LLR of the MAP equalizer.

Within the iterative receiver, the mutual information evolves as the SISO equalizer and SISO decoder exchange the LLR. During the first iteration, the SISO equalizer provides the LLR with zero input LLR from the decoder. This corresponds to the first vertical arrow in Fig. 6.13, which results in a mutual information of 0.52. The SISO decoding is performed with the LLR from the SISO equalizer and provides a better mutual information, 0.58 (corresponding to the first horizontal arrow), and the output LLR becomes the input LLR to the equalizer during the next iteration. The iteration continues and the mutual information can converge to the point where the two EXIT charts meet.

As shown in Fig. 6.13, the two EXIT charts build a tunnel. The EXIT chart of the SISO equalizer depends on the LLR from the decoder as well as the SNR. If the SNR is lower, the EXIT chart will move down since the mutual information of the output LLR of the SISO equalizer decreases. Hence, when the SNR is sufficiently low, the tunnel can be closed and the two EXIT charts will meet at a low mutual information. In this case, the performance of

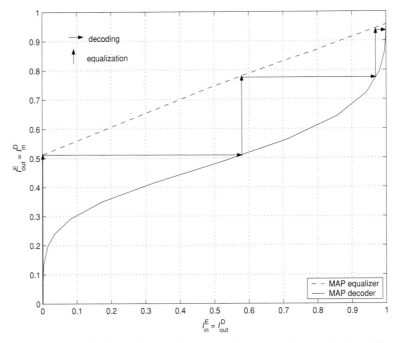

Figure 6.13. EXIT charts of the iterative receiver with an MAP equalizer and an MAP decoder. (For channel coding, a rate-half convolutional code with generator polynomial $(7, 5)$ in octal is used.) I_{in}^{E}, I_{in}^{D} = input extrinsic information to the equalizer and decoder, respectively; $I_{\text{out}}^{\text{E}}$, $I_{\text{out}}^{\text{D}}$ = output extrinsic information to the equalizer and decoder, respectively.

the iterative receiver is not satisfactory. On the other hand, if the SNR is higher, the EXIT chart will move up and the tunnel becomes open.

6.4.2 Approaches taken to construct EXIT charts

So far, we have not addressed how the mutual information can be obtained for the output of the channel decoder. The mutual information in Eq. (6.14) can be obtained by histograms from realizations. From the sequence of output LLR, $\{\bar{z}_m\}$, a histogram may be obtained to approximate the pdf $f(\bar{z})$. With known $\{b_m\}$, we can also find two histograms for the conditional pdfs $f(\bar{z}|b = 1)$ and $f(\bar{z}|b = -1)$. Denote by $\hat{f}(\bar{z})$ and $\hat{f}(\bar{z}|b = \pm 1)$ the empirical pdf of $\bar{z}$ and empirical conditional pdfs of $\bar{z}$ given b. Then, the mutual information can be found as in Eq. (6.14). In general, the histogram-based approach to obtain the mutual information requires a number of samples for a precise result.

The histogram-based approach is a nonparametric estimation method used to estimate the mutual information. A parametric estimation method is available under the Gaussian assumption. With the consistency condition, let $2\varphi = \sigma_e^2 = E[|e_l|^2]$. Then, the likelihood function of $\bar{z}_l$ given b_l is given by

$$f\left(\bar{z}_l | b_l; \sigma_e^2\right) = \frac{1}{\sqrt{2\pi \sigma_e^2}} \exp\left(-\frac{1}{2\sigma_e^2}\left(\bar{z}_l - \frac{\sigma_e^2}{2} b_l\right)^2\right). \tag{6.55}$$

For convenience, let $x_l = \bar{z}_l b_l$ and $v = \sigma_e^2$. Then, the ML estimation problem to estimate v is written as follows:

$$\hat{v}_{\text{ml}} = \arg\max_{v \geq 0} \prod_{l=0}^{L} f(x_l; v)$$

$$= \arg\max_{v \geq 0} \left\{ -\frac{1}{2v} \sum_{l=0}^{L-1} \left(x_l - \frac{v}{2} \right)^2 - \frac{L}{2} \log v \right\}, \tag{6.56}$$

where

$$f(x_l; v) = \frac{1}{\sqrt{2\pi v}} \exp\left(-\frac{1}{2v} \left(x_l - \frac{v}{2} \right)^2 \right). \tag{6.57}$$

Since

$$-\frac{1}{2v} \left(\sum_{l=0}^{L-1} x_l^2 - \sum_{l=0}^{L-1} x_l v + L \frac{v^2}{4} \right) = -\frac{S_x}{2v} + \frac{M_x}{2} - \frac{L}{8} v,$$

where $S_x = \sum_{l=0}^{L-1} x_l^2$ and $M_x = \sum_{l=0}^{L-1} x_l$, the ML estimate of v maximizes the following cost function:

$$C(v) = -\frac{S_x}{v} - \frac{L}{4} v - L \log v$$

$$= L \left(-\frac{\bar{S}_x}{v} - \frac{v}{4} - \log v \right), \quad v \geq 0, \tag{6.58}$$

where $\bar{S}_x = S_x / L$. This function has two extreme points, and they can be found by taking the derivative with respect to v as follows:

$$\frac{dC(v)}{dv} = L \left(\frac{\bar{S}_x}{v^2} - \frac{1}{4} - \frac{1}{v} \right).$$

It can be readily shown that one root is less than zero and the other is greater than zero. The root which is greater than zero is the maximum point and the ML estimate of v:

$$\hat{v}_{\text{ml}} = 2 \left(\sqrt{1 + \bar{S}_x} - 1 \right). \tag{6.59}$$

The advantage of this parametric approach over the nonparametric approach (i.e. the histogram-based approach) is that a precise EXIT chart can be obtained with a small number of samples.

Figure 6.14 shows the EXIT charts. The two different approaches are used with a coded sequence of length $L = 1024$. The results with solid lines are obtained by the parametric approach using the ML estimate, while the results with dashed lines are obtained by the nonparametric approach based on histograms. Two different numbers (10 and 10 000) of realizations are used. We confirm that the parametric approach can provide a consistent result. Note that the parametric approach is valid only when the pdf (as a function) is known but the values of its parameters are unknown. If the pdf is different from the assumed one, the parametric approach could not provide a satisfactory result. On the other hand, since

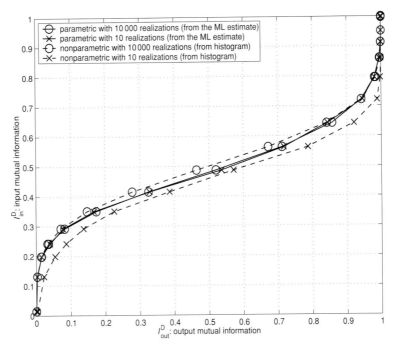

Figure 6.14. EXIT charts of the MAP decoder for a rate-half convolutional code with generator polynomial $(7, 5)$ in octal.

the nonparametric approach is not based on any assumption of the pdf, it can be robust and can provide a reasonable performance even if the pdf is not Gaussian.

The same parametric approach taken to build the EXIT chart can be considered for the MAP equalizer. However, the LLRs of the MAP equalizer do not follow the Gaussian pdf. To show this, we obtain a histogram of the output LLRs; this is shown in Fig. 6.15 for the CIR in Eq. (6.52) when the SNR, $1/N_0$, is 4 dB. We assume that there is no prior information. That is, the b_l's are assumed to be equally likely. For comparison purposes, the Gaussian pdf with the same mean and variance as the empirical pdf of the LLRs is also presented in Fig. 6.15. It is shown that the (empirical) pdf of the LLRs is different from the Gaussian pdf and has a heavy tail. In this case, we cannot use the parametric approach with the Gaussian assumption to build the EXIT chart. However, other pdfs can be used to model the pdf of the LLRs in order to use the parametric approach. Since the nonparametric approach does not assume any pdf, it can be used to build the EXIT chart.

6.4.3   Analytical approach taken to construct the EXIT chart for the MMSE-SC equalizer

To construct the EXIT chart for the MMSE-SC equalizer, the pdf of the LLR in Eq. (6.51) is required. A histogram can be used to estimate the pdf. As mentioned earlier, a large number

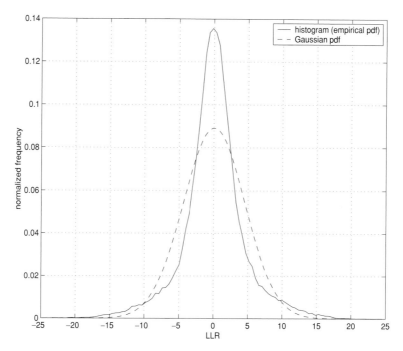

Figure 6.15. Histogram of the LLRs from the MAP equalizer when SNR = 4 dB and there is no prior information.

of samples are required to obtain a good estimate of the pdf. In this subsection, we introduce an analytical approach to find the pdf of the output LLR of the MMSE-SC equalizer.

As shown in Eq. (6.51), the mean $\mu_{\mathrm{mmse},l}$ of the LLR varies over time. As a result, the LLR is not a stationary random sequence. Hence, the mutual information of the LLR cannot be well defined. To avoid this problem, we assume that $\mathrm{LLR}(b_l, \hat{b}_{\mathrm{mmse},l})$ is a stationary random process approximated by

$$\mathrm{LLR}(b_l; \hat{b}_{\mathrm{mmse},l}) = \bar{\mu}_{\mathrm{mmse}} b_l + \bar{\eta}_{\mathrm{mmse},l}, \tag{6.60}$$

where $\bar{\sigma}_\eta^2 = E[|\bar{\eta}_{\mathrm{mmse},l}|^2] = 2\bar{\mu}_{\mathrm{mmse}}$ and $\bar{\mu}_{\mathrm{mmse}}$ is the average of $\mu_{\mathrm{mmse},l}$. Once $\bar{\mu}_{\mathrm{mmse}}$ is found, the pdf of the LLR can be obtained using the Gaussian assumption. Then, the EXIT chart can be easily found from Eq. (6.18).

From Eqs (6.48) and (6.50), we can show that

$$\bar{\mu}_{\mathrm{mmse}} = E[\mu_{\mathrm{mmse},l}] = E\left[\frac{2\mu_l}{1 - \mu_l}\right]. \tag{6.61}$$

Let $\mu_l = \bar{\mu} + \epsilon_l$, where $\bar{\mu} = E[\mu_l]$. Using a Taylor series, we can show that

$$\mu_{\mathrm{mmse},l} = 2\left(\frac{\bar{\mu}}{1 - \bar{\mu}} - \frac{\epsilon_l}{1 - \bar{\mu}} - \frac{\epsilon_l \bar{\mu}}{(1 - \bar{\mu})^2} + \frac{\epsilon_l^2}{(1 - \bar{\mu})^2}\right) + O(\epsilon_l^3) \tag{6.62}$$

and $E[\epsilon_l] = 0$. Then, a second-order approximation is given by

$$E[\mu_{\mathrm{mmse},l}] \simeq \frac{2\bar{\mu}}{1 - \bar{\mu}} + \frac{2E[\epsilon_l^2]}{(1 - \bar{\mu})^2}. \tag{6.63}$$

We need to find $E[\bar{\mu}]$ and $E[\epsilon_l^2]$.

Firstly, we consider $\bar{\mu}$. From Eq. (6.50), it follows that

$$\mu_l = \mathbf{h}^\mathrm{T} \left(\mathbf{h}\mathbf{h}^\mathrm{T} + \mathbf{H}_+\mathbf{Q}_{l,+}\mathbf{H}_+^\mathrm{T} + \mathbf{H}_-\mathbf{Q}_{l,-}\mathbf{H}_-^\mathrm{T} + \frac{N_0}{2}\mathbf{I} \right)^{-1} \mathbf{h}.$$

Let $\mathbf{Q}_+ = E[\mathbf{Q}_{l,+}]$ and $\mathbf{Q}_- = E[\mathbf{Q}_{l,+}]$. Then, it follows that

$$\begin{aligned} \mu_l &= \mathbf{h}^\mathrm{T} \left(\mathbf{h}\mathbf{h}^\mathrm{T} + \mathbf{H}_+\mathbf{Q}_+\mathbf{H}_+^\mathrm{T} + \mathbf{H}_-\mathbf{Q}_-\mathbf{H}_-^\mathrm{T} + \frac{N_0}{2}\mathbf{I} + \boldsymbol{\Delta} \right)^{-1} \mathbf{h} \\ &= \mu_o - \mathbf{w}_o^\mathrm{T}\boldsymbol{\Delta}\mathbf{w}_o + O(\|\boldsymbol{\Delta}\|^2), \end{aligned} \tag{6.64}$$

where

$$\begin{aligned} \boldsymbol{\Delta} &= \mathbf{H}_+(\mathbf{Q}_{l,+} - \mathbf{Q}_+)\mathbf{H}_+^\mathrm{T} + \mathbf{H}_-(\mathbf{Q}_{l,-} - \mathbf{Q}_-)\mathbf{H}_-^\mathrm{T}, \\ \mathbf{w}_o &= \left(\mathbf{h}\mathbf{h}^\mathrm{T} + \mathbf{H}_+\mathbf{Q}_+\mathbf{H}_+^\mathrm{T} + \mathbf{H}_-\mathbf{Q}_-\mathbf{H}_-^\mathrm{T} + \frac{N_0}{2}\mathbf{I} \right)^{-1} \mathbf{h}, \\ \mu_o &= \mathbf{w}_o^\mathrm{T}\mathbf{h}. \end{aligned} \tag{6.65}$$

Since, $\mu_l = \bar{\mu} + \epsilon_l$, we can show that $\bar{\mu} = \mu_o$ and $\epsilon_l \simeq \mathbf{w}_o^\mathrm{T}\boldsymbol{\Delta}\mathbf{w}_o$. To obtain μ_o or $\mathbf{w}_o$, we need to find $\mathbf{Q}_+ = E[\mathbf{Q}_{l,+}]$ and $\mathbf{Q}_- = E[\mathbf{Q}_{l,-}]$ using the LLR from the channel decoder, $\bar{z}_l$. In the MMSE-SC equalizer, the mean and variance of the ISI terms are obtained as follows:

$$\begin{aligned} \bar{b}_l &= E[b_l] = \tanh(\bar{z}_l/2), \\ \sigma_l^2 &= Var(b_l) = 1 - \tanh^2(\bar{z}_l/2), \end{aligned} \tag{6.66}$$

using $\bar{z}_l$. From Eq. (6.54), the averaged σ_l^2 can be obtained as follows:

$$\begin{aligned} \bar{\sigma}^2 &= E[\sigma_l^2] \\ &= \int \left(1 - \tanh^2(\bar{z}/2)\right) f(\bar{z})\,\mathrm{d}\bar{z} \\ &= 1 - \frac{1}{2} \int f(\bar{z}|b = +1) \tanh^2(\bar{z}/2)(1 + \mathrm{e}^{-\bar{z}})\,\mathrm{d}\bar{z}, \end{aligned} \tag{6.67}$$

where

$$f(\bar{z}|b) = \frac{1}{\sqrt{2\pi}\bar{\sigma}_\eta} \exp\left(-\frac{1}{2\bar{\sigma}_\eta^2} \left(\bar{z} - \frac{\bar{\sigma}_\eta^2}{2} b \right)^2 \right).$$

By definition, we readily find that

$$\mathbf{Q}_+ = \mathbf{Q}_- = \bar{\sigma}^2\mathbf{I}.$$

From this and Eqs (6.65), $\mathbf{w}_o$ and $\mu_o = \bar{\mu}$ can be found.

To find the second-order term $E[\epsilon_l^2]$ in Eq. (6.63), let $\mathbf{H}_{\text{isi}} = [\mathbf{H}_- \ \mathbf{H}_+]$. Then, we can show that

$$\mathbf{\Delta} = \mathbf{H}_{\text{isi}} \mathbf{V} \mathbf{H}_{\text{isi}}^{\text{T}},$$

where

$$\mathbf{V} = \text{Diag}\left(\left(\sigma_{l-P+1}^2 - \bar{\sigma}^2 \right), \ldots, \left(\sigma_{l-1}^2 - \bar{\sigma}^2 \right), \left(\sigma_{l+1}^2 - \bar{\sigma}^2 \right), \ldots, \left(\sigma_{l+P-1}^2 - \bar{\sigma}^2 \right) \right).$$

Let $\mathbf{u} = \mathbf{H}_{\text{isi}}^{\text{T}} \mathbf{w}_0$. Then, if the (q, q)th element of $\mathbf{V}$, denoted by v_q, is uncorrelated with the other (diagonal) elements, it can be shown that

$$\begin{aligned} E\left[\epsilon_l^2\right] &\simeq E\left[\left|\mathbf{w}_0^{\text{T}} \mathbf{\Delta} \mathbf{w}_0\right|^2\right] \\ &= E\left[\left(\sum_q u_q v_q u_q\right)^2\right] \\ &= \sum_q |u_q|^4 E[|v_q|^2], \end{aligned} \tag{6.68}$$

where u_q represents the qth element of $\mathbf{u}$ and we can find $E[|v_q|^2]$ as follows:

$$E[|v_q|^2] = E\left[\left(\sigma_l^2 - \bar{\sigma}^2\right)^2\right] = \int \left(1 - \tanh^2(\bar{z}/2) - \bar{\sigma}^2\right)^2 f(\bar{z}) \, d\bar{z}.$$

Using the analytical approach, for which we need to find the mean in Eq. (6.63), an EXIT chart for the MMSE-SC equalizer is found with the following CIR:

$$\mathbf{h} = [0.227 \ \ 0.460 \ \ 0.688 \ \ 0.460 \ \ 0.227]^{\text{T}},$$

shown in Fig. 6.16 with an EXIT chart obtained using the histogram-based approach. It is shown that they are close to each other. In particular, at the two extreme points corresponding to input mutual information of 0 and 1, the two approaches provide the same result. In both cases, $\mathbf{\Delta} = \mathbf{0}$ or $\mu_l = \bar{\mu}$ for all l because $\mathbf{Q}_{l,+} = \mathbf{Q}_+$ and $\mathbf{Q}_{l,-} = \mathbf{Q}_-$. This implies that the approximation in Eq. (6.63) becomes exact at the two extreme points.

The EXIT chart for the MMSE-SC equalizer depends on the SNR. Thus, to understand the convergence behavior of the iterative receiver with the MMSE-SC equalizer, we need the EXIT charts for the MMSE-SC equalizer for various values of SNR. Using the analytical approach, we can easily build EXIT charts for various values of SNR without extensive simulations for each SNR. On the other hand, if the histogram-based approach is employed, we need to run simulations for each SNR, and it becomes a computationally expensive option.

Figure 6.17 shows the EXIT charts for the iterative receiver with the MMSE-SC equalizer; these charts are analytically obtained from Eq. (6.63). We can see that the iterative receiver can provide a satisfactory performance after several iterations if the SNR is equal to or higher than 4 dB; at an SNR of 4 dB, seven iterations are required to converge. However, if the SNR is 3 dB, a large number of iterations are required (therefore, it results in an undesirably long delay). For a lower SNR (< 3 dB), the iterative receiver provides a poor performance.

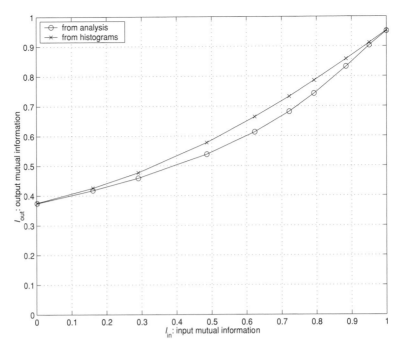

Figure 6.16. EXIT charts for the MMSE-SC equalizer with the CIR $\mathbf{h} = [0.227 \; 0.460 \; 0.688 \; 0.460 \; 0.227]^T$.

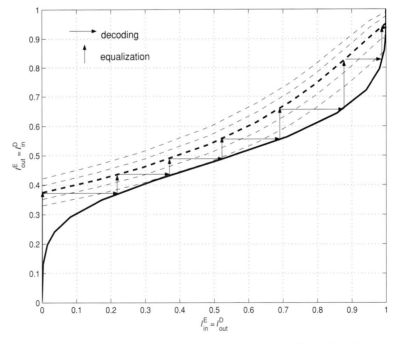

Figure 6.17. EXIT charts of the iterative receiver with the MMSE filter with SC and a rate-half convolutional code with generator polynomial $(7, 5)$ in octal. Dashed curves: the EXIT charts for the MMSE-SC with values of the SNR from 2 to 6 dB (the lowest one corresponding to 2 dB and the highest one corresponding to 6 dB); solid curve: the EXIT chart for the convolutional code.

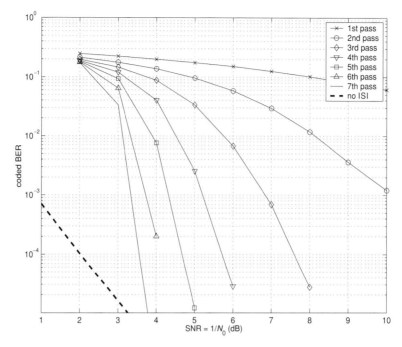

Figure 6.18. Bit error rate performance of the iterative receiver with the MMSE-SC equalizer when $L = 2 \times 10^5$.

The convergence behavior observed by the EXIT charts can be confirmed by BER results. Coded BERs are obtained through simulations, and the results are shown in Fig. 6.18. We can see that a convergence (to a low BER) is achieved after seven iterations at an SNR of 4 dB. It is also shown that a number of iterations would be required to converge at a SNR of 3 dB.

Note that the EXIT charts can be used to understand the convergence behavior when the length of the coded sequence is sufficiently long. In addition, we also note that the iterative receiver provides a better performance if the coded sequence is long, because the performance of convolutional codes can be improved and the random bit interleaver can ideally break the correlation for a long sequence.

6.5 Summary and notes

In this chapter, we briefly reviewed information theory and convolutional codes that are necessary to understand the iterative receiver. The structure and operation of the iterative receiver were studied with different SISO equalizers. For performance analysis, the EXIT chart was introduced.

As the iterative receiver can provide excellent performance, it has been extensively studied (Laot *et al.*, 2001; Tuchler *et al.*, 2002a,b). An overview of the iterative receiver can also be found in Koetter *et al.* (2004).

There are other topics pertaining to the iterative receiver that are not addressed in this chapter. For example, precoding issues can be found in Lee (2001) and Narayanan (2001), and bit-labeling issues with a higher-order modulation can be found in Dejonghe and Vandendorphe (2004). In this chapter, it is assumed that the CIR is known. Since the CIR must be estimated, the iterative receiver with channel estimation is an important issue. This is discussed in Chapter 7.

7 Iterative receivers under time-varying channel conditions

It was shown in Chapter 6 that the iterative receiver can provide an excellent performance through iterations with known CIR. In practice, the CIR should be estimated and the channel estimation error can degrade the performance of the receiver. Various approaches taken to design the iterative receiver with channel estimation can be considered. It is desirable that the channel estimation can be improved through iterations by taking advantage of the iterative processing (i.e., the availability of prior (or extrinsic) information from the channel decoder).

This chapter discusses iterative receiver design for unknown time-varying channels taking two fundamentally different approaches. The first approach is an extension of the iterative receiver with the MAP symbol detection studied in Chapter 6. To deal with unknown and time-varying channels, the MAP symbol detection is generalized to incorporate the channel estimation. Since a straightforward implementation requires high computational complexity, an approximation will be introduced.

The second approach is a channel estimation oriented approach. The main focus of this approach is to find the MAP estimate of a time-varying channel process rather than the MAP data symbols. The detection/decoding is carried out as part of the iterative channel estimation based on the EM algorithm. Some approximations will also be studied to reduce the complexity.

7.1 Detection/decoding for unknown time-varying channels

In this section, we will discuss two different approaches taken to perform the detection/decoding and channel estimation for unknown time-varying random channels. The first is the detection/decoding oriented approach. Suppose that $\mathbf{b}$ denotes a coded symbol vector (or sequence) that is transmitted over a time-varying ISI channel. It is desirable to detect the transmitted coded vector $\mathbf{b}$ from the received signal sequence, denoted by $\mathbf{y}$. In principle, the MLSD or MAP symbol detection can be employed. For example, the MAP symbol detection can be formulated to find the *a posteriori* probability of symbol b_l as follows:

$$
\begin{aligned}
\Pr(b_l|\mathbf{y}) &= \sum_{\mathbf{b} \in \mathcal{B}_{\text{code}}; b_l} \Pr(\mathbf{b}|\mathbf{y}) \\
&\propto \sum_{\mathbf{b} \in \mathcal{B}_{\text{code}}; b_l} f(\mathbf{y}|\mathbf{b}) \Pr(\mathbf{b}),
\end{aligned} \tag{7.1}
$$

with a code constraint that $\mathbf{b} \in \mathcal{B}_{\text{code}}$, where $\mathcal{B}_{\text{code}}$ stands for the set of coded symbol vectors. Here, $f(\mathbf{y}|\mathbf{b})$ is the likelihood function and $\Pr(\mathbf{b})$ is the *a priori* probability of $\mathbf{b}$. As shown in Chapter 4, in this case channel estimation by the Kalman filter is embedded in computing $f(\mathbf{y}|\mathbf{b})$. This differs from the approach taken in Chapter 4 in that the signal vector $\mathbf{b}$ is a coded sequence. This kind of approach will be referred to as the *explicit* detection/decoding–*implicit* channel estimation (ExDD–ImC) approach.

The second approach is the channel estimation oriented one. The main purpose of this approach is to estimate time-varying channels. It is obvious that a good estimate of the CIR is necessary not to degrade the performance of detection/decoding. Let $\mathbf{h}$ denote the time-varying channel vector ($\mathbf{h}$ will be defined later). Then, the MAP channel estimation becomes

$$\hat{\mathbf{h}}_{\text{map}} = \arg\max_{\mathbf{h}} f(\mathbf{h}|\mathbf{y}),$$

where $f(\mathbf{h}|\mathbf{y})$ represents the *a posteriori* pdf of $\mathbf{h}$ given $\mathbf{y}$. In the MAP channel estimation, it is often necessary to perform data detection and decoding implicitly. This approach will be referred to as the *explicit* channel estimation–*implicit* data detection/decoding (ExC–ImDD) approach.

In the following sections, we will derive iterative receivers based on the two approaches.

7.2   Iterative receiver based on ExDD–ImC approach

In this section, we focus on the MAP symbol detection for time-varying channels. An iterative receiver that includes the MAP symbol detector is derived.

7.2.1   *System model and structure of iterative receivers*

Suppose that a symbol sequence or packet, $\{b_m\}$, consists of coded symbols from a channel encoder[†] and pilot symbols. Assume that pilot symbols are periodically inserted. Let

$$\mathbf{b} = [b_0 \ b_1 \ \cdots \ b_{L-1}]^{\mathrm{T}},$$

where L is the length of the symbol sequence. After bit interleaving of a coded sequence, pilot symbols are inserted as shown in Fig. 7.1. The resulting modulation will be referred to as the pilot symbol aided bit interleaved coded modulation (PSA-BICM). It is also assumed that $b_l \in \{-1, +1\}$.

Recall the received signal over a time-varying ISI channel. The lth received signal is given by

$$y_l = \sum_{p=0}^{P-1} h_{p,l} b_{l-p} + n_l, \tag{7.2}$$

[†]   It is assumed that a convolutional encoder is used throughout this chapter.

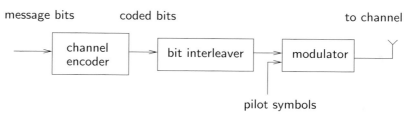

Figure 7.1. Block diagram for the transmitter with the PSA-BICM.

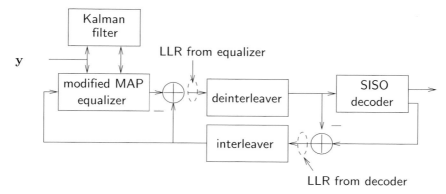

Figure 7.2. Block diagram for the iterative receiver with Kalman filtering to track time-varying channels.

where $\{h_{p,l}\}$ is the complex-valued time-varying CIR of length P at time l and n_l is the background white noise. It is assumed that n_l is a circular complex Gaussian random variable with zero mean and variance $E[|n_l|^2] = N_0$.

It is possible to perform the MAP decoding in Eq. (7.1) based on an exhaustive search. Again, the complexity grows exponentially with L. To reduce the complexity, trellis structures from the convolutional encoder and ISI channel can be used. However, since the number of memory elements of both the ISI channel and the convolutional encoder can be large, it would not be helpful to reduce the complexity significantly. As shown in Chapter 6, the turbo principle can be adopted to reduce the complexity by developing an iterative receiver.

Figure 7.2 shows the structure of the iterative receiver, which consists mainly of two building blocks: the MAP symbol detector and the MAP decoder. Within the iterative receiver, the LLR obtained from the MAP symbol detector becomes the input to the MAP decoder. After decoding, the LLR from the channel decoder becomes the input as the extrinsic information to the MAP symbol detector. Through this iteration which exchanges the LLR, it is generally expected that the performance of the iterative receiver approaches an ideal one.

In Section 7.2.2, we study the MAP symbol detector for time-varying random channels. Since the channel is unknown, we need to modify the MAP symbol detector for known channels in Chapter 5.

7.2.2   MAP symbol detection for random channels

The MAP symbol detection is to provide the LAPP of symbol b_l:

$$L(b_l) = \log \frac{\Pr(b_l = +1 | \mathbf{y})}{\Pr(b_l = -1 | \mathbf{y})}$$

$$= \log \frac{\sum_{\mathbf{b} \in \mathcal{B}_l^+} \Pr(\mathbf{y}|\mathbf{b}) \Pr(\mathbf{b})}{\sum_{\mathbf{b} \in \mathcal{B}_l^-} \Pr(\mathbf{y}|\mathbf{b}) \Pr(\mathbf{b})}, \qquad (7.3)$$

where $\mathcal{B}_l^+$ and $\mathcal{B}_l^-$ are the sets of binary vectors which are defined by $\mathcal{B}_l^+ = \{[b_0 \ b_1 \ \cdots \ b_{L-1}]^{\mathrm{T}} \mid b_l = +1, b_m \in \{+1, -1\}, \forall m \neq l\}$ and $\mathcal{B}_l^- = \{[b_0 \ b_1 \ \cdots \ b_{L-1}]^{\mathrm{T}} \mid b_l = -1, b_m \in \{+1, -1\}, \forall m \neq l\}$, respectively. Since we consider the iterative receiver, the code structure is not imposed in the MAP symbol detection in Eq. (7.3).

Assume that the b_l's are independent due to bit interleaving. Then, the *a posteriori* probability is given by

$$f(\mathbf{y}|\mathbf{b}) \Pr(\mathbf{b}) = f\left(\mathbf{y}_0^{L-1} | \mathbf{b}_0^{L-1}\right) \Pr\left(\mathbf{b}_0^{L-1}\right)$$

$$\propto \prod_{l=0}^{L-1} f\left(y_l | \mathbf{y}_0^{l-1}, \mathbf{b}_0^l\right) \Pr(b_l), \qquad (7.4)$$

where $\Pr(b_l)$ is the *a priori* probability of b_l (which is available from the extrinsic bit information from the MAP decoder) and $f(y_l|\mathbf{y}_0^{l-1}, \mathbf{b}_0^l)$ is the likelihood function of $\mathbf{b}_0^l$. Here, $\mathbf{y}_0^l$ and $\mathbf{b}_0^l$ are given by

$$\mathbf{y}_0^l = \{y_0, \quad y_1, \quad \ldots, \quad y_l\},$$
$$\mathbf{b}_0^l = \{b_0, \quad b_1, \quad \ldots, \quad b_l\}.$$

When a state-space model for time-varying channels is available, as discussed in Chapter 4, the Kalman filter can be used to find the individual likelihood functions as follows:

$$f\left(y_l|\mathbf{y}_0^{l-1}, \mathbf{b}_0^l\right), \quad l = 0, 1, \ldots, L - 1.$$

Unfortunately, although it is possible to find the LAPP, its complexity is prohibitively high since there are 2^L **b**'s.

Due to the high complexity, it is necessary to devise less complex approximations. To reduce the complexity, we can introduce a forced folding to the MAP symbol detection for random channels. Suppose that

$$f\left(y_l|\mathbf{y}_0^{l-1}, \mathbf{b}_0^l\right) \simeq f\left(y_l|\mathbf{y}_0^{l-1}, \mathbf{b}_{l-\bar{P}+1}^l\right), \quad l = 0, 1, \ldots, L - 1, \qquad (7.5)$$

where $\bar{P}$ is the folding memory. At time l, define state $\mathcal{S}_l$ with the $(\bar{P} - 1)$ past symbols as follows:

$$\{b_{l-1}, b_{l-2}, \ldots, b_{l-\bar{P}+1}\}.$$

Then, the BCJR algorithm can be used to find the LAPP.

It can be shown that

$$\Pr(\mathcal{S}_l \to \mathcal{S}_{l+1} | \{y_m\}) = \frac{f(\mathcal{S}_l \to \mathcal{S}_{l+1}, \{y_m\})}{f(\{y_m\})}. \tag{7.6}$$

Then, we can show that

$$\Pr(b_l = +1 | \{y_m\}) = \sum_{ST_{+1}} \Pr(\mathcal{S}_l \to \mathcal{S}_{l+1} | \{y_m\})$$

$$\Pr(b_l = -1 | \{y_m\}) = \sum_{ST_{-1}} \Pr(\mathcal{S}_l \to \mathcal{S}_{l+1} | \{y_m\}),$$

where ST_{+1} and ST_{-1} represent the sets of the state transitions, $\mathcal{S}_l \to \mathcal{S}_{l+1}$, that are caused by the inputs $b_l = +1$ and $b_l = -1$, respectively, and $\sum_{ST_{+1}}$ and $\sum_{ST_{-1}}$ stand for the summations taken over the sets ST_{+1} and ST_{-1}, respectively. Then, it follows that

$$L(b_l) = \log \frac{\sum_{ST_{+1}} f(\mathcal{S}_l \to \mathcal{S}_{l+1}, \{y_m\})}{\sum_{ST_{-1}} f(\mathcal{S}_l \to \mathcal{S}_{l+1}, \{y_m\})}. \tag{7.7}$$

The BCJR algorithm finds $f(\mathcal{S}_l \to \mathcal{S}_{l+1}, \{y_m\})$ using the backward and forward recursions. In Chapter 5, the derivation of the BCJR algorithm is studied with known channels. However, since the channel is unknown, the BCJR algorithm needs to be modified.

Using Bayes' rule, we can show that

$$\begin{aligned}
f(\mathcal{S}_l \to \mathcal{S}_{l+1}, \{y_m\}) &= f(\mathcal{S}_l, \mathcal{S}_{l+1}, \mathbf{y}_{l,+}, \mathbf{y}_{l,-}, y_l) \\
&= f(\mathbf{y}_{l,+} | \mathcal{S}_l, \mathcal{S}_{l+1}, \mathbf{y}_{l,-}, y_l) f(\mathcal{S}_l, \mathcal{S}_{l+1}, \mathbf{y}_{l,-}, y_l) \\
&= f(\mathbf{y}_{l,+} | \mathcal{S}_{l+1}, \mathbf{y}_{l+1,-}) f(\mathcal{S}_l, \mathcal{S}_{l+1}, \mathbf{y}_{l,-}, y_l) \\
&= f(\mathbf{y}_{l,+} | \mathcal{S}_{l+1}, \mathbf{y}_{l+1,-}) f(\mathcal{S}_{l+1}, y_l | \mathcal{S}_l, \mathbf{y}_{l,-}) f(\mathcal{S}_l, \mathbf{y}_{l,-}), \tag{7.8}
\end{aligned}$$

where

$$\mathbf{y}_{l,-} = \{y_0, \quad y_1, \quad \ldots, \quad y_{l-1}\},$$
$$\mathbf{y}_{l,+} = \{y_{l+1}, \quad y_{l+2}, \quad \ldots, \quad y_{L-1}\}.$$

Since the channel vector is random and unknown, we note that

$$f(\mathbf{y}_{l,+} | \mathcal{S}_{l+1}, \mathbf{y}_{l+1,-}) \neq f(\mathbf{y}_{l,+} | \mathcal{S}_{l+1}).$$

Define

$$\begin{aligned}
\gamma(\mathcal{S}_l \to \mathcal{S}_{l+1}) &= f(\mathcal{S}_{l+1}, y_l | \mathcal{S}_l, \mathbf{y}_{l,-}), \\
\alpha(\mathcal{S}_l) &= f(\mathcal{S}_l, \mathbf{y}_{l,-}), \tag{7.9} \\
\beta(\mathcal{S}_{l+1}) &= f(\mathbf{y}_{l,+} | \mathcal{S}_{l+1}, \mathbf{y}_{l+1,-}).
\end{aligned}$$

Then, it is readily shown that

$$f(\mathcal{S}_l \to \mathcal{S}_{l+1}, \{y_m\}) = \gamma(\mathcal{S}_l \to \mathcal{S}_{l+1}) \alpha(\mathcal{S}_l) \beta(\mathcal{S}_{l+1}).$$

The backward recursion to find $\beta(\mathcal{S}_l)$ from $\beta(\mathcal{S}_{l+1})$ is similar to the original backward recursion in Chapter 5. It can be shown that

$$
\begin{aligned}
\beta(\mathcal{S}_l) &= f(\mathbf{y}_{l-1,+}|\mathcal{S}_l, \mathbf{y}_{l,-}) \\
&= f(y_l, \mathbf{y}_{l,+}|\mathcal{S}_l, \mathbf{y}_{l,-}) \\
&= \sum_{s_k \in \bar{\mathbf{S}}} f(y_l, \mathbf{y}_{l,+}, \mathcal{S}_{l+1} = s_k|\mathcal{S}_l, \mathbf{y}_{l,-}) \\
&= \sum_{s_k \in \bar{\mathbf{S}}} \frac{f(y_l, \mathbf{y}_{l,+}, \mathcal{S}_{l+1} = s_k, \mathcal{S}_l, \mathbf{y}_{l,-})}{f(\mathcal{S}_l, \mathbf{y}_{l,-})}.
\end{aligned}
\tag{7.10}
$$

The numerator in Eq. (7.10) can be rewritten as follows:

$$
\begin{aligned}
f(y_l, &\mathbf{y}_{l,+}, \mathcal{S}_{l+1} = s_k, \mathcal{S}_l, \mathbf{y}_{l,-}) \\
&= f(\mathbf{y}_{l,+}, \mathcal{S}_{l+1} = s_k, \mathcal{S}_l, \mathbf{y}_{l+1,-}) \\
&= f(\mathbf{y}_{l,+}, \mathcal{S}_{l+1} = s_k, \mathbf{y}_{l+1,-}) f(\mathcal{S}_l|\mathbf{y}_{l,+}, \mathcal{S}_{l+1} = s_k, \mathbf{y}_{l+1,-}) \\
&= f(\mathbf{y}_{l,+}|\mathcal{S}_{l+1} = s_k, \mathbf{y}_{l+1,-}) f(\mathcal{S}_{l+1} = s_k, \mathbf{y}_{l+1,-}) f(\mathcal{S}_l|\mathbf{y}_{l,+}, \mathcal{S}_{l+1} = s_k, \mathbf{y}_{l+1,-}).
\end{aligned}
\tag{7.11}
$$

Noting that $\beta(\mathcal{S}_{l+1} = s_k) = f(\mathbf{y}_{l,+}|\mathcal{S}_{l+1} = s_k, \mathbf{y}_{l+1,-})$ and substituting

$$
f(\mathcal{S}_l|\mathbf{y}_{l,+}, \mathcal{S}_{l+1} = s_k, \mathbf{y}_{l+1,-}) = f(\mathcal{S}_l|\mathcal{S}_{l+1} = s_k, \mathbf{y}_{l+1,-})
$$

into Eq. (7.11), we have

$$
f(y_l, \mathbf{y}_{l,+}, \mathcal{S}_{l+1} = s_k, \mathcal{S}_l, \mathbf{y}_{l,-}) = \beta(\mathcal{S}_{l+1} = s_k) f(\mathcal{S}_l, \mathcal{S}_{l+1} = s_k, \mathbf{y}_{l+1,-}).
\tag{7.12}
$$

Finally, substituting Eq. (7.12) into Eq. (7.10) yields

$$
\begin{aligned}
\beta(\mathcal{S}_l) &= \sum_{s_k \in \bar{\mathbf{S}}} \frac{\beta(\mathcal{S}_{l+1} = s_k) f(\mathcal{S}_l, \mathcal{S}_{l+1} = s_k, \mathbf{y}_{l+1,-})}{f(\mathcal{S}_l, \mathbf{y}_{l,-})} \\
&= \sum_{s_k \in \bar{\mathbf{S}}} \beta(\mathcal{S}_{l+1} = s_k) \gamma(\mathcal{S}_l \to \mathcal{S}_{l+1} = s_k),
\end{aligned}
\tag{7.13}
$$

where

$$
\gamma(\mathcal{S}_l \to \mathcal{S}_{l+1} = s_k) = f(\mathcal{S}_{l+1} = s_k, y_l|\mathcal{S}_l, \mathbf{y}_{l,-}).
$$

As shown in Eq. (7.13), the backward recursion is identical to that with known CIR. However, the definitions of the β's and γ's are slightly different. It is straightforward to derive the forward recursion for α's. Actually, the forward recursion is also identical to that for known CIR.

The quantity $\gamma(\mathcal{S}_l \to \mathcal{S}_{l+1})$ is given by

$$
\gamma(\mathcal{S}_l \to \mathcal{S}_{l+1}) = \begin{cases} \Pr(b_l) f(y_l|\mathcal{S}_l, b_l, \mathbf{y}_{l,-}), & \text{for valid transition;} \\ 0, & \text{for invalid transition.} \end{cases}
$$

Since

$$
f(y_l|\mathcal{S}_l, b_l, \mathbf{y}_{l,-}) = f\left(y_l|\mathbf{y}_0^{l-1}, \mathbf{b}_{l-\bar{P}+1}^l\right),
$$

$\gamma(\mathcal{S}_l \to \mathcal{S}_{l+1})$ can be found in terms of the mean and variance of y_l conditioned on $\mathbf{y}_0^{l-1}$ and $\mathbf{b}_{l-\bar{P}+1}^l$. This is similar to the MLSD for unknown CIR in Chapter 4.

Once the LAPP is found, we obtain the LLR as follows:

$$\text{LLR}(b_l) = L(b_l) - L_{\text{pi}}(b_l), \qquad (7.14)$$

where

$$L_{\text{pi}}(b_l) = \log \frac{\Pr(b_l = +1)}{\Pr(b_l = -1)}.$$

Since the LAPP is obtained with a forced folding condition, it is an approximation. Thus, the performance would vary depending on the folding memory.

Simulations are carried out to observe the performance of the iterative receiver based on the ExDD–ImC approach. For the channel tracking, the LMS filter is used instead of the Kalman filter with the optimal value of μ in Eq. (4.44). Fading channel processes are assumed to have the Clarke's spectrum with fading rate $f_d T = 0.005$ and $P = 4$. The autocorrelation of fading channel processes is given by

$$E[h_{p,l} h_{p',l'}^*] = \frac{1}{P} J_0(2\pi f_d T |l - l'|) \delta_{p,p'}.$$

A rate-half convolutional code with generator polynomial $(5, 7)$ in octal and a random bit interleaver are used. Two different structures of transmission block (or packet) are considered as follows.

- Block structure A: 7 subblocks of 7 pilot symbols and 7 subblocks of 85 data symbols and the last subblock of 5 data symbols (there are 600 data symbols and 49 pilot symbols in a transmission block).
- Block structure B: 14 subblocks of 7 pilot symbols and 14 subblocks of 42 data symbols and the last subblock of 12 data symbols (there are 600 data symbols and 98 pilot symbols in a transmission block).

Pilot subblocks are uniformly distributed over a packet. Block structure A is suitable for a fading rate up to $f_d T = 0.005$. Since a pilot subblock is inserted every 85 data symbols, the pilot subblock interval becomes $M_{\text{ps}} = 7 + 85 = 92$ and

$$M_{\text{ps}} = 92 < \frac{1}{2 f_d T} = 100.$$

Thus, the Nyquist condition is satisfied (see Eq. (4.20)). For block structure B, we have $M_{\text{ps}} = 49$. Hence, both block structures are suitable for a fading rate of $f_d T = 0.005$.

Figure 7.3 shows simulation results with $\bar{P} = P = 4$. The SNR is given by

$$\text{SNR} = \frac{E\left[\sum_{p=0}^{P-1} |h_{p,l}|^2\right]}{N_0} = \frac{1}{N_0}.$$

For both transmission blocks, the performance is improved through iterations. In addition, we observe that the performance can be significantly improved if more pilot symbols are transmitted.

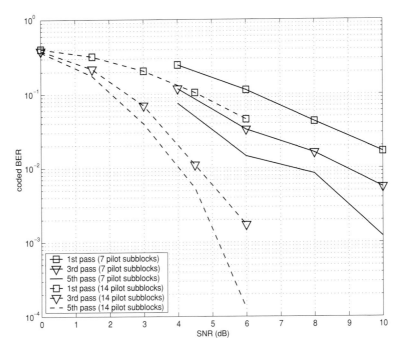

Figure 7.3. Performance of the iterative receiver based on the ExDD–ImC approach; $f_{\mathrm{d}}T = 0.005$.

However, the increase of pilot symbols results in two undesirable consequences: a decrease of data throughput and an increase of transmission energy (since more pilot symbols are transmitted).

For a fair comparison in terms of bit energy, consider the energy ratios per bit as follows:

$$E R_A = \frac{600 + 49}{600} = 0.7850 \ (\mathrm{dB}),$$

$$E R_B = \frac{600 + 98}{600} = 1.5129 \ (\mathrm{dB}).$$

For block structure A, 0.7850 dB more energy per bit is used compared with the case that no pilot symbol is transmitted. For block structure B, 1.5129 dB more energy per bit is assigned. The bit energy difference between block structures A and B is 0.7279 dB (i.e. less than 1 dB). Even though we compensate 0.7279 dB, we can see that block structure B, which transmits more pilot symbols, can provide a better performance than block structure A.

As pointed out in Chapter 4, it is desirable to have a larger folding memory for a better performance. In Fig. 7.4, simulation results are presented under the same conditions as in Fig. 7.3, except for a different folding memory $\bar{P}$. We can see that a better performance can be achieved for a larger folding memory. Since more candidate data sequences are taken into account to approximate the LAPP in the MAP symbol detection for a larger $\bar{P}$, a better performance is expected. However, since the number of states becomes $2^6 = 64$ when

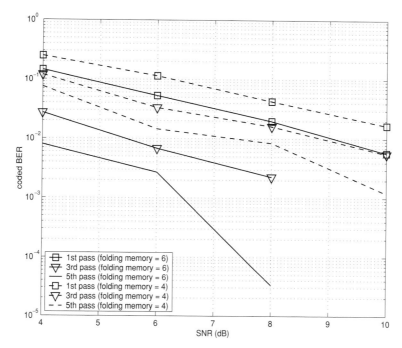

Figure 7.4. Performance of the iterative receiver based on the ExDD–ImC approach with different folding memory $\bar{P}$; $f_\mathrm{d}T = 0.005$, seven pilot subblocks.

$\bar{P} = 6$, the complexity becomes higher than in the case of $\bar{P} = 4$, in which the number of states is $2^4 = 16$.

7.3 Iterative receiver based on ExC–ImDD approach

The other method of treating the detection/decoding and channel estimation for unknown random channels in this chapter is the ExC–ImDD approach, in which the channel estimation is the main target. As one of the ExC–ImDD-based methods, the MAP channel estimation is studied in this section with the EM algorithm.

7.3.1 MAP channel estimation

From Eq. (7.2), the received signal is rewritten as follows:

$$y_l = \mathbf{b}_l^\mathsf{T}\mathbf{h}_l + n_l, \quad l = 0, 1, \ldots, L - 1, \tag{7.15}$$

where $\mathbf{h}_l = [h_{0,l} \; h_{1,l} \; \cdots \; h_{P-1,l}]^\mathsf{T}$ and $\mathbf{b}_l = [b_l \; b_{l-1} \; \cdots \; b_{l-P+1}]^\mathsf{T}$. Define $\mathbf{y} = [y_0 \; y_1 \; \cdots \; y_{L-1}]^\mathsf{T}$ and $\mathbf{n} = [n_0 \; n_1 \; \cdots \; n_{L-1}]^\mathsf{T}$. Then, we have

$$\mathbf{y} = \mathbf{X}^\mathsf{T}\mathbf{h} + \mathbf{n}, \tag{7.16}$$

where

$$\mathbf{X}^{\mathrm{T}} = \begin{bmatrix} \mathbf{b}_0^{\mathrm{T}} & \mathbf{0} & \cdots & \mathbf{0} \\ \mathbf{0} & \mathbf{b}_1^{\mathrm{T}} & \cdots & \mathbf{0} \\ \vdots & \vdots & \ddots & \vdots \\ \mathbf{0} & \mathbf{0} & \cdots & \mathbf{b}_{L-1}^{\mathrm{T}} \end{bmatrix}, \tag{7.17a}$$

$$\mathbf{h} = \begin{bmatrix} \mathbf{h}_0^{\mathrm{T}} & \mathbf{h}_1^{\mathrm{T}} & \cdots & \mathbf{h}_{L-1}^{\mathrm{T}} \end{bmatrix}^{\mathrm{T}}. \tag{7.17b}$$

The ExC–ImDD approach aims to estimate $\mathbf{h}$ from $\mathbf{y}$.

There are PL elements in $\mathbf{h}$, while the size of $\mathbf{y}$ is $L \times 1$. If $\mathbf{h}$ is completely unknown and random, the estimation of $\mathbf{h}$ becomes very difficult and impractical, because there are less observations than there are parameters to be estimated. On the other hand, if the channel is static, there are P parameters as the $\mathbf{h}_l$'s are the same for all l. Thus, a better estimate is available for a larger L. The channel estimation problem of interest in this section is to estimate temporally correlated fading channels. Although the CIR is time-varying and random, a good channel estimate would be available by exploiting the temporal correlation.

Suppose that the *a priori* pdf of the channel vector $\mathbf{h}$ is available. The temporal correlation of the channel can be imposed in the *a priori* pdf of $\mathbf{h}$. The MAP channel estimation problem is given by

$$\begin{aligned} \hat{\mathbf{h}}_{\mathrm{map}} &= \arg\max_{\mathbf{h}} f(\mathbf{h}|\mathbf{y}) \\ &= \arg\max_{\mathbf{h}} \{ \log f(\mathbf{y}|\mathbf{h}) + \log f(\mathbf{h}) \}, \end{aligned} \tag{7.18}$$

where $f(\mathbf{h}|\mathbf{y})$, $f(\mathbf{y}|\mathbf{h})$, and $f(\mathbf{h})$ are the *a posteriori* pdf, the likelihood function, and the *a priori* pdf of $\mathbf{h}$, respectively. Note that the MAP channel estimation in Eq. (7.18) is able to use pilot symbols as well as data symbols to find a better channel estimate. Throughout this section, we assume that $\mathbf{h}$ is a complex Gaussian random vector with zero mean and covariance matrix $\mathbf{R_h} = E[\mathbf{h}\mathbf{h}^{\mathrm{H}}]$. Thus, the temporal correlation of the channel is represented through $\mathbf{R_h}$.

Since $\mathbf{n}$ is the AWGN, for given $\mathbf{h}$ and $\mathbf{X}$, we have

$$f(\mathbf{y}|\mathbf{h}, \mathbf{X}) = C \exp\left(-\frac{1}{N_0} ||\mathbf{y} - \mathbf{X}^{\mathrm{T}}\mathbf{h}||^2 \right),$$

where C is a normalizing constant. Using the marginalization, the likelihood function can be obtained as follows:

$$\begin{aligned} f(\mathbf{y}|\mathbf{h}) &= \sum_{\mathbf{X}} f(\mathbf{y}|\mathbf{h}, \mathbf{X}) \Pr(\mathbf{X}|\mathbf{h}) \\ &= C \sum_{\mathbf{X}} \exp\left(-\frac{1}{N_0} ||\mathbf{y} - \mathbf{X}^{\mathrm{T}}\mathbf{h}||^2 \right) \Pr(\mathbf{X}), \end{aligned} \tag{7.19}$$

where $\sum_{\mathbf{X}}$ represents the summation over all the possible $\mathbf{X}$'s. Note that since $\mathbf{h}$ and $\mathbf{X}$ are independent, we have $\Pr(\mathbf{X}|\mathbf{h}) = \Pr(\mathbf{X})$. In finding the MAP estimate of $\mathbf{h}$, there are two major difficulties. One is due to a high computational complexity to obtain the likelihood function since there are 2^L different $\mathbf{X}$'s for the summation, as shown in Eq. (7.19). The other is due to a highly nonlinear form of the likelihood function $f(\mathbf{y}|\mathbf{h})$ (neither a convex nor a concave function). However, the EM algorithm can assist in finding the estimate numerically through iterations to overcome these difficulties.

7.3.2 EM algorithm for the MAP channel estimation

To apply the EM algorithm, we define the complete data as $\{\mathbf{X}, \mathbf{y}\}$. In this case, $\{\mathbf{X}\}$ becomes the missing data, while $\{\mathbf{y}\}$ is the incomplete data. The EM algorithm is an iterative algorithm, and each iteration consists of two steps: the E-step and the M-step. A better estimate (in terms of the likelihood) is available during every EM iteration consisting of the E-step and M-step.

Denote by $\hat{\mathbf{h}}^{(q)}$ the qth estimate of $\mathbf{h}$. Assume that the initial estimate denoted by $\hat{\mathbf{h}}^{(0)}$ is available. The E-step for the qth iteration is given by

$$
\begin{aligned}
Q\big(\mathbf{h}|\hat{\mathbf{h}}^{(q)}\big) &= E\big[\log f(\mathbf{y}|\mathbf{h}, \mathbf{X})|\mathbf{y}, \hat{\mathbf{h}}^{(q)}\big] + E\big[\log f(\mathbf{h})|\mathbf{y}, \hat{\mathbf{h}}^{(q)}\big] \\
&= E\big[\log f(\mathbf{y}|\mathbf{h}, \mathbf{X})|\mathbf{y}, \hat{\mathbf{h}}^{(q)}\big] + \log f(\mathbf{h}) \\
&= \log C - \frac{1}{N_0}\sum_{\mathbf{X}}||\mathbf{y} - \mathbf{X}^{\mathrm{T}}\mathbf{h}||^2 \Pr\big(\mathbf{X}|\mathbf{y}, \hat{\mathbf{h}}^{(q)}\big) + \log f(\mathbf{h}).
\end{aligned}
\tag{7.20}
$$

Since the (*a priori*) pdf of $\mathbf{h}$ is given by

$$
f(\mathbf{h}) = C'\exp\big(-\mathbf{h}^{\mathrm{H}}\mathbf{R}_{\mathbf{h}}^{-1}\mathbf{h}\big),
\tag{7.21}
$$

where C' is a normailzing constant, we can show that

$$
Q\big(\mathbf{h}|\hat{\mathbf{h}}^{(q)}\big) = C'' + \frac{2}{N_0}\Re\big(\mathbf{y}^{\mathrm{H}}(\bar{\mathbf{X}}^{(q)})^{\mathrm{T}}\mathbf{h}\big) - \mathbf{h}^{\mathrm{H}}\big(\mathbf{C}^{(q)} + \mathbf{R}_{\mathbf{h}}^{-1}\big)\mathbf{h}.
\tag{7.22}
$$

Here, C'' is a constant and $\bar{\mathbf{X}}^{(q)}$ and $\mathbf{C}^{(q)}$ are given by

$$
\begin{aligned}
\bar{\mathbf{X}}^{(q)} &= E\big[\mathbf{X}|\mathbf{y}, \hat{\mathbf{h}}^{(q)}\big] = \sum_{\mathbf{X}}\mathbf{X}\Pr\big(\mathbf{X}|\mathbf{y}, \hat{\mathbf{h}}^{(q)}\big), \\
\mathbf{C}^{(q)} &= E\big[\mathbf{X}^*\mathbf{X}^{\mathrm{T}}|\mathbf{y}, \hat{\mathbf{h}}^{(q)}\big] = \sum_{\mathbf{X}}\mathbf{X}^*\mathbf{X}^{\mathrm{T}}\Pr\big(\mathbf{X}|\mathbf{y}, \hat{\mathbf{h}}^{(q)}\big),
\end{aligned}
\tag{7.23}
$$

respectively. From the structure of $\mathbf{X}$ shown in Eq. (7.17a), it is straightforward to obtain the quantities $\bar{\mathbf{X}}^{(q)}$ and $\mathbf{C}^{(q)}$ without the summations over all the possible $\mathbf{X}$'s.

The M-step to find the vector $\mathbf{h}$ which maximizes $Q(\mathbf{h}|\hat{\mathbf{h}}^{(q)})$ can be written as follows:

$$
\begin{aligned}
\hat{\mathbf{h}}^{(q+1)} &= \arg\max_{\mathbf{h}} Q\big(\mathbf{h}|\hat{\mathbf{h}}^{(q)}\big) \\
&= \frac{1}{N_0}\big(\mathbf{R}_{\mathbf{h}}^{-1} + \mathbf{C}^{(q)}\big)^{-1}\big(\bar{\mathbf{X}}^{(q)}\big)^*\mathbf{y}.
\end{aligned}
\tag{7.24}
$$

As shown in Eq. (7.24), the complexity required to perform the M-step is $O(L^3 P^3)$ because of the matrix inversion. The estimate of $\mathbf{h}$ in Eq. (7.24) can be seen as the MAP estimate of $\mathbf{h}$ when $\Pr(\mathbf{X}|\mathbf{y}, \hat{\mathbf{h}}^{(q)})$ becomes the distribution of $\mathbf{X}$, i.e. $\hat{\mathbf{h}}^{(q+1)}$ is the MAP estimate of $\mathbf{h}$ conditioned on $\mathbf{y}$ and $\hat{\mathbf{h}}^{(q)}$.

Through EM iterations, we expect to obtain the MAP channel estimate after convergence. To perform the EM algorithm, we need to know the conditional probability $\Pr(\mathbf{X}|\mathbf{y}, \hat{\mathbf{h}}^{(q)})$, as shown in Eq. (7.23). This conditional probability can be obtained by the joint MAP

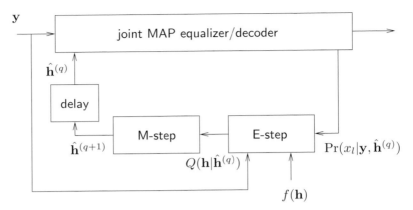

Figure 7.5. Block diagram for the EM algorithm to find the MAP channel estimate.

equalization/decoding with the channel estimate $\hat{\mathbf{h}}^{(q)}$. Thus, the EM-based MAP channel estimation method needs to include the joint MAP equalization/decoding. In general, with the channel estimate $\hat{\mathbf{h}}^{(q)}$, the joint MAP equalizer/decoder can provide the *a posteriori* probability of b_l, i.e. $\Pr(b_l|\mathbf{y}, \hat{\mathbf{h}}^{(q)})$. If we assume the b_l's are independent conditioned on $\mathbf{y}$ and $\hat{\mathbf{h}}^{(q)}$ (this is generally reasonable with a random bit interleaver), it is straightforward to obtain $\Pr(\mathbf{X}|\mathbf{y}, \hat{\mathbf{h}}^{(q)})$ from $\Pr(b_l|\mathbf{y}, \hat{\mathbf{h}}^{(q)})$. This further reduces the complexity to obtain $\mathbf{C}^{(q)}$ in Eqs (7.23). Figure 7.5 shows a block diagram for the EM algorithm to find the MAP channel estimate with the joint MAP equalizer/decoder.

7.3.3 Approximations to reduce complexity

In general, the complexity of the joint MAP equalization/decoding is prohibitively high because of the bit interleaver and the memories of the ISI channel and convolutional encoder. The iterative equalizer/decoder (IED) can replace the joint MAP equalizer/decoder to reduce the complexity. This approach results in a doubly iterative receiver, which will be discussed in this subsection.

Approximation I: use of IED

It is interesting to note that the EM-based MAP channel estimation method forms an iterative receiver that can provide the MAP channel estimate, $\hat{\mathbf{h}}^{(q)} = \hat{\mathbf{h}}_{map}$, after several EM iterations. Once $\hat{\mathbf{h}}_{map}$ is available, the *a posteriori* probability $\Pr(\mathbf{X}|\mathbf{y}, \hat{\mathbf{h}}_{map})$ can be found for a soft-decision. Since the complexity of the joint MAP equalizer/decoder is prohibitively high, the IED, consisting of the MAP equalizer, MAP channel decoder, bit interleaver, and bit deinterleaver, can replace the joint MAP equalizer/decoder. Consequently, the resulting receiver becomes a doubly iterative receiver, which is depicted in Fig. 7.6. The outer iteration is referred to as the EM iteration and the inner iteration is referred to as the equalization/decoding (ED) iteration. Note that the extrinsic bit information, which is

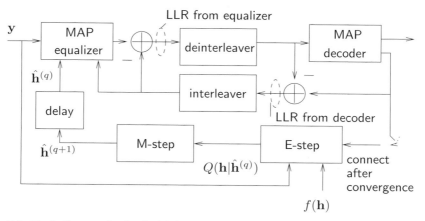

Figure 7.6. Block diagram for the doubly iterative receiver (the output of the MAP decoder in this block diagram is assumed to be the LAPP).

the LLR of bits, is exchanged within the ED iteration, while the *a posteriori* probability, $\Pr(\mathbf{X}|\mathbf{y}, \hat{\mathbf{h}}^{(q)})$, is updated in the EM iteration, as shown in Fig. 7.6.

As the IED replaces the joint MAP equalizer/decoder, several iterations are required for the IED so that the *a posteriori* probability can converge. For convenience, denote by Q_{ED} and Q_{EM} the numbers of ED iterations and EM iterations, respectively. We need to have $Q_{\mathrm{ED}} > 1$ and $Q_{\mathrm{EM}} > 1$ for a good performance. It is difficult to determine the numbers of iterations analytically. In general, three or four EM iterations are enough for the channel estimation to converge (however, in general, it depends on the application). From this, we have $Q_{\mathrm{EM}} = 3$ or 4. The number of iterations for the IED for convergence varies depending on the SNR. Generally, as the SNR increases, less iterations are required.

We can summarize the operation of the doubly iterative receiver as follows.

(S0) Let $q = 0$, where q is the index for the EM iteration. Find an initial channel estimate $\hat{\mathbf{h}}^{(q)} = \hat{\mathbf{h}}^{(0)}$.

(S1) Perform the IED with channel estimate $\hat{\mathbf{h}}^{(q)}$ to obtain (approximate) $\Pr(b_l|\mathbf{y}, \hat{\mathbf{h}}^{(q)})$, $l = 0, 1, \ldots, L - 1$.

(S2) Find the new channel estimate $\hat{\mathbf{h}}^{(q+1)}$ from the E- and M-steps.

(S3) If $||\hat{\mathbf{h}}^{(q+1)} - \hat{\mathbf{h}}^{(q)}||^2 \le \epsilon$ (or $|Q(\hat{\mathbf{h}}^{(q+1)}|\hat{\mathbf{h}}^{(q)}) - Q(\hat{\mathbf{h}}^{(q)}|\hat{\mathbf{h}}^{(q-1)})| < \bar{\epsilon}$), where ϵ and $\bar{\epsilon}$ are small positive constants, stop. Otherwise, let $q = q + 1$ and go to (S1).

Note that (S1) is performed with a new channel estimate for each q. Hence, the LLR as the extrinsic bit information needs to be initialized (i.e. the LLR becomes zero) when the IED starts at each EM iteration.

Approximation II: use of the LMMSE estimation

The M-step requires a prohibitively higher complexity, as shown in Eq. (7.24), for a larger PL. To reduce the complexity, we can consider alternative channel estimation methods, for example the LMMSE channel estimation.

The LMMSE estimate of $\mathbf{h}$ can be found using the *a posteriori* probability of $\mathbf{X}$, i.e. $\Pr(\mathbf{X}|\mathbf{y}, \hat{\mathbf{h}}^{(q)})$. The LMMSE estimator for the qth EM iteration is given by

$$\mathbf{V}^{(q)} = \arg\min_{\mathbf{V}} E_{\mathbf{X},\mathbf{n}}\left[||\mathbf{h} - \mathbf{V}\mathbf{y}||^2 |\mathbf{y}, \hat{\mathbf{h}}^{(q)}\right]. \tag{7.25}$$

To obtain the LMMSE estimator, the following second-order statistics are required:

$$\begin{aligned} E_{\mathbf{X},\mathbf{n}}\left[\mathbf{h}\mathbf{y}^H|\mathbf{y}, \hat{\mathbf{h}}^{(q)}\right] &= \mathbf{R}_{\mathbf{h}} E_{\mathbf{X}}\left[\mathbf{X}^*|\mathbf{y}, \hat{\mathbf{h}}^{(q)}\right] = \mathbf{R}_{\mathbf{h}}\left(\bar{\mathbf{X}}^{(q)}\right)^*, \\ E_{\mathbf{X},\mathbf{n}}\left[\mathbf{y}\mathbf{y}^H|\mathbf{y}, \hat{\mathbf{h}}^{(q)}\right] &= E_{\mathbf{X}}\left[\mathbf{X}^T\mathbf{R}_{\mathbf{h}}\mathbf{X}^*|\mathbf{y}, \hat{\mathbf{h}}^{(q)}\right] + N_0\mathbf{I}. \end{aligned} \tag{7.26}$$

Then, the LMMSE estimator becomes

$$\begin{aligned} \mathbf{V}^{(q)} &= E_{\mathbf{X},\mathbf{n}}\left[\mathbf{h}\mathbf{y}^H|\mathbf{y}, \hat{\mathbf{h}}^{(q)}\right]\left(E_{\mathbf{X},\mathbf{n}}\left[\mathbf{y}\mathbf{y}^H|\mathbf{y}, \hat{\mathbf{h}}^{(q)}\right]\right)^{-1} \\ &= \mathbf{R}_{\mathbf{h}}\left(\bar{\mathbf{X}}^{(q)}\right)^*\left(E_{\mathbf{X}}\left[\mathbf{X}^T\mathbf{R}_{\mathbf{h}}\mathbf{X}^*|\mathbf{y}, \hat{\mathbf{h}}^{(q)}\right] + N_0\mathbf{I}\right)^{-1} \end{aligned} \tag{7.27}$$

and the next channel estimate becomes

$$\hat{\mathbf{h}}^{(q+1)} = \mathbf{V}^{(q)}\mathbf{y}. \tag{7.28}$$

This estimate can replace the estimate obtained by the M-step in Eq. (7.24).

As in Eq. (7.27), a matrix inversion of size $L \times L$ is required to find the LMMSE channel estimator. This means that the complexity required to update the channel estimate becomes $O(L^3)$ given that $E_{\mathbf{X}}[\mathbf{X}^T\mathbf{R}_{\mathbf{h}}\mathbf{X}^*|\mathbf{y}, \hat{\mathbf{h}}^{(q)}]$ is available. On making a certain mild assumption (which will be given later), we will show that $E_{\mathbf{X}}[\mathbf{X}^T\mathbf{R}_{\mathbf{h}}\mathbf{X}^*|\mathbf{y}, \hat{\mathbf{h}}^{(q)}]$ can be readily found using the *a posteriori* probability $\Pr(b_l|\mathbf{y}, \hat{\mathbf{h}}^{(q)})$ (*not* $\Pr(\mathbf{X}|\mathbf{y}, \hat{\mathbf{h}}^{(q)})$).

In general, the LMMSE and MAP estimates obtained from the *a posteriori* probability, $\Pr(\mathbf{X}|\mathbf{y}, \hat{\mathbf{h}}^{(q)})$, are different. However, if the *a posteriori* information of $\mathbf{X}$ is so reliable that $\bar{\mathbf{X}}^{(q)}$ is identical to the hard-decision as follows:

$$\Pr\left(\mathbf{X} = \bar{\mathbf{X}}^{(q)}|\mathbf{y}, \hat{\mathbf{h}}^{(q)}\right) = 1,$$

we can show that

$$E_{\mathbf{X}}\left[\mathbf{X}^T\mathbf{R}_{\mathbf{h}}\mathbf{X}^*|\mathbf{y}, \hat{\mathbf{h}}^{(q)}\right] = \left(\bar{\mathbf{X}}^{(q)}\right)^T\mathbf{R}_{\mathbf{h}}\left(\bar{\mathbf{X}}^{(q)}\right)^*. \tag{7.29}$$

In this case, the LMMSE estimate is identical to the MAP estimate. Thus, the LMMSE estimate can be a good approximation of the MAP estimate if a more reliable *a posteriori* information of $\mathbf{X}$ is available, which can happen after few iterations.

As mentioned above, we can readily find $E_{\mathbf{X}}[\mathbf{X}^T\mathbf{R}_{\mathbf{h}}\mathbf{X}^*|\mathbf{y}, \hat{\mathbf{h}}^{(q)}]$ on making a mild assumption. Assume that the channel coefficients for each tap are uncorrelated. Then, the correlation of the channel coefficients is given by

$$E[h_{p,l}h_{p',l'}^*] = \sigma_{h,p}^2 \delta_{p,p'} \sigma_T^2(l, l'), \tag{7.30}$$

where $\sigma_{h,p}^2$ is the power delay profile and $\sigma_T^2(l, l')$ is the temporal autocorrelation function. For Clarke's fading channel, we have $\sigma_T^2(l, l') = J_0(2\pi f_d T|l - l'|)$. From this, we can show that

$$\mathbf{R}_{\mathbf{h}} = \mathbf{J} \otimes \mathbf{P}, \tag{7.31}$$

where $\otimes$ stands for the Kronecker product[†] and

$$\mathbf{P} = \mathrm{Diag}\big(\sigma_{h,0}^2, \sigma_{h,1}^2, \ldots, \sigma_{h,P-1}^2\big),$$

$$\mathbf{J} = \begin{bmatrix} \sigma_{\mathrm{T}}^2(0,0) & \sigma_{\mathrm{T}}^2(0,1) & \cdots & \sigma_{\mathrm{T}}^2(0,L-1) \\ \sigma_{\mathrm{T}}^2(1,0) & \sigma_{\mathrm{T}}^2(1,1) & \cdots & \sigma_{\mathrm{T}}^2(1,L-1) \\ \vdots & \vdots & \ddots & \vdots \\ \sigma_{\mathrm{T}}^2(L-1,0) & \sigma_{\mathrm{T}}^2(L-1,1) & \cdots & \sigma_{\mathrm{T}}^2(L-1,L-1) \end{bmatrix}.$$

It follows that

$$\big[E\big[\mathbf{X}^{\mathrm{T}}\mathbf{R_h}\mathbf{X}^* | \mathbf{y}, \hat{\mathbf{h}}^{(q)}\big]\big]_{l,l'} = \underbrace{E\big[\mathbf{x}_{l-1}^{\mathrm{T}}\mathbf{P}\mathbf{x}_{l'-1}^* | \mathbf{y}, \hat{\mathbf{h}}^{(q)}\big]}_{=u_{l,l'}^{(q)}} \sigma_{\mathrm{T}}^2(l-1,l'-1). \tag{7.32}$$

Furthermore, we can show that

$$u_{l,l'}^{(q)} = \sum_{p=0}^{P-1} \sigma_{h,p}^2 E\big[b_{l-1-p} b_{l'-1-p}^* | \mathbf{y}, \hat{\mathbf{h}}^{(q)}\big]$$

$$= \begin{cases} \sum_{p=0}^{P-1} \sigma_{h,p}^2, & \text{if } l = l'; \\ \sum_{p=0}^{P-1} \sigma_{h,p}^2 \bar{b}_{l-1-p}^{(q)} \big(\bar{b}_{l'-1-p}^{(q)}\big)^*, & \text{if } l \neq l', \end{cases} \tag{7.33}$$

where

$$\bar{b}_l^{(q)} = E\big[b_l | \mathbf{y}, \hat{\mathbf{h}}^{(q)}\big] = \Pr\big(b_l = 1 | \mathbf{y}, \hat{\mathbf{h}}^{(q)}\big) - \Pr\big(b_l = -1 | \mathbf{y}, \hat{\mathbf{h}}^{(q)}\big).$$

This mean value can be readily obtained with the *a posteriori* probability of b_l from the IED after several ED iterations.

Let $[\mathbf{U}^{(q)}]_{l,l'} = u_{l,l'}^{(q)}$. Then, from Eqs (7.32) and (7.27), we have

$$\mathbf{V}^{(q)} = \mathbf{R_h}\bar{\mathbf{X}}^* \big(\mathbf{U}^{(q)} \odot \mathbf{J} + N_0\mathbf{I}\big)^{-1}, \tag{7.34}$$

where $\odot$ represents the Hadamard product. Here, $(\mathbf{U}^{(q)} \odot \mathbf{J} + N_0\mathbf{I})$ is the conditional covariance matrix of $\mathbf{y}$. That is,

$$E\big[\mathbf{y}\mathbf{y}^{\mathrm{H}} | \mathbf{y}, \hat{\mathbf{h}}^{(q)}\big] = \mathbf{U}^{(q)} \odot \mathbf{J} + N_0\mathbf{I}.$$

The complexity involved in finding the LMMSE estimator becomes $O(L^3)$ as the size of the matrix $\mathbf{U}^{(q)} \odot \mathbf{J} + N_0\mathbf{I}$ is $L \times L$.

In general, the length of the signal blocks, L, is several hundred to a few thousand. Therefore, the complexity involved in finding the LMMSE channel estimate is still high. To reduce the complexity further, we note that $\mathbf{U}^{(q)} \odot \mathbf{J} + N_0\mathbf{I}$ is positive definite and symmetric. The Gauss–Seidel (GS) iteration is applied to find the LMMSE channel estimate with a lower complexity. The GS iteration is explained in detail in Section 7.5.

Let

$$\mathbf{z}^{(q)} = \big(\mathbf{U}^{(q)} \odot \mathbf{J} + N_0\mathbf{I}\big)^{-1} \mathbf{y}.$$

[†] See Section A2.4 in Appendix 2 for the definition of the Kronecker product and its properties.

The GS iteration used to obtain $\mathbf{z}^{(q)}$ is as follows:

$$z_l^{(k+1,q)} = \frac{1}{a_{l,l}^{(q)}} \left(y_l - \sum_{t=1}^{l-1} a_{l,t}^{(q)} z_t^{(k+1,q)} - \sum_{t=l+1}^{L} a_{l,t}^{(q)} z_t^{(k,q)} \right), \quad l = 0, 1, \ldots, L-1, \quad (7.35)$$

where $a_{l,t}^{(q)} = [\mathbf{U}^{(q)} \odot \mathbf{J} + N_0 \mathbf{I}]_{l,t}$ and k denotes the index for the GS iteration. Once the GS iteration converges, we have $\hat{\mathbf{h}}^{(q)} = \mathbf{R}_{\mathbf{h}} \bar{\mathbf{X}}^* \mathbf{z}^{(q)}$. Generally, the number of GS iterations, denoted by Q_{GS}, is not large. Since the complexity per GS iteration is $O(L^2)$, the overall complexity becomes $O(Q_{GS}L^2)$, where $Q_{GS} \ll L$. Note that Q_{GS} would be different for each EM iteration.

In the GS iteration, the converged vector of $\mathbf{z}$ of the previous EM iteration can be used as the initial vector in the current EM iteration. That is, $\mathbf{z}^{(0,q)} = \mathbf{z}^{(\infty,q-1)}$, where $\mathbf{z}^{(\infty,q-1)}$ denotes the converged vector of $\mathbf{z}$ at the $(q-1)$th EM iteration. If $\mathbf{U}^{(q)} \simeq \mathbf{U}^{(q-1)}$, the number of GS iterations for convergence at the qth EM iteration becomes small. Note that when $q = 0$, we have

$$\mathbf{U}^{(0)} = \left(\sum_{p=0}^{P-1} \sigma_{h,p}^2 \right) \mathbf{I}$$

from Eq. (7.33). Hence, we do not need the GS iteration to find $\mathbf{z}^{(0)}$ since

$$\mathbf{z}^{(0)} = \frac{1}{\sum_{p=0}^{P-1} \sigma_{h,p}^2 + N_0} \mathbf{y}.$$

It is noteworthy that the MAP channel estimation requires the independency of b_l conditioned on $\mathbf{y}$ and $\hat{\mathbf{h}}^{(q)}$ so that $\Pr(\mathbf{X}|\mathbf{y}, \hat{\mathbf{h}}^{(q)})$ can be found from $\Pr(b_l|\mathbf{y}, \hat{\mathbf{h}}^{(q)})$. On the other hand, if the LMMSE channel estimation is employed, this assumption is not necessary as the second-order statistics in Eqs (7.26) and (7.32) can be obtained from $\Pr(b_l|\mathbf{y}, \hat{\mathbf{h}}^{(q)})$ directly.

Further approximations

We have discussed two major approximations (the LMMSE channel estimation to replace the MAP channel estimation; the IED to replace the joint equalization/decoding) in order to reduce the complexity. To reduce the complexity further, a few more options are available as follows.

(O1) The MAP equalizer can be replaced by the MMSE-SC equalizer.
(O2) To avoid the double iteration, it is possible to update the channel estimation every ED iteration. The iterative receivers in Berthet, Unal, and Visoz (2001), Nefedov et al. (2003), and Otnes and Tuchler (2004) belong to this case. In terms of the EM algorithm, an approximation of the *a posteriori* probability, $\Pr(\mathbf{X}|\mathbf{y}, \hat{\mathbf{h}}^{(q)})$, is used since the convergence of the IED will not have been achieved before the channel estimate is updated.
(O3) Instead of the LMMSE estimation, adaptive algorithms of low complexity, including the LMS algorithm, can be used. This approximation is employed in Otnes and Tuchler (2004).

We can combine approximations to derive various suboptimal, but less complex, iterative receivers, as follows.

- Type 1 iterative receiver. This is a doubly iterative receiver consisting of the MAP equalizer and the LMMSE channel estimator (no additional options are taken from (O1)–(O3)).
- Type 2 iterative receiver. This is a singly iterative receiver consisting of the MAP equalizer and the LMMSE channel estimator with option (O2).
- Type 3 iterative receiver. This is a singly iterative receiver with options (O1) (i.e. the MAP equalizer is replaced by the MMSE-SC equalizer) and (O2) and the GS iteration for the channel estimation.

Clearly, the Type 3 iterative receiver would be the most computationally efficient. Note that when option (O2) is used, the EM and ED iterations are combined. For convenience, the combined EM and ED iterations will be referred to as the *combined* iteration.

7.3.4 Simulation results

We present simulation results in this subsection. A rate-half convolutional code with generator polynomial (5, 7) in octal and a random bit interleaver are used in the transmitter. The following three data block structures are considered.

- Block structure A: 7 subblocks of 7 pilot symbols and 7 subblocks of 85 data symbols and the last subblock of 5 data symbols (there are 600 data symbols and 49 pilot symbols in a transmission block).
- Block structure B: 14 subblocks of 7 pilot symbols and 14 subblocks of 42 data symbols and the last subblock of 12 data symbols (there are 600 data symbols and 98 pilot symbols in a transmission block).
- Block structure C: 14 subblocks of 4 pilot symbols and 14 subblocks of 42 data symbols and the last subblock of 12 data symbols (there are 600 data symbols and 56 pilot symbols in a transmission block).

Both block structures B and C are suitable for a fading rate up to $f_d T = 0.01$, while block structure A is suitable up to $f_d T = 0.005$. For block structures B and C, we have $M_{ps} = 49$ and 46, respectively, less than $1/2 f_d T = 50$ when $f_d T = 0.01$.

For frequency-selective fading channels, a four-tap channel model is considered with $\sigma_{h,p}^2 = 1/P$ for all p. Each channel tap is independent and time-varying according to Clarke's fading channel model.

Figure 7.7 shows simulation results with a fading rate $f_d T = 0.005$. Note that a Type 3 iterative receiver is an ExC–ImDD based receiver, while the iterative receiver introduced in Section 7.2 (referred to as the PSP-LMS based iterative receiver) is an ExDD–ImC based receiver. It is shown that the Type 3 iterative receiver outperforms the PSP-LMS based iterative receiver. This shows that the channel estimation is very important for signal detection/decoding over fading channels. Since the Type 3 iterative receiver can fully exploit the statistical properties of fading channels (e.g. the *a priori* pdf of **h**), a better channel estimate is available, which results in a better BER performance.

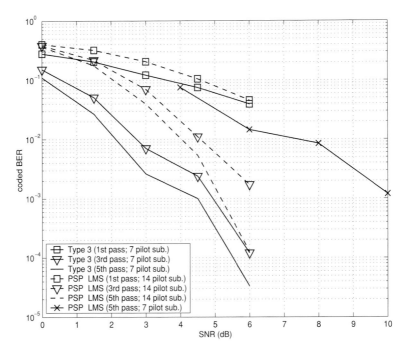

Figure 7.7. Bit error rate simulation results for a Type 3 iterative receiver and a PSP-LMS based iterative receiver; $f_d T = 0.005$.

Figure 7.8 shows the BER of a Type 3 iterative receiver for different data block structures and illustrates the impact of the number of pilot symbols when $f_d T = 0.01$. For convenience, denote by R_{p-d} the pilot-symbol to data-symbol ratio. For block structure C, $R_{p-d} = 56/600 \simeq 1/11$, while $R_{p-d} = 98/600 \simeq 1/6$ for block structure B. It can be confirmed that more pilot symbols can increase performance.

In the Type 3 iterative receiver, the GS iteration is used for the LMMSE channel estimation. If the number of GS iterations is large, the Type 3 iterative receiver would not be computationally efficient. Figure 7.9 shows the number of GS iterations for each combined iteration through simulations with block structure B. We observe that the number of GS iterations decreases with the number of combined iterations. Since $\mathbf{U}^{(q)} \simeq \mathbf{U}^{(q+1)}$ after four or five combined iterations, fewer GS iterations are required as mentioned earlier. It is observed that the largest number of GS iterations (about 20) happens when $q = 2$. However, $Q_{GS} = 20$ is much smaller than $L = 698$. Thus, the GS iteration based LMMSE channel estimation would be much more computationally efficient.

7.4 Summary and notes

In this chapter, we have studied two different approaches to the design of the iterative receiver for random fading channels: the ExDD–ImC and ExC–ImDD approaches. Several

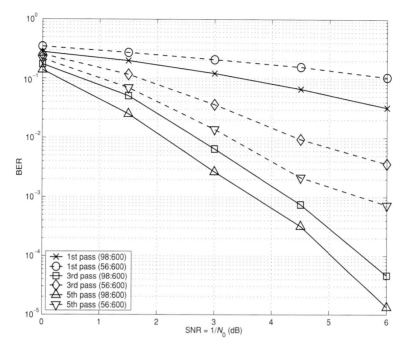

Figure 7.8. Performance of Type 3 iterative receiver. In the inset, $(x : y)$ stands for the numbers of pilot symbols (denoted by x) and data symbols (denoted by y) in a block.

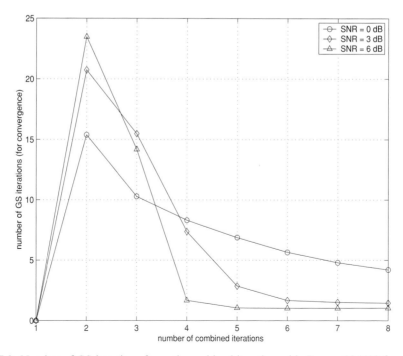

Figure 7.9. Number of GS iterations for each combined iteration with $R_{p-d} = 98/600$ for a Type 3 interative receiver.

approximations made to reduce the complexity further were discussed, in an attempt to make the resulting iterative receivers more computationally affordable.

More ExDD–ImC based iterative receivers have been investigated than ExC–ImDD based ones since the signal detection/decoding is the major task of the receiver. In Anastasopoulos and Chugg (2000), MAP symbol detection algorithms are discussed for the iterative receiver based on the ExDD–ImC approach. A factor graph based approach is considered in Worthen and Stark (2001) to design iterative receivers for flat-fading channels. Only a few iterative receivers are based on the ExC–ImDD approach. (There are other approaches; see Baccarelli and Cusani (1998), Cozzo and Hughes (2003), Garcia-Frias and Villasenor (2003), and Iltis (1992).) However, the iterative receivers in Berthet *et al.* (2001), Nefedov *et al.* (2003), and Otnes and Tuchler (2004) can be considered to be ExC–ImDD based receivers.

The performance analysis of the iterative receiver for random fading channels has not been well addressed due to difficulties. The EXIT chart analysis would be applicable; however, as the CIR is random and time-varying, the EXIT chart for the MAP symbol detection becomes random. Hence, an averaged EXIT chart is necessary.

7.5   Appendix to Chapter 7: Gauss–Seidel iteration

The Gauss–Seidel (GS) iteration can solve linear systems. Consider a linear system $\mathbf{Az} = \mathbf{y}$, where $\mathbf{A}$ is square and full-rank. Given $\mathbf{y}$ and $\mathbf{A}$, $\mathbf{z}$ is to be found. A direct approach to find the solution $\mathbf{z} = \mathbf{A}^{-1}\mathbf{y}$ requires the inverse of $\mathbf{A}$. However, if $\mathbf{A}$ is large, this direct approach is prohibitive because of a high computational complexity. Fortunately, there are iterative methods, including the GS iteration, that can obtain $\mathbf{z}$ without finding $\mathbf{A}^{-1}$; see Axelsson (1994) and Golub and Van Loan (1983) for detailed accounts of iterative methods. In this appendix, we focus only on the GS iteration.

Assume that the size of $\mathbf{A}$ is $N \times N$. To solve the linear system $\mathbf{Az} = \mathbf{y}$, consider a recursion as follows:

$$\mathbf{Bz}^{(k+1)} = \mathbf{Cz}^{(k)} + \mathbf{y}, \tag{7.36}$$

where $\mathbf{B} - \mathbf{C} = \mathbf{A}$. Suppose that the iteration in Eq. (7.36) converges. Let $\lim_{k \to \infty} \mathbf{z}^{(k)} = \bar{\mathbf{z}}$. Then, it follows that

$$\mathbf{B}\bar{\mathbf{z}} = \mathbf{C}\bar{\mathbf{z}} + \mathbf{y}$$

or $\bar{\mathbf{z}} = (\mathbf{B} - \mathbf{C})^{-1}\mathbf{y} = \mathbf{A}^{-1}\mathbf{y} = \mathbf{z}$. Thus, we can see that the recursion in Eq. (7.36) can find $\mathbf{z}$. In the GS iteration,

$$[\mathbf{B}]_{n,m} = \begin{cases} [\mathbf{A}]_{n,m}, & \text{if } n \geq m; \\ 0, & \text{otherwise} \end{cases} \tag{7.37}$$

and

$$[\mathbf{C}]_{n,m} = \begin{cases} 0, & \text{if } n \geq m; \\ [\mathbf{A}]_{n,m}, & \text{otherwise.} \end{cases} \tag{7.38}$$

Thus, $\mathbf{B}$ is lower triangle and $\mathbf{C}$ is strictly upper triangle. The GS iteration is given by

$$z_n^{(k+1)} = \frac{1}{a_{nn}} \left(y_n - \sum_{m=1}^{n-1} a_{nm} z_m^{(k+1)} - \sum_{m=n+1}^{N} a_{nm} z_m^{(k)} \right) \qquad (7.39)$$

for $n = 1, 2, \ldots, N$. The GS iteration has $O(N^2)$ complexity per iteration as shown in Eq. (7.39). If the number of iterations is smaller than N, the GS iteration can be more computationally efficient than the direct approach to find $\mathbf{z}$.

The convergence depends on $\mathbf{A}$ in the GS iteration. Let $\tilde{\mathbf{z}}^{(k)} = \mathbf{z}^{(k)} - \mathbf{z}$. From Eq. (7.36) and using $\mathbf{Bz} = \mathbf{Cz} + \mathbf{y}$, we can show that

$$\tilde{\mathbf{z}}^{(k+1)} = \mathbf{B}^{-1}\mathbf{C}\tilde{\mathbf{z}}^{(k)}$$

or

$$\tilde{\mathbf{z}}^{(k+1)} = \mathbf{D}^k\tilde{\mathbf{z}}^{(k)},$$

where $\mathbf{D} = \mathbf{B}^{-1}\mathbf{C}$. From this, we have

$$\lim_{k \to \infty} \tilde{\mathbf{z}}^{(k)} = \mathbf{0}, \quad \text{if} \quad \lim_{k \to \infty} \mathbf{D}^k = \mathbf{0}.$$

This shows that the spectral radius of $\mathbf{D}$ decides the convergence.

We state the following important theorem from Golub and Van Loan (1983).

Theorem 7.5.1 *If $\mathbf{A}$ is positive definite and symmetric, the spectral radius of $\mathbf{D} = \mathbf{B}^{-1}\mathbf{C}$ is less than 1 or*

$$\lim_{k \to \infty} (\mathbf{B}^{-1}\mathbf{C})^k = \lim_{k \to \infty} \mathbf{D}^k = \mathbf{0}.$$

Thus, if $\mathbf{A}$ is a covariance matrix, the GS iteration converges.

III Other interference-limited systems

8 CDMA systems and multiuser detection

In a multiuser communication system, a multiple access scheme is required to share a common channel resource, e.g. the frequency band. For mobile multiuser communication systems, especially cellular systems, a code division multiple access (CDMA) scheme is widely employed due to several advantages over other schemes. In CDMA systems, users can transmit signals simultaneously over the same frequency band. This makes CDMA systems interference-limited since interfering signals (i.e. other users' signals) degrade the performance. To improve the performance of CDMA systems, multiuser detection has been investigated. Multiuser detection attempts to detect the desired signal as well as other signals (this implies a joint detection†) for a better performance. Indeed, this is an analogy of the MLSD, in which a symbol sequence (i.e. multiple symbols), rather than individual symbols, is detected. There are also a number of different approaches taken to mitigate or suppress interfering signals for multiuser detection. In this chapter, we introduce CDMA systems and multiuser detectors; adaptive multiuser detectors are also introduced.

8.1 Overview of cellular systems

Cellular systems are wireless communication systems that can provide services over a large area. As the range of radio signals is limited, a cellular structure is employed to cover a large area. A cellular structure is shown in Fig. 8.1, in which each hexagonal area is called a cell. There is a basestation at the center of each cell. The role of the basestation is to communicate with mobile subscribers (i.e. users) within a cell. Hence, the size of the cell is closely related to the range of the radio signals between the basestation and users.

When a user wants to call another user in a different cell, the two users have to communicate with their basestations. Then, the basestations communicate with each other to relay signals. A cellular system is depicted in Fig. 8.2. The basestations are connected through a wired network (i.e. a backbone network).

In a cellular system, there are two different types of channels: downlink and uplink channels. A downlink channel is from a basestation to a user, while an uplink channel is from a user to a basestation. If each signal needs a bandwidth of B Hz, a total bandwidth

---

† It is from this that the term multiuser detection was coined.

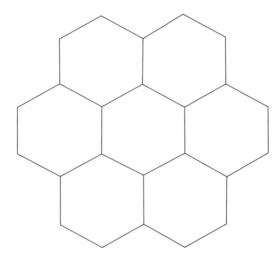

Figure 8.1. Cell structure.

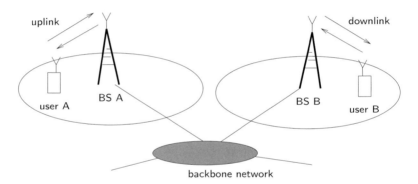

Figure 8.2. Cellular system (BS = basestation).

of $2KB$ Hz is required to support K users in a cell. Here, we have assumed that different frequency bands are assigned for each channel to avoid interference from the other users. This method of allocating channels is called frequency division multiple access (FDMA). As the number of cells increases, the total bandwidth for a cellular system increases. This leads to an unrealistic demand of bandwidth if a cellular system attempts to cover a large area. To avoid this problem, the concept of frequency reuse was introduced.

The power of a radio signal decays as follows:

$$P(d) \propto P_0 d^{-n}, \tag{8.1}$$

where P_0 is the transmitted power at a reference location and $P(d)$ is the power measured at a location distance d from the reference location. Here, n is the path loss exponent that varies from 3 to 4 in outdoor communications ($n = 2$ for free space communications); see Rappaport (1996) for details. Equation (8.1) shows that a higher power is required for longer distance communication to keep a certain SNR at the receiver. If two cells are sufficiently

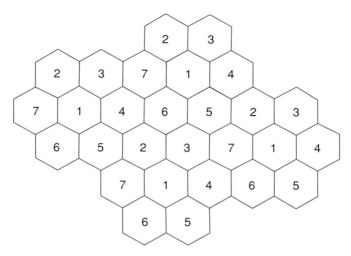

Figure 8.3. Seven-cell reuse pattern (each number corresponds to a different frequency band).

far apart, they can use the same frequency band without significant interference because signals from the first cell would not interfere with the second cell due to the propagation loss in Eq. (8.1). This is the frequency reuse. In Fig. 8.3, we show a repeated cluster of seven cells that use different frequency bands covering a large area. Cells that use the same frequency band are sufficiently far apart. The total bandwidth becomes $7 \times 2KB$ Hz if a cell supports K users. The number of cells in a cluster is called the *frequency reuse factor*. As the frequency reuse factor increases, the total bandwidth increases. However, since the effect of interference becomes lower, a more reliable channel is expected.

Instead of FDMA, time division multiple access (TDMA) can provide multiple channels for the users in a cell. A time frame is divided into multiple time slots, and for each time frame a user can use a time slot to communicate with the basestation. Hence, the maximum number of users in a cell is the same as the number of slots in a frame. In both FDMA and TDMA, the signals within a cell do not interfere with each other as the channels are orthogonal. However, there are interfering signals from other cells due to a finite frequency reuse factor.

Since the bandwidth or data rate per user is limited by the frequency reuse factor for a given total bandwidth, it is desirable to have a lower frequency reuse factor to support higher data rate services. Comparing with FDMA and TDMA, CDMA-based cellular systems can easily have frequency reuse factor of unity. This becomes an important advantage of CDMA.

8.2 Introduction to CDMA systems

In this section, we use an example to explain CDMA systems. A CDMA system over the AWGN channel is considered in order to simplify the presentation. Furthermore, we assume that all the users' signals are synchronized.

In CDMA systems, multiple users can share a common frequency band at the same time by using different signature waveforms. To illustrate this idea, let us consider an example. Suppose that there are K users and that user 1 transmits the following signal:

$$x_1(t) = s_1(t)b_1, \quad 0 \le t \le T, \quad b_1 \in \{-1, +1\}, \tag{8.2}$$

where b_1 is the symbol that user 1 transmits, $s_1(t)$ is the signature waveform or spreading waveform for user 1, and T is the symbol period. The symbol b_1 is modulated by $s_1(t)$ and $x_1(t)$ and is called the spread signal. If there is only one user, the received signal is given by

$$r(t) = s_1(t)b_1 + n(t), \quad 0 \le t \le T, \tag{8.3}$$

where $n(t)$ is the background noise that is often modeled as the AWGN. Generally, we assume that $E[n(t)] = 0$ and $E[n(t)n(\tau)] = (N_0/2)\delta(t - \tau)$, where $N_0/2$ is the double-sided spectral density and $\delta(t)$ is the Dirac delta function. Since the background noise is Gaussian, the likelihood function is given by

$$f(r(t)|b_1) = C \exp\left(-\frac{1}{N_0} \int_0^T |r(t) - s_1(t)b_1|^2 \, dt\right), \tag{8.4}$$

where C is a normalizing constant. The ML detector chooses b_1 that maximizes the likelihood function. Since

$$\int_0^T |r(t) - s_1(t)b_1|^2 \, dt = \int_0^T r^2(t) \, dt + \int_0^T s_1^2(t) \, dt - 2\int_0^T r(t)s_1(t)b_1 \, dt,$$

we can see that

$$f(r(t)|b_1) = C' \exp\left(-\frac{1}{N_0} \int_0^T r(t)s_1(t)b_1 \, dt\right),$$

where C' is a constant independent of b_1. The ML detection is based on the output of the correlator with the signature waveform $\int_0^T r(t)s_1(t) \, dt$, and the correlator detector becomes the ML detector. The output of the correlator detector is given by

$$\hat{b}_1 = \frac{1}{\int_0^T s_1^2(t) \, dt} \int_0^T r(t)s_1(t) \, dt, \tag{8.5}$$

where the denominator normalizes the output of the correlator so that $E[\hat{b}_1|b_1] = b_1$. Then, a hard-decision of b_1 becomes $\text{Sign}(\hat{b}_1)$.

With a different signature waveform, say $s_2(t)$, user 2 can also transmit its signal as follows:

$$x_2(t) = s_2(t)b_2, \quad 0 \le t \le T, \quad b_2 \in \{-1, +1\},$$

where b_2 is the symbol that user 2 transmits. A different signature waveform has to be assigned to each user so that the receiver can distinguish one from another. Given that

$K = 2$, if the two users transmit signals simultaneously, the received signal is given by

$$r(t) = x_1(t) + x_2(t) + n(t)$$
$$= s_1(t)b_1 + s_2(t)b_2 + n(t), \quad 0 \le t \le T.$$

If the two signature waveforms are orthogonal, i.e. $\int_0^T s_1(t)s_2(t)\,dt = 0$, the receiver can detect both symbols without the interference from the other user's signal. That is, we can have

$$\hat{b}_1 = \frac{1}{\int_0^T s_1^2(t)\,dt} \int_0^T s_1(t)r(t)\,dt$$

$$= b_1 + \frac{1}{\int_0^T s_1^2(t)\,dt} \int_0^T s_1(t)n(t)\,dt;$$

$$\hat{b}_2 = \frac{1}{\int_0^T s_2^2(t)\,dt} \int_0^T s_2(t)r(t)\,dt$$

$$= b_2 + \frac{1}{\int_0^T s_2^2(t)\,dt} \int_0^T s_2(t)n(t)\,dt.$$

However, if $s_1(t)$ and $s_2(t)$ are not orthogonal, interference occurs. The interference from the other users' signals is called the multiuser interference (MUI). To effectively deal with the MUI, we can jointly detect all the signals; this results in a multiuser detector, which will be discussed in Section 8.3.

The bandwidth of signature waveforms cannot be arbitrarily wide due to the limited availability of the spectrum. In addition, to accommodate a large number of users, the signature waveforms need to be designed carefully. One approach taken to generate signature waveforms is as follows:

$$s_k(t) = \sum_{m=0}^{N-1} c_{m,k} p(t - mT_c), \quad 0 \le t \le T, \tag{8.6}$$

where $s_k(t)$ and $\{c_{0,k}, c_{1,k}, \ldots, c_{N-1,k}\}$ are the signature waveform, the spreading sequence (or code) of user k, $p(t)$, is the chip waveform, and T_c is the chip interval. In addition, we assume that $c_{m,k} \in \{-1/\sqrt{N}, 1/\sqrt{N}\}$ so that $\sum_{m=0}^{N-1} |c_{m,k}|^2 = 1$ for normalization purposes. It is generally assumed that

$$T = NT_c,$$

where N is called the processing gain. Spreading sequences should be different to generate different signature waveforms. In addition, we assume that $p(t)$ satisfies

$$\int_{-\infty}^{\infty} p(t - nT_c)p(t - mT_c)\,dt = E_c \delta_{n,m}, \tag{8.7}$$

where E_c is the energy of $p(t)$. For normalization purposes, we assume that $E_c = 1$ throughout this chapter.

The received signal from K users over the AWGN channel is given by

$$r(t) = \sum_{k=1}^{K} A_k s_k(t)b_k + n(t), \tag{8.8}$$

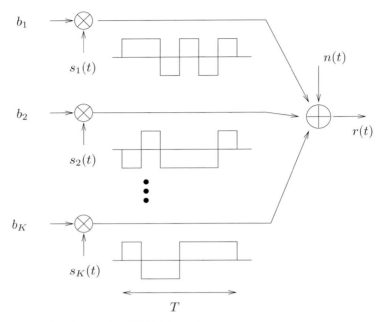

Figure 8.4. Spread signals over the AWGN channel.

where A_k is the amplitude of the kth user signal. A block diagram for multiuser trans-missions with signature waveforms is shown in Fig. 8.4. According to Eq. (8.6), it can be shown that the spectrum of the users' signals is determined by $p(t)$. In general, the bandwidth of $p(t)$ is wider than $1/T$, which can be considered as the bandwidth of the unspread signals. If $p(t)$ is a rectangular pulse with pulse width $T_c = T/N$, the bandwidth of $p(t)$ becomes $1/T_c = N/T$ and the spread signal has an N times wider bandwidth than the unspread signal. Thus, we can see that the signal has spread in the frequency domain.

A conventional approach for detecting the desired signal assumes that the sum of the other users' signals as well as the background noise is white Gaussian noise. This leads to the correlator detector. Suppose that the desired user is user 1. Then, the received signal can be rewritten as follows:

$$r(t) = A_1 s_1(t) b_1 + u_1(t), \tag{8.9}$$

where $u_1(t) = \sum_{k=2}^{K} A_k s_k(t) b_k + n(t)$. We can see that $u_1(t)$, which includes the other users' signals and the noise, $n(t)$, becomes the interfering signal. Since $u_1(t)$ is the sum of many spread signals, we can simply assume that $u_1(t)$ is white and Gaussian according to the central limit theorem. Then, under the Gaussian assumption, the optimal detector for b_1 (based on the ML detection) is the correlator detector which provides the detector output as follows:

$$\hat{b}_1 = \int_0^T s_1(t) r(t) \, dt. \tag{8.10}$$

The sign of $\hat{b}_1$ becomes the decision of b_1. Even though this detector seems optimal,[†] it does not actually exploit the *structure* of interfering signals. Thus, there would be better approaches to detect signals by exploiting the structure of interfering signals.

Example 8.2.1 The output of the correlator detector has the desired signal as well as interfering signals. Hence, the signal-to-interference-plus-noise ratio (SINR) can be used to measure the performance. Suppose there are K spread signals. It is assumed that the spreading sequences are iid sequences with $\Pr(c_{m,k} = +1/\sqrt{N}) = \Pr(c_{m,k} = -1/\sqrt{N}) = 1/2$. The output of the correlator for the kth signal is given by

$$\hat{b}_k = \int_0^T s_k(t) r(t) \, dt$$

$$= A_k b_k + \sum_{q \neq k} \int_0^T A_q s_q(t) b_q s_k(t) \, dt + \int_0^T n(t) s_k(t) \, dt. \tag{8.11}$$

The second term on the right hand side of Eq. (8.11) is due to the interference. The mean is zero if b_q is equally likely. To find the variance of the second term, we need to show that

$$E\left[\int s_k(t) s_q(t) \, dt\right] = E\left[\int \sum_m c_{m,k} p(t - mT_c) \sum_i c_{i,q} p(t - iT_c) \, dt\right]$$

$$= \sum_m \sum_i E[c_{m,k} c_{i,q}] \int p(t - mT_c) p(t - iT_c) \, dt$$

$$= \sum_m E[c_{m,k} c_{m,q}]$$

$$= \begin{cases} 1, & \text{if } k = q; \\ 0, & \text{otherwise.} \end{cases} \tag{8.12}$$

Using this, the variance of the interference can be obtained as follows:

$$\sigma_I^2 = E\left[\left(\sum_{q \neq k} \int A_q s_q(t) s_k(t) \, dt\right)\left(\sum_{q' \neq k} \int A_{q'} s_{q'}(t) s_k(t) \, dt\right)\right]$$

$$= \sum_{q \neq k} \sum_{q' \neq k} A_q A_{q'} E\left[\int s_q(t) s_{q'}(t) \underbrace{s_k^2(t)}_{=1/N} \, dt\right]$$

$$= \frac{1}{N} \sum_{q \neq k} \sum_{q' \neq k} A_q A_{q'} \underbrace{E\left[\int s_q(t) s_{q'}(t) \, dt\right]}_{=\delta_{q,q'}}$$

$$= \frac{1}{N} \sum_{q \neq k} A_q^2. \tag{8.13}$$

[†] The correlator detector is optimal if the interference including the background noise is white Gaussian. However, for a finite number of users, the interference cannot be approximated by a white Gaussian noise and, thereby, the correlator detector cannot be optimal.

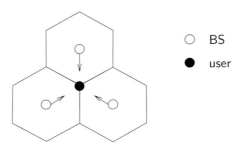

$\bigcirc$ BS

$\bullet$ user

Figure 8.5. Interference from three cells in a CDMA downlink.

The variance of the third term of the right hand side of Eq. (8.11), which is the noise, can be obtained as follows:

$$\sigma_N^2 = E\left[\left(\int n(t)s_k(t)\,dt\right)^2\right]$$

$$= \frac{N_0}{2}. \tag{8.14}$$

The SINR of Eq. (8.11) is given by

$$SINR_k = \frac{A_k^2}{\sigma_I^2 + \sigma_N^2}$$

$$= \frac{A_k^2}{(1/N)\sum_{q\neq k} A_q^2 + (N_0/2)}. \tag{8.15}$$

If $A_k = 1$ for all k, the SINR expression is simplified as follows:

$$SINR_k = \frac{1}{(K-1)/N + (N_0/2)}.$$

Equation (8.15) indicates that the CDMA system is interference-limited. Even though N_0 decreases, the SINR would not increase much due to the interfering signals. To reduce the impact of the interfering signals on the performance, the processing gain N can be increased. However, this results in an increase of bandwidth.

When CDMA is used for a cellular system, the frequency reuse factor can be unity. That is, all cells can use a common frequency band. In this case, a user at a boundary location of three cells, as shown in Fig. 8.5, will experience severe interference in the downlink channel. Other cells can be ignored as they are far enough away from the boundary location. If there are K users in each cell and $A_k = 1$ for all k, the SINR at the boundary location is given by

$$SINR_{worst} = \frac{1}{(3K-1)/N + (N_0/2)} \simeq \frac{N}{3K-1}. \tag{8.16}$$

In Eq. (8.16), the noise is ignored as the interference dominates. If $SINR_{worst} \geq 4\,(\simeq 6\,dB)$ is required, the maximum number of users per cell would be $N/12$.

In voice communications, half the time could be silent (consider a conversation) and no transmission of signal is necessary. Accounting for this voice activity factor, the number of effective users can be $K/2$. In this case, the number of maximum users per cell doubles to $N/6$. Compared with CDMA, TDMA and FDMA cannot easily accommodate the voice activity factor because a dynamic slot assignment is required.

8.3 Multiuser detection

By exploiting the structure of the interfering signals in CDMA systems, a conventional correlator detector's performance may be surpassed. Multiuser detection exploits the structure of the interfering signals. Multiuser detectors are joint detectors in the sense of detecting multiple users' signals simultaneously.

For multiuser detection, it is convenient to consider the following chip-rate sampled signals:[†]

$$
\begin{aligned}
y_m &= \int_{-\infty}^{\infty} r(t)p(t - mT_c)\,\mathrm{d}t \\
&= \int_{-\infty}^{\infty} \left(\sum_{k=1}^{K} A_k s_k(t)b_k \right) p(t - mT_c)\,\mathrm{d}t + \int_{-\infty}^{\infty} n(t)p(t - mT_c)\,\mathrm{d}t \\
&= \sum_{k=1}^{K} A_k b_k c_{m,k} + n_m, \quad m = 0, 1, \ldots, N-1,
\end{aligned}
\tag{8.17}
$$

where $n_m = \int_{-\infty}^{\infty} n(t)p(t - mT_c)\,\mathrm{d}t$. In Eq. (8.17), we use Eq. (8.7) to derive the third equality from the second one. Then, by stacking y_m's, we can show that

$$
\begin{aligned}
\mathbf{y} &= [y_0 \ y_1 \ \cdots \ y_{N-1}]^{\mathrm{T}} \\
&= \mathbf{CAb} + \mathbf{n},
\end{aligned}
\tag{8.18}
$$

where $\mathbf{b} = [b_1 \ b_2 \ \cdots \ b_K]^{\mathrm{T}}$, $\mathbf{n} = [n_0 \ n_1 \ \cdots \ n_{N-1}]^{\mathrm{T}}$, $\mathbf{A} = \mathrm{Diag}(A_1, A_2, \ldots, A_K)$, and

$$
\mathbf{C} = \begin{bmatrix}
c_{0,1} & c_{0,2} & \cdots & c_{0,K} \\
c_{1,1} & c_{1,2} & \cdots & c_{1,K} \\
\vdots & \vdots & \ddots & \vdots \\
c_{N-1,1} & c_{N-1,2} & \cdots & c_{N-1,K}
\end{bmatrix}.
$$

The column vectors of $\mathbf{C}$ are the users' spreading codes. That is, $\mathbf{C} = [\mathbf{c}_1 \ \mathbf{c}_2 \ \cdots \ \mathbf{c}_K]$. Note that

$$
\begin{aligned}
E[n_m n_q] &= \int_{-\infty}^{\infty} \int_{-\infty}^{\infty} E[n(t)n(\tau)]p(t - mT_c)p(\tau - qT_c)\,\mathrm{d}t\,\mathrm{d}\tau \\
&= \frac{N_0}{2}\delta_{m,q}.
\end{aligned}
$$

Hence, $\mathbf{n}$ is a Gaussian random noise vector of zero mean and covariance matrix $E[\mathbf{nn}^{\mathrm{T}}] = (N_0/2)\mathbf{I}$.

[†] A bank of matched filters can be applied to obtain a sufficient statistic for all the users' data symbols; see Verdu (1998) for details.

In Eq. (8.18), the binary vector $\mathbf{b}$ is the signal to be detected or estimated. This is the main difference from the single-user detector, in which only the user's signal, say b_1, is to be estimated.

8.3.1 ML multiuser detector

For the MLSD, the binary vector $\mathbf{b}$ that maximizes the likelihood function can be found. This approach leads to the ML multiuser detector. With a known amplitude matrix, $\mathbf{A}$, and a spreading code matrix, $\mathbf{C}$, the likelihood function is given by

$$f(\mathbf{y}|\mathbf{b}) = C \exp\left(-\frac{1}{N_0}\|\mathbf{y} - \mathbf{CAb}\|^2\right),$$

where C is a normalizing constant. An exhaustive search for the ML vector requires a complexity of $O(2^K)$:

$$\begin{aligned}
\mathbf{b}_{\mathrm{ml}} &= \arg\max_{\mathbf{b}} f(\mathbf{y}|\mathbf{b}) \\
&= \arg\min_{\mathbf{b}} \|\mathbf{y} - \mathbf{CAb}\|^2,
\end{aligned} \tag{8.19}$$

where $\mathbf{b}_{\mathrm{ml}}$ stands for the ML vector that maximizes the likelihood function.

Example 8.3.1 Suppose that there are two users ($K = 2$) with the same unit power $A_1 = A_2 = 1$. Let

$$\mathbf{c}_1 = [+1 \ \ +1 \ \ +1]^{\mathrm{T}}; \ \ \mathbf{c}_2 = [+1 \ \ -1 \ \ +1]^{\mathrm{T}}.$$

When the received signal vector is $\mathbf{y} = [2 \ 2 \ 0]^{\mathrm{T}}$, the ML vector can be found by an exhaustive search.

There are four possible cases for $\mathbf{b}$. The values of the negative log-likelihood for each case are given by

$$\begin{aligned}
\mathbf{b} = [+1 \ \ +1]^{\mathrm{T}} &\Rightarrow \|\mathbf{y} - \mathbf{Cb}\|^2 = 8, \\
\mathbf{b} = [+1 \ \ -1]^{\mathrm{T}} &\Rightarrow \|\mathbf{y} - \mathbf{Cb}\|^2 = 4, \\
\mathbf{b} = [-1 \ \ +1]^{\mathrm{T}} &\Rightarrow \|\mathbf{y} - \mathbf{Cb}\|^2 = 20, \\
\mathbf{b} = [-1 \ \ -1]^{\mathrm{T}} &\Rightarrow \|\mathbf{y} - \mathbf{Cb}\|^2 = 24.
\end{aligned}$$

Hence, the ML vector is given by $\mathbf{b} = [+1 \ \ -1]^{\mathrm{T}}$.

8.3.2 Decorrelating detector

The ML detector may be impractical due to a high complexity. Therefore, suboptimal but less complex approaches are considered. The decorrelating detector is a suboptimal detector that suppresses interference using a linear transform (Lupas and Verdu, 1989).

As for a single-user detector using matched filtering, consider the outputs of the matched filters:

$$z_k = \mathbf{c}_k^{\mathrm{T}}\mathbf{y}, \quad k = 1, 2, \ldots, K. \tag{8.20}$$

We have the desired user's output as well as the interfering signals from the other users. Stacking the z_k's, we have

$$\begin{aligned}
\mathbf{z} &= [z_1 \ z_2 \ \cdots \ z_K]^\mathsf{T} \\
&= \mathbf{C}^\mathsf{T}\mathbf{y} \\
&= \mathbf{C}^\mathsf{T}\mathbf{CAb} + \mathbf{C}^\mathsf{T}\mathbf{n}.
\end{aligned} \tag{8.21}$$

To suppress the MUI, a linear transform,

$$\mathbf{R}^{-1} = (\mathbf{C}^\mathsf{T}\mathbf{C})^{-1},$$

results in a zero-MUI such as

$$\begin{aligned}
\mathbf{d} &= \mathbf{R}^{-1}\mathbf{z} \\
&= \mathbf{Ab} + \mathbf{R}^{-1}\mathbf{C}^\mathsf{T}\mathbf{n}.
\end{aligned} \tag{8.22}$$

If the desired user is user k, the kth element of $\mathbf{d}$ becomes the signal of interest:

$$d_k = [\mathbf{d}]_k = A_k b_k + [\mathbf{R}^{-1}\mathbf{C}^\mathsf{T}\mathbf{n}]_k. \tag{8.23}$$

Even though no MUI exists, the decorrelating detector has the following shortcomings:

- all users' spreading codes should be known;
- the noise can be enhanced if there are spreading codes of high cross-correlation. This can be seen through $\mathbf{R}^{-1}$.

To see the impact of the spreading codes' cross-correlation, let us derive the SNR of the signal in Eq. (8.23) as follows:

$$\mathrm{SNR}_{\mathrm{dc},k} = \frac{A_k^2}{\frac{N_0}{2}[\mathbf{R}^{-1}]_{k,k}} \tag{8.24}$$

since

$$\begin{aligned}
E\left[\left([\mathbf{R}^{-1}\mathbf{C}^\mathsf{T}\mathbf{n}]_k\right)^2\right] &= \left[E\left[(\mathbf{R}^{-1}\mathbf{C}^\mathsf{T}\mathbf{n})(\mathbf{R}^{-1}\mathbf{C}^\mathsf{T}\mathbf{n})^\mathsf{T}\right]\right]_{k,k} \\
&= \left[\mathbf{R}^{-1}\mathbf{C}^\mathsf{T}E[\mathbf{nn}^\mathsf{T}]\mathbf{CR}^{-1}\right]_{k,k} \\
&= \frac{N_0}{2}[\mathbf{R}^{-1}]_{k,k}.
\end{aligned}$$

Indeed, the decorrelating detector is an analogy of the zero-forcing equalizer. The MMSE approach can alleviate the noise enhancement problem.

8.3.3 LMMSE detector

The LMMSE detector can be found from Eq. (8.18). We assume that the detector is linear and is to estimate the signal vector $\mathbf{Ab}$. Then, the LMMSE detector is given by

$$\begin{aligned}
\mathbf{L}_{\mathrm{lm}} &= \arg\min_{\mathbf{L}} E[\|\mathbf{Ab} - \mathbf{Ly}\|^2] \\
&= \mathbf{R}_{\mathbf{Ab},\mathbf{y}}\mathbf{R}_{\mathbf{y}}^{-1},
\end{aligned} \tag{8.25}$$

where $\mathbf{R_y} = E[\mathbf{yy}^T]$ is the covariance matrix of $\mathbf{y}$ and $\mathbf{R}_{\mathbf{Ab},y} = E[\mathbf{Aby}^T]$. Assuming that the elements of $\mathbf{b}$ are iid binary random variables and independent of $\mathbf{n}$, we can show that

$$\mathbf{R}_{\mathbf{Ab},y} = \mathbf{A}^2\mathbf{C}^T,$$

$$\mathbf{R_y} = \mathbf{CA}^2\mathbf{C}^T + \frac{N_0}{2}\mathbf{I}.$$

Thus, the LMMSE detector is given by

$$\mathbf{L}_{lm} = \mathbf{A}^2\mathbf{C}^T\left(\mathbf{CA}^2\mathbf{C}^T + \frac{N_0}{2}\mathbf{I}\right)^{-1}. \tag{8.26}$$

The correlation matrix of the error, $\mathbf{Ab} - \mathbf{L}_{lm}\mathbf{y}$, is given by

$$\begin{aligned}
\mathbf{R}_{lm} &= E[(\mathbf{Ab} - \mathbf{L}_{lm}\mathbf{y})(\mathbf{Ab} - \mathbf{L}_{lm}\mathbf{y})^T] \\
&= \mathbf{A}^2 - \mathbf{A}^2\mathbf{C}^T\left(\mathbf{CA}^2\mathbf{C}^T + \frac{N_0}{2}\mathbf{I}\right)^{-1}\mathbf{CA}^2 \\
&= \mathbf{A}^2\left(\mathbf{I} - \mathbf{A}^2\mathbf{C}^T\left(\mathbf{CA}^2\mathbf{C}^T + \frac{N_0}{2}\mathbf{I}\right)^{-1}\mathbf{C}\right) \\
&= \left[\mathbf{A}^{-2} + \left(\frac{N_0}{2}\right)^{-1}\mathbf{C}^T\mathbf{C}\right]^{-1}, \tag{8.27}
\end{aligned}$$

where the last equality is obtained by using the matrix inversion lemma. If $\mathbf{A} = \mathbf{I}$, $\mathbf{R}_{lm}$ is simplified as follows:

$$\begin{aligned}
\mathbf{R}_{lm} &= \mathbf{I} - \mathbf{C}^T\left(\mathbf{CC}^T + \frac{N_0}{2}\mathbf{I}\right)^{-1}\mathbf{C} \\
&= \left[\mathbf{I} + \left(\frac{N_0}{2}\right)^{-1}\mathbf{C}^T\mathbf{C}\right]^{-1}. \tag{8.28}
\end{aligned}$$

Note that the LMMSE detector can be considered as $\mathbf{b}$ being the desired signal vector rather than $\mathbf{Ab}$; in this case, the LMMSE detector becomes

$$\bar{\mathbf{L}}_{lm} = \mathbf{AC}^T\left(\mathbf{CA}^2\mathbf{C}^T + \frac{N_0}{2}\mathbf{I}\right)^{-1} = \mathbf{A}^{-1}\mathbf{L}_{lm} \tag{8.29}$$

and the correlation matrix of the error vector, $\mathbf{b} - \mathbf{L}_{lm}\mathbf{y}$, is given by

$$\begin{aligned}
\bar{\mathbf{R}}_{lm} &= \left[\mathbf{I} + \left(\frac{N_0}{2}\right)^{-1}\mathbf{AC}^T\mathbf{CA}\right]^{-1} \\
&= \mathbf{A}^{-1}\mathbf{R}_{lm}\mathbf{A}^{-1}. \tag{8.30}
\end{aligned}$$

There is no difference between $\mathbf{L}_{lm}$ and $\bar{\mathbf{L}}_{lm}$ except scaling by the amplitude matrix, $\mathbf{A}$.

From Eq. (8.29), the LMMSE estimate of b_k becomes

$$
\begin{aligned}
\hat{b}_{\text{lm},k} &= \mathbf{w}_k^{\text{T}} \mathbf{y} \\
&= A_k \mathbf{c}_k^{\text{T}} \mathbf{R}_{\mathbf{y}}^{-1} \mathbf{y} \\
&= A_k^2 \mathbf{c}_k^{\text{T}} \mathbf{R}_{\mathbf{y}}^{-1} \mathbf{c}_k b_k + A_k \mathbf{c}_k^{\text{T}} \mathbf{R}_{\mathbf{y}}^{-1} \left(\sum_{q \neq k} A_q \mathbf{c}_q b_q + \mathbf{n} \right),
\end{aligned} \tag{8.31}
$$

where $\mathbf{w}_k = \mathbf{R}_{\mathbf{y}}^{-1} \mathbf{c}_k A_k$ is the LMMSE weighting vector for user k. From Eq. (8.30), the MSE of $\hat{b}_{\text{lm},k}$ is given by

$$
\text{MSE}_k = 1 - A_k^2 \mathbf{c}_k^{\text{T}} \mathbf{R}_{\mathbf{y}}^{-1} \mathbf{c}_k. \tag{8.32}
$$

Then, the SINR of $b_{\text{lm},k}$ can be found in terms of MSE_k as follows:

$$
\begin{aligned}
\text{SINR}_k &= \frac{\left| A_k^2 \mathbf{c}_k^{\text{T}} \mathbf{R}_{\mathbf{y}}^{-1} \mathbf{c}_k \right|^2}{A_k^2 \mathbf{c}_k^{\text{T}} \mathbf{R}_{\mathbf{y}}^{-1} (\mathbf{R}_{\mathbf{y}} - A_k^2 \mathbf{c}_k \mathbf{c}_k^{\text{T}}) \mathbf{R}_{\mathbf{y}}^{-1} \mathbf{c}_k} \\
&= \frac{A_k^2 \mathbf{c}_k^{\text{T}} \mathbf{R}_{\mathbf{y}}^{-1} \mathbf{c}_k}{1 - A_k^2 \mathbf{c}_k^{\text{T}} \mathbf{R}_{\mathbf{y}}^{-1} \mathbf{c}_k} \\
&= \frac{1 - \text{MSE}_k}{\text{MSE}_k}.
\end{aligned} \tag{8.33}
$$

Example 8.3.2 Consider $K = 8$ users with processing gain $N = 12$. We assume that $A_k = 1$ for all users. The spreading codes are given as follows:

$$
\mathbf{C} = \frac{1}{\sqrt{12}}
\begin{bmatrix}
-1 & 1 & -1 & 1 & 1 & -1 & 1 & 1 \\
1 & 1 & -1 & -1 & -1 & -1 & 1 & 1 \\
1 & 1 & 1 & 1 & -1 & 1 & 1 & -1 \\
1 & -1 & -1 & 1 & -1 & 1 & 1 & -1 \\
-1 & -1 & 1 & -1 & -1 & -1 & -1 & -1 \\
1 & -1 & 1 & 1 & -1 & -1 & -1 & 1 \\
1 & 1 & 1 & -1 & 1 & 1 & -1 & 1 \\
1 & -1 & 1 & -1 & 1 & 1 & -1 & 1 \\
-1 & 1 & -1 & 1 & -1 & 1 & -1 & 1 \\
-1 & -1 & 1 & -1 & 1 & -1 & -1 & 1 \\
1 & 1 & -1 & 1 & 1 & -1 & -1 & -1 \\
-1 & -1 & 1 & 1 & -1 & -1 & 1 & -1
\end{bmatrix} . \tag{8.34}
$$

Each column represents a spreading code. Figure 8.6 shows the simulation results. The ML detector provides the best performance; the MMSE and decorrelating detectors provide good performance; and the correlator detector has an error floor. This is mainly due to the interference. When the noise variance approaches zero, the interference becomes dominant and limits the performance of the correlator detector, while the multiuser detectors' BER curves approach zero as they suppress the interference.

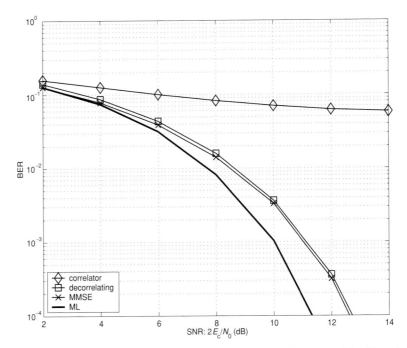

Figure 8.6. Bit error rate performance of the single-user correlator detector and (multiuser) decorrelating, MMSE, and ML detectors in a CDMA system.

8.3.4   Near–far effect

The near–far problem is one of the major drawbacks of CDMA systems. In uplink channels (from mobile users to the basestation), each user signal can be received with a different signal energy. For example, the signals from the users close to the basestation can have much higher energies than the signals from the users far away from the basestation if they transmit signals with the same power. Thus, depending on the users' locations, signals can be strong or weak. This problem is called the near–far problem. To alleviate this problem, the transmit power control can be employed (Viterbi, 1995). However, if the transmit power control cannot be employed in some cases, the multiuser detector can be used as it has a near–far resistance. In particular, the LMMSE detector has a near–far resistance at the expense of the dimension reduction.

To understand the near–far problem and dimension reduction, consider the following received signal vector:

$$\mathbf{y} = \mathbf{CAb} + \mathbf{n}$$

$$= \sum_{k=1}^{K} A_k \mathbf{c}_k b_k + \mathbf{n}. \tag{8.35}$$

Suppose that user k ($k \neq K$) is the desired user and that user K transmits the signal with a much higher power than the others. Let $A_k^2 = P_k$. The covariance matrix of $\mathbf{y}$ is given by

$$\mathbf{R}_\mathbf{y} = \sum_{k=1}^{K} P_k \mathbf{c}_k \mathbf{c}_k^\mathrm{T} + \frac{N_0}{2}\mathbf{I}$$

$$= \underbrace{\sum_{k=1}^{K-1} P_k \mathbf{c}_k \mathbf{c}_k^\mathrm{T} + \frac{N_0}{2}\mathbf{I}}_{=\mathbf{R}_K} + P_K \mathbf{c}_K \mathbf{c}_K^\mathrm{T}.$$

Then, using the matrix inversion lemma, we have

$$\mathbf{R}_\mathbf{y}^{-1} = \mathbf{R}_K^{-1} - \frac{P_K}{1 + P_K \mathbf{c}_K^\mathrm{T} \mathbf{R}_K^{-1} \mathbf{c}_K} \mathbf{R}_K^{-1} \mathbf{c}_K \mathbf{c}_K^\mathrm{T} \mathbf{R}_K^{-1}.$$

The LMMSE estimate of b_k becomes

$$\hat{b}_{\mathrm{lm},k} = \mathbf{w}_k^\mathrm{T} \mathbf{y}$$

$$= A_k \mathbf{c}_k^\mathrm{T} \mathbf{R}_\mathbf{y}^{-1} \mathbf{y}$$

$$= A_k \mathbf{c}_k^\mathrm{T} \left(\mathbf{R}_K^{-1} - \frac{P_K}{1 + P_K \mathbf{c}_K^\mathrm{T} \mathbf{R}_K^{-1} \mathbf{c}_K} \mathbf{R}_K^{-1} \mathbf{c}_K \mathbf{c}_K^\mathrm{T} \mathbf{R}_K^{-1} \right) \mathbf{y}. \qquad (8.36)$$

We can see an asymptotic behavior of the LMMSE estimate as $P_K \to \infty$:

$$\lim_{P_K \to \infty} \hat{b}_{\mathrm{lm},k} = A_k \mathbf{c}_k^\mathrm{T} \left(\mathbf{R}_K^{-1} - \frac{1}{\mathbf{c}_K^\mathrm{T} \mathbf{R}_K^{-1} \mathbf{c}_K} \mathbf{R}_K^{-1} \mathbf{c}_K \mathbf{c}_K^\mathrm{T} \mathbf{R}_K^{-1} \right) \mathbf{y}. \qquad (8.37)$$

We can show that

$$\left(\mathbf{R}_K^{-1} - \frac{1}{\mathbf{c}_K^\mathrm{T} \mathbf{R}_K^{-1} \mathbf{c}_K} \mathbf{R}_K^{-1} \mathbf{c}_K \mathbf{c}_K^\mathrm{T} \mathbf{R}_K^{-1} \right) \mathbf{c}_K = \mathbf{0}.$$

This implies that

$$\lim_{P_K \to \infty} \mathbf{w}_k \perp \mathbf{c}_K = \mathbf{0}, \quad k = 1, 2, \ldots, K-1.$$

As user K becomes a strong interference, the LMMSE weight vector becomes orthogonal to the spreading code of user K to reject its impact.

As the number of strong interfering signals increases, the LMMSE weight vector becomes orthogonal to more spreading codes. Suppose that $P_k = \bar{P} \to \infty$, $k = \bar{K}+1, \bar{K}+2, \ldots, K$. In this case, after repeating the same procedure above, we can conclude that

$$\lim_{\bar{P} \to \infty} \mathbf{w}_k \perp \mathbf{c}_{k'}, \quad k \in \{1, 2, \ldots, \bar{K}\}, k' \in \{\bar{K}+1, \bar{K}+2, \ldots, K\}$$

or

$$\lim_{\bar{P} \to \infty} \mathbf{w}_k \perp \mathrm{Span}\{\mathbf{c}_{\bar{K}+1}, \mathbf{c}_{\bar{K}+2}, \ldots, \mathbf{c}_K\}, \quad k \in \{1, 2, \ldots, \bar{K}\}.$$

As the number of strong interfering signals increases, the LMMSE detector exploits less dimension; this is the so-called dimension reduction. The LMMSE estimate, $\mathbf{w}_k^\mathrm{T} \mathbf{y}$, becomes a combined signal from an $(N - \bar{K})$-dimensional vector space rather than an N-dimensional vector space. Since less dimension is utilized, the SINR would be low, even though strong interfering signals are suppressed. The worst case happens if $\mathbf{c}_k$ belongs to the subspace Span$\{\mathbf{c}_{\bar{K}+1}, \mathbf{c}_{\bar{K}+2}, \ldots, \mathbf{c}_K\}$. In this case, since $\mathbf{c}_k \perp \mathbf{w}_k$ or $\mathbf{w}_k^T \mathbf{c}_k = 0$, the LMMSE estimate $\hat{b}_{\text{lm},k}$ does not have the desired signal:

$$\hat{b}_{\text{lm},k} = \mathbf{w}_k^T \mathbf{y}$$

$$= \mathbf{w}_k^T \left(A_k \mathbf{c}_k b_k + \sum_{k' \neq k}^{K} A_{k'} \mathbf{c}_{k'} b_{k'} + \mathbf{n} \right)$$

$$= \mathbf{w}_k^T \left(\sum_{k' \neq k}^{K} A_{k'} \mathbf{c}_{k'} b_{k'} + \mathbf{n} \right).$$

We can see that the near–far problem still exists in the LMMSE detector, although it is not as severe as in the conventional correlator detector.

8.4 Adaptive detectors

In the LMMSE detector, we need to know all the users' spreading codes or $\mathbf{C}$. In some cases, this could be impractical. To avoid this problem (i.e. to avoid the requirement of $\mathbf{C}$), adaptive techniques can be used.

8.4.1 Adaptive LMMSE detector

Before we start discussing the adaptive LMMSE detector, we need to consider the decomposition of the MSE in Eq. (8.25) into the MSEs for each user. The squared error is given by

$$\|\mathbf{Ab} - \mathbf{Ly}\|^2 = \sum_{k=1}^{K} \left| A_k b_k - \mathbf{g}_k^T \mathbf{y} \right|^2,$$

where $\mathbf{g}_k^T$ represents the kth row vector of $\mathbf{L}$. It follows that

$$\min_{\mathbf{L}} E[\|\mathbf{Ab} - \mathbf{Ly}\|^2] = \min_{\{\mathbf{g}_k\}} E \left[\sum_{k=1}^{K} \left| A_k b_k - \mathbf{g}_k^T \mathbf{y} \right|^2 \right]$$

$$= \sum_{k=1}^{K} \min_{\mathbf{g}_k} E \left[\left| A_k b_k - \mathbf{g}_k^T \mathbf{y} \right|^2 \right]. \tag{8.38}$$

This implies that the LMMSE detector minimizing the overall MSE can be decomposed into the K LMMSE detectors that minimize the MSE for each user *without* any joint processing. If the kth user is the desired user, we only need to find $\mathbf{g}_k$ that minimizes $E[|A_k b_k - \mathbf{g}_k^T \mathbf{y}|^2]$ without attempting to find the other users' LMMSE detectors. In other words, if we determine the following vector:

$$\mathbf{g}_{\text{lm},k} = \arg\min_{\mathbf{g}_k} E\left[\left| A_k b_k - \mathbf{g}_k^T \mathbf{y} \right|^2 \right], \tag{8.39}$$

then $\mathbf{g}_{\text{lm},k}^{\text{T}}$ is the kth row vector of the optimal transform $\mathbf{L}_{\text{lm}}$ and $\mathbf{g}_{\text{lm},k}^{\text{T}}\mathbf{y}$ is the LMMSE estimate of $A_k b_k$. We can use this observation to construct an adaptive LMMSE detector which does not require the spreading codes of the other users.

The LMMSE detector in Eq. (8.39) is used to derive the following adaptive algorithms. We assume that the binary data sequences are transmitted from K users. From Eq. (8.18), the received signal vector sequence is now given by

$$\mathbf{y}_l = \mathbf{CAb}_l + \mathbf{n}_l, \quad l = 0, 1, \ldots, \tag{8.40}$$

where l stands for the discrete-time index, $\mathbf{b}_l = [b_{l,1} \; b_{l,2} \; \cdots \; b_{l,K}]^{\text{T}}$, and $\mathbf{n}_l = [n_{l,0} \; n_{l,1} \; \cdots \; n_{l,N-1}]^{\text{T}}$. Here, $b_{l,k}$ stands for the lth symbol of the kth user. Then, the LMS and RLS algorithms for the MMSE problem in Eq. (8.39) are given by

$$\text{(LMS)} \quad \mathbf{g}_k^{(l)} = \mathbf{g}_k^{(l-1)} + 2\mu \mathbf{y}_l \left(A_k b_{l,k} - \mathbf{y}_l^{\text{T}} \mathbf{g}_k^{(l-1)} \right),$$

$$\text{(RLS)} \quad \mathbf{g}_k^{(l)} = \mathbf{g}_k^{(l-1)} + \mathbf{m}_{l,k} \left(A_k b_{l,k} - \mathbf{y}_l^{\text{T}} \mathbf{g}_k^{(l-1)} \right)$$

$$= \mathbf{g}_k^{(l-1)} + \beta_{l,k} \mathbf{\Phi}_{l-1,k} \mathbf{y}_l \left(A_k b_{l,k} - \mathbf{y}_l^{\text{T}} \mathbf{g}_k^{(l-1)} \right), \tag{8.41}$$

where

$$\beta_{l,k} = \frac{\lambda^{-1}}{1 + \lambda^{-1} \mathbf{y}_l^{\text{T}} \mathbf{\Phi}_{l-1,k} \mathbf{y}_l},$$

$$\mathbf{m}_{l,k} = \beta_{l,k} \mathbf{\Phi}_{l-1,k} \mathbf{y}_l,$$

$$\mathbf{\Phi}_{l,k} = \lambda^{-1} \left(\mathbf{\Phi}_{l-1,k} - \mathbf{m}_{l,k} \mathbf{y}_l^{\text{T}} \mathbf{\Phi}_{l-1,k} \right).$$

As shown in Eqs (8.41), we only need to know the desired user's signal for the LMMSE detector based on Eq. (8.39). When the training sequence is available, the adaptive algorithms can find the vector $\mathbf{g}_k^{(l)}$ to suppress the MUI. It is interesting that the spreading code of the desired user is not required in the adaptive LMMSE detector.

Using the same spreading codes of processing gain 12 (i.e. $N = 12$) in Eq. (8.34) for $K = 8$ users, the LMS and RLS algorithms are applied for the adaptive MMSE receiver. The resulting MSE curves, which are obtained by averaging 100 realizations, are shown in Fig. 8.7. For the LMS algorithm, we have the adaptation gain $\mu = 0.001$, with the forgetting factor $\lambda = 0.99$ and the initial matrix $\mathbf{\Phi}_{0,k} = 0.01\mathbf{I}$. The first user is assumed to be the desired user, i.e. $k = 1$. We can see that both LMS and RLS algorithms converge to the MMSE (which is shown by a dash-dot line), while the RLS algorithm has a faster convergence rate than the LMS algorithm.

8.4.2 Adaptive blind detectors

Both LMS and RLS algorithms require a training or reference sequence to adapt the filtering vector. However, it is often desirable to adapt the detector during the transmission of data sequences, because the channel condition can be changed. For example, new interfering signals can arise and/or some existing interfering signals can drop out. Therefore, the blind adaptation (during the transmission of a data sequence) can be important for the multiuser detection to avoid performance degradation due to changing channel conditions.

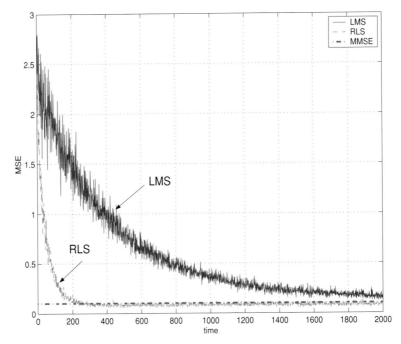

Figure 8.7. The MSE performance of the LMS and RLS algorithms for the adaptive MMSE receiver in a CDMA system.

Minimum variance distortionless detector

The minimum variance distortionless (MVDR) detector uses the MVDR criterion, which has been successfully applied for adaptive arrays (Capon, 1969); see Johnson and Dudgen (1993) and Monzingo and Miller (1980) for detailed accounts of adaptive arrays. A linear detector is used below and the output of the detector is given by

$$d_1 = \mathbf{a}_1^{\mathrm{T}} \mathbf{y}, \tag{8.42}$$

where $\mathbf{a}_1$ is the weight vector that detects the signal, given that user 1 is the desired user. In the conventional correlator detector, $\mathbf{a}_1 = \mathbf{c}_1$ to perform matched filtering. In the MVDR detector, the variance of the output of the linear detector, $E[|d_1|^2]$, is to be minimized without distorting the desired signal by imposing the following constraint:

$$\mathbf{a}_1^{\mathrm{T}} \mathbf{c}_1 = 1. \tag{8.43}$$

Then, it follows that

$$
\begin{aligned}
d_1 &= \mathbf{a}_1^{\mathrm{T}} \mathbf{y} \\
&= \mathbf{a}_1^{\mathrm{T}} A_1 \mathbf{c}_1 b_1 + \mathbf{a}_1^{\mathrm{T}} \left(\sum_{k=2}^{K} A_k \mathbf{c}_k b_k + \mathbf{n} \right) \\
&= \underbrace{A_1 b_1}_{\text{desired signal}} + \mathbf{a}_1^{\mathrm{T}} \left(\sum_{k=2}^{K} A_k \mathbf{c}_k b_k + \mathbf{n} \right).
\end{aligned} \tag{8.44}
$$

The minimization of the output variance minimizes the effect of the interfering signals, while the constraint keeps the desired signal component unchanged. Hence, the optimal MVDR detector is given by

$$\mathbf{a}_{mvdr,1} = \arg\min_{\mathbf{a}_1} E\left[|\mathbf{a}_1^T\mathbf{y}|^2\right]$$
$$\text{subject to} \quad \mathbf{a}_1^T\mathbf{c}_1 = 1. \tag{8.45}$$

The optimal solution of the MVDR detector can be found using the Lagrangian multiplier. The constrained optimization problem in Eq. (8.45) becomes an unconstrained optimization problem using the Lagrangian multiplier as follows:

$$\mathbf{a}_{mvdr,1}(\lambda) = \arg\min_{\mathbf{a}_1}\left\{E\left[|\mathbf{a}_1^T\mathbf{y}|^2\right] + \lambda\mathbf{a}_1^T\mathbf{c}_1\right\}, \tag{8.46}$$

where λ is the Lagrangian multiplier. The optimal solution of Eq. (8.46) depends on λ and is given by

$$\mathbf{a}_{mvdr,1}(\lambda) = -\frac{\lambda}{2}\mathbf{R}_y^{-1}\mathbf{c}_1. \tag{8.47}$$

To find the λ that satisfies the constraint in Eq. (8.45), we write

$$1 = \mathbf{a}_{mvdr,1}^T(\lambda)\mathbf{c}_1 = -\frac{\lambda}{2}\mathbf{c}_1^T\mathbf{R}_y^{-1}\mathbf{c}_1$$

or

$$\lambda = -\frac{2}{\mathbf{c}_1^T\mathbf{R}_y^{-1}\mathbf{c}_1}.$$

This provides the optimal solution of the MVDR detector:

$$\mathbf{a}_{mvdr,1} = \frac{1}{\mathbf{c}_1^T\mathbf{R}_y^{-1}\mathbf{c}_1}\mathbf{R}_y^{-1}\mathbf{c}_1. \tag{8.48}$$

Example 8.4.1 The linear filter $\mathbf{a}_1$ can be found to maximize the SINR or minimize the MSE.

According to Eq. (8.31), the linear vector that maximizes the MSE is given by

$$\mathbf{a}_{lm,1} = A_1\mathbf{R}_y^{-1}\mathbf{c}_1. \tag{8.49}$$

From Eq. (8.48), we find that

$$\mathbf{a}_{mvdr,1} \propto \mathbf{a}_{lm,1}.$$

The two linear filters are identical except for the scaling.

From Eq. (8.44), the SINR of the output of the linear filter $\mathbf{a}_1$ with no constraint is given by

$$\text{SINR}_1 = \frac{A_1^2\mathbf{a}_1^T\mathbf{c}_1\mathbf{c}_1^T\mathbf{a}_1}{\mathbf{a}_1^T(\mathbf{R}_y - A_1^2\mathbf{c}_1\mathbf{c}_1^T)\mathbf{a}_1}.$$

To find a vector that maximizes the SINR, assume that $\mathbf{c}_1^T\mathbf{a}_1 = \alpha_1$, where α_1 is a constant. To maximize the SINR, we need to minimize

$$\mathbf{a}_1^T(\mathbf{R}_y - A_1^2\mathbf{c}_1\mathbf{c}_1^T)\mathbf{a}_1$$

or $\mathbf{a}_1$ should satisfy

$$0 = \left(\mathbf{R_y} - A_1^2 \mathbf{c}_1 \mathbf{c}_1^T\right)\mathbf{a}_1.$$

Since $\mathbf{c}_1^T \mathbf{a}_1 = \alpha_1$, we can show that

$$\mathbf{a}_1 = \alpha_1 A_1^2 \mathbf{R_y}^{-1} \mathbf{c}_1,$$

where α_1 can be any constant except 0. If $\alpha_1 = (A_1^2 \mathbf{c}_1^T \mathbf{R_y}^{-1} \mathbf{c}_1)^{-1}$, the MVDR solution is obtained. On the other hand, if $\alpha_1 = A_1^{-1}$, the MMSE solution is obtained. This shows that both MVDR and MMSE solutions are the maximum SINR solution.

Various adaptive implementations for the MVDR detector are proposed. In particular, the Frost algorithm is well known in antenna arrays (Frost, 1972). To obtain the Frost algorithm, we first consider the cost function of the MVDR detector in Eq. (8.46) in a slightly different form:

$$\mathbf{a}_1^T \mathbf{R_y} \mathbf{a}_1 + \lambda \mathbf{a}_1^T \mathbf{c}_1.$$

We assume that $\mathbf{R_y}$ is *a priori* known. The steepest descent (SD) algorithm can be applied and it provides the recursion with the adaptation gain μ as follows:

$$\mathbf{a}_1^{(n+1)} = \mathbf{a}_1^{(n)} - \mu \nabla_{\mathbf{a}_1} \left(\mathbf{a}_1^T \mathbf{R_y} \mathbf{a}_1 + \lambda \mathbf{a}_1^T \mathbf{c}_1\right)\Bigg|_{\mathbf{a}_1 = \mathbf{a}_1^{(n)}}$$

$$= \mathbf{a}_1^{(n)} - \mu \left(2\mathbf{R_y} \mathbf{a}_1^{(n)} + \lambda \mathbf{c}_1\right). \tag{8.50}$$

The Lagrangian multiplier has to be determined to satisfy the constraint as follows:

$$1 = \mathbf{c}_1^T \mathbf{a}^{(n+1)} = \mathbf{c}_1^T \mathbf{a}^{(n)} - 2\mu \mathbf{c}_1^T \mathbf{R_y} \mathbf{a}^{(n)} - \mu \lambda \|\mathbf{c}_1\|^2.$$

From this, we can find λ, and the recursion is rewritten as follows:

$$\mathbf{a}_1^{(n+1)} = \mathbf{a}_1^{(n)} - 2\mu \left(\mathbf{I} - \frac{1}{\|\mathbf{c}_1\|^2} \mathbf{c}_1 \mathbf{c}_1^T\right) \mathbf{R_y} \mathbf{a}_1^{(n)} + \frac{1}{\|\mathbf{c}_1\|^2} \mathbf{c}_1 \left(1 - \mathbf{c}_1^T \mathbf{a}_1^{(n)}\right). \tag{8.51}$$

Since the covariance matrix $\mathbf{R_y}$ is not available, we replace this matrix by its estimate $\hat{\mathbf{R}}_\mathbf{y}$ in the adaptive implementation from actual received signal vectors. As in the LMS algorithm, consider the received signal vector sequence in Eq. (8.40). For the lth epoch, we have

$$\hat{\mathbf{R}}_\mathbf{y} = \mathbf{y}_l \mathbf{y}_l^T$$

for the lth updating. The resulting adaptive algorithm is given by

$$\mathbf{a}_1^{(l+1)} = \mathbf{a}_1^{(l)} - 2\mu \left(\mathbf{I} - \frac{1}{\|\mathbf{c}_1\|^2} \mathbf{c}_1 \mathbf{c}_1^T\right) \mathbf{y}_l \mathbf{y}_l^T \mathbf{a}_1^{(l)} + \frac{1}{\|\mathbf{c}_1\|^2} \mathbf{c}_1 \left(1 - \mathbf{c}_1^T \mathbf{a}_1^{(l)}\right). \tag{8.52}$$

Let

$$\mathbf{P}_1^\perp = \mathbf{I} - \frac{1}{\|\mathbf{c}_1\|^2} \mathbf{c}_1 \mathbf{c}_1^T. \tag{8.53}$$

This is the orthogonal projection matrix. Then, the adaptive algorithm is rewritten as follows:

$$\mathbf{a}_1^{(l+1)} = \frac{1}{\|\mathbf{c}_1\|^2} \mathbf{c}_1 + \mathbf{P}_1^\perp \left(\mathbf{I} - 2\mu \mathbf{y}_l \mathbf{y}_l^T\right) \mathbf{a}_1^{(l)}.$$

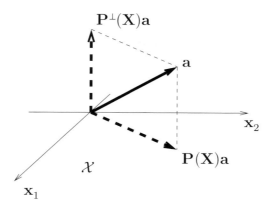

Figure 8.8. Projection of the vector $\mathbf{a}$ onto $\mathcal{X}$, where $\mathbf{X} = [\mathbf{x}_1 \ \mathbf{x}_2]$.

Since the spreading code, $\mathbf{c}_1$, is normalized to yield unit norm, we obtain finally

$$\mathbf{a}_1^{(l+1)} = \mathbf{c}_1 + \mathbf{P}_1^{\perp}\left(\mathbf{I} - 2\mu\mathbf{y}_l\mathbf{y}_l^{\mathrm{T}}\right)\mathbf{a}_1^{(l)}, \tag{8.54}$$

where $\mathbf{P}_1^{\perp} = \mathbf{I} - \mathbf{c}_1\mathbf{c}_1^{\mathrm{T}}$.

Before we move on to another adaptive blind detector, we will review the properties of the orthogonal projection matrix in Eq. (8.53).

(i) $\mathbf{P}^{\perp}(\mathbf{X})\mathbf{X} = \mathbf{0}$ for any tall matrix $\mathbf{X}$ (i.e. one with more rows than columns), where

$$\mathbf{P}^{\perp}(\mathbf{X}) \overset{\triangle}{=} \mathbf{I} - \mathbf{X}(\mathbf{X}^{\mathrm{T}}\mathbf{X})^{-1}\mathbf{X}^{\mathrm{T}}.$$

If $\mathbf{X}^{\mathrm{T}}\mathbf{X}$ does not have an inverse, its pseudo-inverse can replace it.

(ii) $\mathbf{P}^{\perp}(\mathbf{X})\mathbf{P}^{\perp}(\mathbf{X}) = \mathbf{P}^{\perp}(\mathbf{X})$ and $(\mathbf{P}^{\perp}(\mathbf{X}))^{\mathrm{T}} = \mathbf{P}^{\perp}(\mathbf{X})$.

(iii) $\mathbf{P}^{\perp}(\mathbf{X})(\mathbf{I} - \mathbf{P}^{\perp}(\mathbf{X})) = \mathbf{0}$.

(iv) Let $\mathbf{P}(\mathbf{X}) = (\mathbf{I} - \mathbf{P}^{\perp}(\mathbf{X}))$. Then, $\mathbf{P}(\mathbf{X})\mathbf{a}$ is the orthogonal projection of the vector $\mathbf{a}$ onto the subspace of $\mathbf{X}$, $\mathcal{X} = \{\mathbf{x} \mid \mathbf{x} = \mathbf{Xu} \text{ for any } \mathbf{u}\}$.

(v) $\mathbf{P}(\mathbf{X})\mathbf{a}$ and $\mathbf{P}^{\perp}(\mathbf{X})\mathbf{a}$ are orthogonal and

$$\mathbf{P}(\mathbf{X})\mathbf{a} + \mathbf{P}^{\perp}(\mathbf{X})\mathbf{a} = \mathbf{a}.$$

In order to illustrate the idea of the orthogonal projection matrix, an example is shown with $\mathbf{X} = [\mathbf{x}_1 \ \mathbf{x}_2]$ in Fig. 8.8. The two vectors $\mathbf{P}(\mathbf{X})\mathbf{a}$ and $\mathbf{P}^{\perp}(\mathbf{X})\mathbf{a}$ are orthogonal.

Using the properties of the orthogonal projection matrix, we can easily show that

$$\mathbf{P}_1^{\perp}\mathbf{c}_1 = 0.$$

From this, we can verify that the weight vector for the adaptive MVDR detector in Eq. (8.54) satisfies the constraint $\mathbf{c}_1^{\mathrm{T}}\mathbf{a}_1^{(n+1)} = 1$ as follows:

$$\begin{aligned}
\mathbf{c}_1^{\mathrm{T}}\mathbf{a}_1^{(n+1)} &= \mathbf{c}_1^{\mathrm{T}}\mathbf{c}_1 + \mathbf{c}_1^{\mathrm{T}}\mathbf{P}_1^{\perp}\left(\mathbf{I} - 2\mu\mathbf{y}_n\mathbf{y}_n^{\mathrm{T}}\right)\mathbf{a}_1^{(n)} \\
&= 1.
\end{aligned}$$

Minimum mean output energy detector

The minimum mean output energy (MMOE) detector minimizes the output energy of the linear detector by constraining the linear filter (Honig, Madhow, and Verdu, 1995). Assuming that the desired user is user 1, the linear detector is given by

$$\mathbf{a}_{\text{mmoe},1} = \arg\min_{\mathbf{a}_1} E\big[|\mathbf{a}_1^\mathsf{T}\mathbf{y}|^2\big]$$

$$\text{subject to} \quad \mathbf{a}_1 = \mathbf{c}_1 + \mathbf{x}_1, \tag{8.55}$$

where $\mathbf{x}_1$ is orthogonal to $\mathbf{c}_1$. This approach is similar to the MVDR detector, as the desired signal is not distorted by linear filtering. Due to the constraint in Eq. (8.55), we have

$$\mathbf{a}_1^\mathsf{T}\mathbf{y} = (\mathbf{c}_1 + \mathbf{x}_1)^\mathsf{T}\left(A_1\mathbf{c}_1 b_1 + \sum_{k=2}^{K} A_k\mathbf{c}_k b_k + \mathbf{n}\right)$$

$$= A_1 b_1 + (\mathbf{c}_1 + \mathbf{x}_1)^\mathsf{T}\left(\sum_{k=2}^{K} A_k\mathbf{c}_k b_k + \mathbf{n}\right).$$

As shown above, the desired signal $A_1 b_1$ is preserved since $\mathbf{x}_1^\mathsf{T}\mathbf{c}_1 = 0$.

The mean output energy (MOE) is closely related to the MSE. With the constraint in Eq. (8.55), the MSE is given by

$$\text{MSE}(\mathbf{x}_1) = E[|(\mathbf{c}_1 + \mathbf{x}_1)^\mathsf{T}\mathbf{y} - A_1 b_1|^2]$$

$$= E\big[|(\mathbf{c}_1 + \mathbf{x}_1)^\mathsf{T}\mathbf{y}|^2\big] + A_1^2 - 2A_1 E\big[(\mathbf{c}_1 + \mathbf{x}_1)^\mathsf{T}\mathbf{y}b_1|^2\big].$$

Since $E[\mathbf{y}b_1] = A_1\mathbf{c}_1$ and $\mathbf{x}_1^\mathsf{T}\mathbf{c}_1 = 0$, the MSE is rewritten as follows:

$$\text{MSE}(\mathbf{x}_1) = E[|(\mathbf{c}_1 + \mathbf{x}_1)^\mathsf{T}\mathbf{y}|^2] + A_1^2 - 2A_1^2 = E[|(\mathbf{c}_1 + \mathbf{x}_1)^\mathsf{T}\mathbf{y}|^2] - A_1^2,$$

where $\mathbf{c}_1^\mathsf{T}\mathbf{c}_1 = 1$. Hence, we have

$$\text{MSE}(\mathbf{x}_1) = \text{MOE}(\mathbf{x}_1) - A_1^2,$$

where $\text{MOE}(\mathbf{x}_1) = E[|(\mathbf{c}_1 + \mathbf{x}_1)^\mathsf{T}\mathbf{y}|^2]$.

We can derive an LMS-like adaptive algorithm using the MMOE criterion. The gradient vector is given by

$$\nabla_{\mathbf{x}_1} E[|(\mathbf{c}_1 + \mathbf{x}_1)^\mathsf{T}\mathbf{y}|^2] = 2E[\mathbf{y}\mathbf{y}^\mathsf{T}(\mathbf{c}_1 + \mathbf{x}_1)].$$

Since $\mathbf{x}_1$ is orthogonal to $\mathbf{c}_1$, the orthogonal projection must apply; it gives

$$\mathbf{P}_1^\perp \nabla_{\mathbf{x}_1} E[|(\mathbf{c}_1 + \mathbf{x}_1)^\mathsf{T}\mathbf{y}|^2] = 2E\big[\mathbf{P}_1^\perp \mathbf{y}\mathbf{y}^\mathsf{T}(\mathbf{c}_1 + \mathbf{x}_1)\big]$$

$$= 2E\big[(\mathbf{y} - (\mathbf{c}_1^\mathsf{T}\mathbf{y})\mathbf{c}_1)\mathbf{y}^\mathsf{T}(\mathbf{c}_1 + \mathbf{x}_1)\big]. \tag{8.56}$$

Hence, the SD algorithm for the MMOE detector is given by

$$\mathbf{x}^{(n+1)} = \mathbf{x}^{(n)} - 2\mu E\Big[\Big(\mathbf{y} - \big(\mathbf{c}_1^\mathsf{T}\mathbf{y}\big)\mathbf{c}_1\Big)\mathbf{y}^\mathsf{T}\Big(\mathbf{c}_1 + \mathbf{x}_1^{(n)}\Big)\Big].$$

After removing the expection, the LMS-like adaptive agorithm for the received signal vector

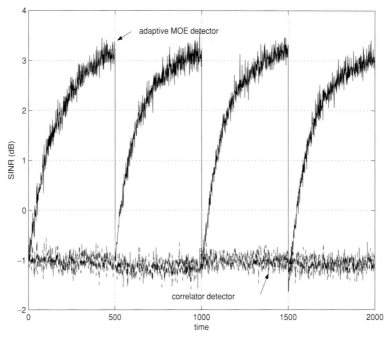

Figure 8.9. Bit error rate performance of the adaptive MOE detector with five users. The fifth user has four times higher power than the other users (i.e. $A_1 = \cdots = A_4 = 1$ and $A_5 = 2$) and changes its spreading code every 500 symbols.

sequence in Eq. (8.40) is given by

$$\mathbf{x}^{(l+1)} = \mathbf{x}^{(l)} - 2\mu(\mathbf{y}_l - z_l\mathbf{c}_1)\mathbf{y}_l^{\mathrm{T}}\left(\mathbf{c}_1 + \mathbf{x}_1^{(l)}\right), \tag{8.57}$$

where $z_l = \mathbf{c}_1^{\mathrm{T}}\mathbf{y}_l$ is the output of the matched filter with $\mathbf{c}_1$.

Simulation results for the adaptive MOE detector are shown in Fig. 8.9 with $\mu = 10^{-3}$ and SNR $= 1/N_0 = 10$ dB. It is assumed that there are five users with $N = 5$. The first four users have the following spreading codes:

$$[\mathbf{c}_1 \ \ \mathbf{c}_2 \ \ \mathbf{c}_3 \ \ \mathbf{c}_4] = \frac{1}{\sqrt{5}}\begin{bmatrix} 1 & 1 & 1 & 1 \\ 1 & -1 & -1 & -1 \\ 1 & -1 & 1 & -1 \\ 1 & -1 & -1 & 1 \\ 1 & -1 & -1 & -1 \end{bmatrix}.$$

The fifth user has a higher power than the other users as $A_5 = 2$ (while $A_1 = \cdots = A_4 = 1$) and changes its spreading code randomly every 500 symbols. The fifth user can be considered to be a new user who is active and passive every 500 symbols. As shown in Fig. 8.9, the adaptive MOE detector exhibits a poor performance when the fifth user changes its spreading code. However, the performance can be improved through adaptation. The correlator detector cannot provide a satisfactory performance as there are five users with $N = 5$ (a full load).

8.5   Summary and notes

CDMA systems and multiuser detection were studied in this chapter. Adaptive methods were also discussed for the multiuser detection.

For cellular systems, CDMA has a pre-eminent role. Several standards for cellular communications adopt CDMA as a key access scheme. A comprehensive treatment for cellular CDMA can be found in Viterbi (1995).

Multiuser detection has been a fascinating research area since Verdu (1986) was published. There have been many attempts to design and analyze multiuser detectors; see Verdu (1998) for details. Comphrehensive tutorials can also be found in Duel-Hallen, Holtzman, and Zvonar (1995) and Moshavi (1996). An overview of adaptive techniques for multiuser detection is presented in Honig and Trutsanis (2000).

We have ignored multipaths in this chapter. It should be noted, however, that in practical situations multipaths should be taken into account. See Xu and Tsatsanis (2001) for an investigation of adaptive algorithms based on the MVDR criterion for multipath channels. An adaptive CDMA receiver over multipath channels incorporating a decison feedback structure is derived in Choi, Kim, and Lim (2004).

9 Iterative CDMA receivers

In CDMA systems, the MAI degrades the performance. Various approaches may be teken to mitigate the MAI. For example, in linear multiuser detection, a linear transform or filtering is employed to mitigate the MAI. In this case, a dimension reduction occurs to suppress interfering signals in a vector space. Thus, if there are many interfering signals, linear multiuser detection becomes inefficient. Alternatively, the cancelation approach can be adopted. If it is possible to estimate interfering signals, they can be canceled without any dimension reduction. This approach becomes effective when the estimate of interfering signals is precise.

Generally, for coded CDMA systems, an optimal receiver has a prohibitively high computational complexity since joint detection and channel decoding is required. To avoid joint detection and channel decoding without siginificant performance degradation, an iterative receiver can be considered. In general, an iterative CDMA receiver consists of a multiuser detector and channel decoders. For multiuser detection, the MAP symbol detector can be an ideal choice. However, it would be impractical for a large number of users due to the high complexity (the complexity grows exponentially with the number of users). Therefore, a computationally efficient suboptimal multiuser detector is desirable. A cancelation-based approach would be preferable. Since decoded signals are more reliable, cancelation can effectively eliminate interfering signals whilst ensuring low complexity.

In this chapter, we study coded CDMA systems and iterative receivers with computationally efficient multiuser detectors. We also discuss multiuser detection with unknown interferers in the context of the iterative receiver.

9.1 Structure of iterative receivers

The iterative receiver may be considered for coded signals as it achieves good performance at a relatively low complexity. A block diagram of coded CDMA systems is shown in Fig. 9.1. Each user transmits an (interleaved) coded sequence with a different spreading code. At a receiver, the received signal becomes a sum of the spread signals and the background white Gaussian noise.

Figure 9.2 shows a block diagram of the iterative CDMA receiver. A multiuser detector provides soft-decisions of all users' signals. In general, it is desirable to exploit the *a priori* probabilities of symbols (from the extrinsic information obtained by the channel decoder)

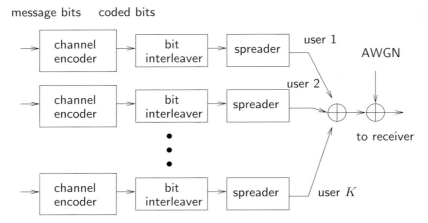

Figure 9.1. Coded CDMA systems.

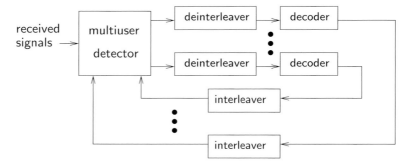

Figure 9.2. Block diagram of the iterative receiver with a multiuser detector.

for better soft-decisions through iterations. In general, the operation of the iterative receiver is the same as that for ISI channels; see Chapter 6 for details.

In Chapter 8, we discussed various linear multiuser detectors. Unfortunately, these linear multiuser detectors, including the decorrelating and MMSE detectors, are not suitable for the iterative receiver because they cannot take advantage of the feedback from channel decoders. A simple but effective approach to exploit the feedback from channel decoders is cancelation, which does not introduce dimension reduction. Furthermore, cancelation is a low complexity operation. Thus, the multiuser detector for the iterative receiver is based primarily on the cancelation, as we will see later.

9.2   MAP detection in iterative receivers

In this section, we discuss an optimal detector, the MAP (symbol) detector in CDMA. The MAP detector finds the LAPP or LLR. In Section 9.3, we approximate the LLR to derive suboptimal but much less complex detectors.

Recall the received signal after chip rate sampling. The lth received signal vector is given by

$$\mathbf{y}_l = \mathbf{CAb}_l + \mathbf{n}_l, \quad l = 0, 1, \ldots, L - 1, \tag{9.1}$$

where $\mathbf{C}$ is a matrix whose column vectors are spreading codes, $\mathbf{A}$ is a diagonal matrix of symbol amplitudes, L is the length of a coded symbol sequence, and $\mathbf{b}_l$ is the lth symbol vector. Here, $\mathbf{n}_l$ is a white Gaussian noise vector with $E[\mathbf{n}_l] = \mathbf{0}$ and $E[\mathbf{n}_l \mathbf{n}_l^\mathsf{T}] = (N_0/2)\mathbf{I}$. It is assumed that there are K users and that the spreading gain is N for all users. Hence, the sizes of matrices $\mathbf{C}$ and $\mathbf{A}$ are $N \times K$ and $K \times K$, respectively, and the sizes of vectors $\mathbf{y}_l$, $\mathbf{b}_l$, and $\mathbf{n}_l$ are $N \times 1$, $K \times 1$, and $N \times 1$, respectively. We assume that $b_{k,l}$, denoting the lth symbol of user k, is a coded symbol and $b_{k,l} \in \{-1, +1\}$.

The MAP detection finds the LAPP as follows:

$$L(b_{k,l}) = \log \frac{\Pr(b_{k,l} = +1|\mathbf{y}_l)}{\Pr(b_{k,l} = -1|\mathbf{y}_l)}. \tag{9.2}$$

For convenience, we omit the time index l. The *a posteriori* probability $\Pr(b_k|\mathbf{y})$ is given by

$$\Pr(\mathbf{b}|\mathbf{y}) = \frac{f(\mathbf{y}|\mathbf{b})\Pr(\mathbf{b})}{f(\mathbf{y})},$$
$$\propto f(\mathbf{y}|\mathbf{b})\Pr(\mathbf{b}), \tag{9.3}$$

where $f(\mathbf{y}|\mathbf{b})$ is the likelihood function and $\Pr(\mathbf{b})$ is the *a priori* probability of $\mathbf{b}$. Using the LLR from the MAP decoder as the extrinsic information, the *a priori* probability of $\mathbf{b}$ can be obtained within the iterative receiver. From the *a posteriori* probability of $\mathbf{b}$, we can find the *a posteriori* probability of b_k as follows:

$$\Pr(b_k = +1|\mathbf{y}) = \sum_{\mathbf{b} \in \mathcal{B}_k^+} \Pr(\mathbf{b}|\mathbf{y}),$$
$$\Pr(b_k = -1|\mathbf{y}) = \sum_{\mathbf{b} \in \mathcal{B}_k^-} \Pr(\mathbf{b}|\mathbf{y}), \tag{9.4}$$

where $\mathcal{B}_k^+$ and $\mathcal{B}_k^-$ are the sets of the binary vectors defined by $\mathcal{B}_k^+ = \{[b_1\ b_2\ \cdots\ b_K]^\mathsf{T} \mid b_k = 1, b_m \in \{+1, -1\}, \forall m \neq k\}$ and $\mathcal{B}_k^- = \{[b_1\ b_2\ \cdots\ b_K]^\mathsf{T} \mid b_k = -1, b_m \in \{+1, -1\}, \forall m \neq k\}$, respectively. Since the likelihood function is given by

$$f(\mathbf{y}|\mathbf{b}) = C \exp\left(-\frac{1}{N_0}||\mathbf{y} - \mathbf{CAb}||^2\right), \tag{9.5}$$

we can show that

$$\Pr(b_k = +1|\mathbf{y}) \propto \sum_{\mathbf{b} \in \mathcal{B}_k^+} \exp\left(-\frac{1}{N_0}||\mathbf{y} - \mathbf{CAb}||^2\right) \Pr(\mathbf{b}),$$
$$\Pr(b_k = -1|\mathbf{y}) \propto \sum_{\mathbf{b} \in \mathcal{B}_k^-} \exp\left(-\frac{1}{N_0}||\mathbf{y} - \mathbf{CAb}||^2\right) \Pr(\mathbf{b}). \tag{9.6}$$

It is assumed that the b_k's are independent (resulting in $\Pr(\mathbf{b}) = \prod_{k=1}^{K} \Pr(b_k)$). Finally, the LAPP is given by

$$L(b_k) = \log \frac{\sum_{\mathbf{b} \in \mathcal{B}_k^+} \exp\left(-\frac{1}{N_0}||\mathbf{y} - \mathbf{CAb}||^2\right) \Pr(\mathbf{b})}{\sum_{\mathbf{b} \in \mathcal{B}_k^-} \exp\left(-\frac{1}{N_0}||\mathbf{y} - \mathbf{CAb}||^2\right) \Pr(\mathbf{b})}. \tag{9.7}$$

The LLR to the channel decoder is given by

$$\mathrm{LLR}(b_k) = L(b_k) - L_{\mathrm{pi}}(b_k), \tag{9.8}$$

where

$$L_{\mathrm{pi}}(b_k) = \log \frac{\Pr(b_k = +1)}{\Pr(b_k = -1)}.$$

Since there are 2^{K-1} binary vectors in $\mathcal{B}_k^+$ or $\mathcal{B}_k^-$, the resulting complexity in obtaining the *a posteriori* probability of b_k is $O(2^K)$. Obviously, the complexity is prohibitively high as K increases. Approximations that reduce the computational complexity are therefore required.

Note that instead of computing the summation in Eq. (9.7), the maximum can be used and it results in an approximation of the LAPP as follows:

$$L_{\max}(b_k) = \log \frac{\max_{\mathbf{b} \in \mathcal{B}_k^+} \exp\left(-\frac{1}{N_0}||\mathbf{y} - \mathbf{CAb}||^2\right) \Pr(\mathbf{b})}{\max_{\mathbf{b} \in \mathcal{B}_k^-} \exp\left(-\frac{1}{N_0}||\mathbf{y} - \mathbf{CAb}||^2\right) \Pr(\mathbf{b})}. \tag{9.9}$$

Even though this can reduce the complexity in comparison with Eq. (9.7), the order of the computational complexity is the same since the binary vector that maximizes the *a posteriori* probability is found by an exhaustive search with a set of 2^{K-1} binary vectors for each case.

Example 9.2.1   A numerical example is considered to show how the LLR can be found.

Suppose there are $K = 3$ users and that the processing gain N is 3. The spreading codes are given by

$$\mathbf{C} = [\mathbf{c}_1 \ \mathbf{c}_2 \ \mathbf{c}_3] = \frac{1}{\sqrt{3}} \begin{bmatrix} 1 & 1 & -1 \\ 1 & -1 & -1 \\ 1 & 1 & 1 \end{bmatrix}.$$

Assume that $A_k = 1$, $\forall k$, and $N_0 = 1$. The *a priori* probabilities are given by

$$\Pr(b_1 = 1) = 0.4, \ \Pr(b_2 = 1) = 0.6, \ \Pr(b_3 = 1) = 0.1.$$

For all the symbol combinations, it is necessary to find the joint *a priori* probability and the likelihood. They are shown in Table 9.1.

Table 9.1. *The joint* a priori *probability and the squared distance,* $||\mathbf{y} - \mathbf{Cb}||^2$, *for likelihood*

$\{b_1, b_2, b_3\}$	$||\mathbf{y} - \mathbf{Cb}||^2$	$\Pr(\mathbf{b})$
$\{+1, +1, +1\}$	2.7385	0.0240
$\{+1, +1, -1\}$	2.7385	0.2160
$\{+1, -1, +1\}$	9.3094	0.0160
$\{+1, -1, -1\}$	11.9761	0.1440
$\{-1, +1, +1\}$	7.3573	0.0360
$\{-1, +1, -1\}$	4.6906	0.3240
$\{-1, -1, +1\}$	16.5949	0.0240
$\{-1, -1, -1\}$	16.5949	0.2160

The LAPPs are found from Table 9.1 as follows:

$$L(b_1) = 1.6445,$$
$$L(b_2) = 8.9635,$$
$$L(b_3) = -2.3747.$$

The LLRs become

$$\text{LLR}(b_1) = L(b_1) - L_{\text{pi}}(b_1) = 2.0500,$$
$$\text{LLR}(b_2) = L(b_2) - L_{\text{pi}}(b_2) = 8.5581,$$
$$\text{LLR}(b_3) = L(b_3) - L_{\text{pi}}(b_3) = -0.1775.$$

9.3 Approximate MAP detection within iterative receivers

Since the trellis structure does not exist in CDMA channels, the BCJR algorithm cannot be applied to the MAP detection. Therefore, we need to consider approximations which do not rely on the trellis structure. In this section, we will study approximations of the LAPP to derive computationally efficient methods for the MAP detection.

9.3.1 First-order approximation

In this subsection, we derive a first-order approximation of the LLR using the Taylor series.

Let $\mathbf{c}_k$ and A_k denote the kth column vector of $\mathbf{C}$ (i.e. the spreading code for user k) and the (k, k)th element of $\mathbf{A}$ (i.e. the amplitude for user k), respectively. Define

$$\mathbf{y}_k^+ = \mathbf{y} - \mathbf{c}_k A_k$$

and a $(K - 1) \times 1$ binary vector as follows:

$$\mathbf{b}_k = [b_1 \ \cdots \ b_{k-1} \ b_{k+1} \ \cdots \ b_K]^\text{T}.$$

Assume that $b_k = 1$ to find $\Pr(b_k = 1|\mathbf{y})$. In this case, we can show that

$$\mathbf{y} - \mathbf{CAb} = \mathbf{y}_k^+ - \mathbf{C}_k A_k \mathbf{b}_k,$$

where $\mathbf{C}_k$ and $\mathbf{A}_k$ are the submatrices of $\mathbf{C}$ and $\mathbf{A}$ obtained by deleting the kth column vectors, respectively. It follows that

$$
\begin{aligned}
\Pr(b_k = 1|\mathbf{y}) &\propto \sum_{\mathbf{b}\in\mathcal{B}_k^+} \exp\left(-\frac{1}{N_0}||\mathbf{y} - \mathbf{C}\mathbf{A}\mathbf{b}||^2\right) \Pr(\mathbf{b}) \\
&= \sum_{\mathbf{b}\in\mathcal{B}^{K-1}} \exp\left(-\frac{1}{N_0}||\mathbf{y}_k^+ - \mathbf{C}_k\mathbf{A}_k\mathbf{b}_k||^2\right) \Pr(\mathbf{b}_k)\Pr(b_k = 1) \\
&= E_{\mathbf{b}_k}\left[\exp\left(-\frac{1}{N_0}||\mathbf{y}_k^+ - \mathbf{C}_k\mathbf{A}_k\mathbf{b}_k||^2\right)\right]\Pr(b_k = 1), \quad (9.10)
\end{aligned}
$$

where $E_{\mathbf{b}_k}[\cdot]$ represents the expectation with respect to $\mathbf{b}_k$.

Let

$$
v_+(\mathbf{b}_k) = \exp\left(-\frac{1}{N_0}||\mathbf{y}_k^+ - \mathbf{C}_k\mathbf{A}_k\mathbf{b}_k||^2\right).
$$

Then, using a Taylor series, we can show that

$$
v_+(\mathbf{b}_k) = v_+(\check{\mathbf{b}}_k) + \nabla_{v,+}^{\mathrm{T}}(\check{\mathbf{b}}_k)(\mathbf{b}_k - \check{\mathbf{b}}_k) + \cdots, \quad (9.11)
$$

where $\nabla_{v,+}$ is the gradient of $v_+(\mathbf{b}_k)$, which is given by

$$
\nabla_{v,+}(\mathbf{x}) = \left[\frac{\partial v_+(\mathbf{x})}{\partial x_1} \quad \frac{\partial v_+(\mathbf{x})}{\partial x_2} \quad \cdots \quad \frac{\partial v_+(\mathbf{x})}{\partial x_{K-1}}\right]^{\mathrm{T}}
$$

and $\check{\mathbf{b}}_k$ is any $(K-1) \times 1$ vector. If $\check{\mathbf{b}}_k = \bar{\mathbf{b}}_k$, where $E[\mathbf{b}_k] = \bar{\mathbf{b}}_k$, we can show that

$$
\begin{aligned}
E[v_+(\mathbf{b}_k)] &= v_+(\bar{\mathbf{b}}_k) + E[\nabla_{v,+}^{\mathrm{T}}(\bar{\mathbf{b}}_k)(\mathbf{b}_k - \bar{\mathbf{b}}_k)] + \cdots \\
&= v_+(\bar{\mathbf{b}}_k) + 0 + \cdots,
\end{aligned}
$$

where the first-order term becomes zero. This Taylor series leads to the following approximation:

$$
\begin{aligned}
\Pr(b_k = 1|\mathbf{y}) &\simeq C \times v_+(\bar{\mathbf{b}}_k)\Pr(b_k = 1) \\
&= C \times \exp\left(-\frac{1}{N_0}||\mathbf{y}_k^+ - \mathbf{C}_k\mathbf{A}_k\bar{\mathbf{b}}_k||^2\right)\Pr(b_k = 1), \quad (9.12)
\end{aligned}
$$

where C is a constant for all k.

We can approximate $\Pr(b_k = -1|\mathbf{y})$ using the same approach. It results in the following approximate LAPP:

$$
\begin{aligned}
L(b_k) &\simeq \log \frac{\exp\left(-\frac{1}{N_0}||\mathbf{y}_k^+ - \mathbf{C}_k\mathbf{A}_k\bar{\mathbf{b}}_k||^2\right)}{\exp\left(-\frac{1}{N_0}||\mathbf{y}_k^- - \mathbf{C}_k\mathbf{A}_k\bar{\mathbf{b}}_k||^2\right)} + L_{\mathrm{pi}}(b_k) \\
&= \frac{1}{N_0}\left(4A_k\mathbf{c}_k^{\mathrm{T}}\mathbf{y} - 4A_k\mathbf{c}_k^{\mathrm{T}}\mathbf{C}_k\mathbf{A}_k\bar{\mathbf{b}}_k\right) + L_{\mathrm{pi}}(b_k) \\
&= \frac{4A_k}{N_0}\mathbf{c}_k^{\mathrm{T}}(\mathbf{y} - \mathbf{C}_k\mathbf{A}_k\bar{\mathbf{b}}_k) + L_{\mathrm{pi}}(b_k), \quad (9.13)
\end{aligned}
$$

where $\mathbf{y}_k^- = \mathbf{y} + \mathbf{c}_k A_k$. Then, the LLR is approximated by

$$\mathrm{LLR}(b_k) = L(b_k) - L_{\mathrm{pi}}(b_k)$$

$$\simeq \mathrm{LLR}_{\mathrm{mth}}(b_k)$$

$$\overset{\triangle}{=} \frac{4A_k}{N_0} \mathbf{c}_k^{\mathrm{T}} (\mathbf{y} - \mathbf{C}_k A_k \bar{\mathbf{b}}_k). \tag{9.14}$$

Hence, this approximate LLR can be considered as a scaled output of the matched filter with $\mathbf{c}_k$ when the input is $\mathbf{y} - \mathbf{C}_k A_k \bar{\mathbf{b}}_k$.

We can see that $\mathbf{y} - \mathbf{C}_k A_k \bar{\mathbf{b}}_k$ is the output of SC. If the cancelation is perfect, the matched filter with $\mathbf{c}_k$ becomes optimal (to maximize the SNR). The resulting detector will be referred to as the matched filter (MF) detector with SC or the MF-SC detector. The MF-SC detector has a low complexity as it only requires the cancelation and matched filtering. The performance of the MF-SC detector relies on SC. Hence, within the iterative receiver, a good performance is expected after several iterations as a more reliable decoding result would be available.

9.3.2 Second-order approximation

For a better approximation, the second-order term can be included, which results in a second-order approximation of the LLR. From Eq. (9.11), letting $\check{\mathbf{b}}_k = \bar{\mathbf{b}}_k$, we can show that

$$v_+(\mathbf{b}_k) = v_+(\bar{\mathbf{b}}_k) + \frac{1}{2}(\mathbf{b}_k - \bar{\mathbf{b}}_k)^{\mathrm{T}} \mathbf{H}_+(\bar{\mathbf{b}}_k)(\mathbf{b}_k - \bar{\mathbf{b}}_k) + \cdots, \tag{9.15}$$

where $\mathbf{H}_+(\bar{\mathbf{b}}_k)$ is the Hessian matrix at $\mathbf{b}_k = \bar{\mathbf{b}}_k$. For notational convenience, let

$$\mathbf{a} = \mathbf{b}_k,$$
$$\bar{\mathbf{a}} = \bar{\mathbf{b}}_k,$$
$$\mathbf{x}^+ = \mathbf{y}_k^+,$$
$$\mathbf{S} = \mathbf{C}_k A_k.$$

Then, we have

$$v_+(\mathbf{b}_k) = v_+(\mathbf{a}) = \exp\left(-\frac{1}{N_0}||\mathbf{x}^+ - \mathbf{S}\mathbf{a}||^2\right).$$

After some manipulation, the Hessian matrix of $v_+(\mathbf{a})$ can be found as follows:

$$\mathbf{H}_+(\mathbf{a}) = \left(\frac{2}{N_0}\right) v_+(\mathbf{a}) \left[\left(\frac{2}{N_0}\right) \mathbf{u}^+(\mathbf{u}^+)^{\mathrm{T}} - \mathbf{S}^{\mathrm{T}}\mathbf{S}\right], \tag{9.16}$$

where $[\mathbf{H}_+(\mathbf{a})]_{p,q} = (\partial^2/\partial a_p \partial a_q) v_+(\mathbf{a})$ and $\mathbf{u}^+ = \mathbf{S}^{\mathrm{T}}(\mathbf{S}\mathbf{a} - \mathbf{x}^+)$.

With a second-order approximation of $v_+(\mathbf{a})$ from Eq. (9.15), we have

$$E[v_+(\mathbf{a})] \simeq v_+(\bar{\mathbf{a}}) + \frac{1}{2} E[\tilde{\mathbf{a}}^{\mathrm{T}} \mathbf{H}_+(\bar{\mathbf{a}})\tilde{\mathbf{a}}], \tag{9.17}$$

where $\tilde{\mathbf{a}} = \mathbf{b}_k - \bar{\mathbf{b}}_k$. Let $\bar{\mathbf{u}}^+ = \mathbf{S}^{\mathsf{T}}(\mathbf{S}\bar{\mathbf{a}} - \mathbf{x}^+)$. From Eq. (9.16), it follows that

$$E[\tilde{\mathbf{a}}^{\mathsf{T}}\mathbf{H}_+(\bar{\mathbf{a}})\tilde{\mathbf{a}}] = \left(\frac{2}{N_0}\right) v_+(\bar{\mathbf{a}}) \left(\left(\frac{2}{N_0}\right) E[\tilde{\mathbf{a}}^{\mathsf{T}}\bar{\mathbf{u}}^+(\bar{\mathbf{u}}^+)^{\mathsf{T}}\tilde{\mathbf{a}}] - E[\tilde{\mathbf{a}}^{\mathsf{T}}\mathbf{S}^{\mathsf{T}}\mathbf{S}\tilde{\mathbf{a}}]\right).$$

We now need to find $E[\tilde{\mathbf{a}}^{\mathsf{T}}\bar{\mathbf{u}}^+(\bar{\mathbf{u}}^+)^{\mathsf{T}}\tilde{\mathbf{a}}]$ and $E[\tilde{\mathbf{a}}^{\mathsf{T}}\mathbf{S}^{\mathsf{T}}\mathbf{S}\tilde{\mathbf{a}}]$. Since

$$E[\tilde{\mathbf{a}}\tilde{\mathbf{a}}^{\mathsf{T}}] = \mathrm{Diag}\left(E[|\tilde{a}_1|^2], E[|\tilde{a}_2|^2], \ldots, E[|\tilde{a}_{K-1}|^2]\right),$$

where $\tilde{a}_p$ denotes the pth element of $\tilde{\mathbf{a}}$, it can be shown that

$$E[\tilde{\mathbf{a}}^{\mathsf{T}}\bar{\mathbf{u}}^+(\bar{\mathbf{u}}^+)^{\mathsf{T}}\tilde{\mathbf{a}}] = \sum_p E[|\tilde{a}_p|^2]|\bar{u}_p^+|^2, \qquad (9.18)$$

where $\bar{u}_p^+$ denotes the pth element of $\bar{\mathbf{u}}^+$, and

$$E[\tilde{\mathbf{a}}^{\mathsf{T}}\mathbf{S}^{\mathsf{T}}\mathbf{S}\tilde{\mathbf{a}}] = \sum_p E[|\tilde{a}_p|^2][\mathbf{S}^{\mathsf{T}}\mathbf{S}]_{p,p}. \qquad (9.19)$$

Here, $[\mathbf{S}^{\mathsf{T}}\mathbf{S}]_{p,p}$ stands for the (p, p)th element of $\mathbf{S}^{\mathsf{T}}\mathbf{S}$. From Eqs (9.18) and (9.19), we can readily show that

$$E[\tilde{\mathbf{a}}^{\mathsf{T}}\mathbf{H}_+(\bar{\mathbf{a}})\tilde{\mathbf{a}}] = v_+(\bar{\mathbf{a}})\left(\frac{2}{N_0}\right)\phi^+, \qquad (9.20)$$

where

$$\phi^+ = \sum_p E[|\tilde{a}_p|^2]\left(\frac{2}{N_0}|\bar{u}_p^+|^2 - [\mathbf{S}^{\mathsf{T}}\mathbf{S}]_{p,p}\right).$$

Substituting Eq. (9.20) into Eq. (9.17), we have

$$E[v_+(\mathbf{a})] \simeq \bar{v}_{2\mathrm{nd},+} \overset{\triangle}{=} v_+(\bar{\mathbf{a}})\left(1 + \frac{\phi^+}{N_0}\right). \qquad (9.21)$$

With $\mathbf{x}^- = \mathbf{y}_k^-$, the second-order approximation of $E[v_-(\mathbf{b}_k)]$ can also be obtained by the same method. We denote by $\bar{v}_{2\mathrm{nd},-}$ the second-order approximation of $E[v_-(\mathbf{b}_k)]$. Then, a second-order approximate LLR can be written as follows:

$$\begin{aligned} \mathrm{LLR}(b_k) &\simeq \mathrm{LLR}_{2\mathrm{nd}}(b_k) \\ &\overset{\triangle}{=} \log \frac{\bar{v}_{2\mathrm{nd},+}}{\bar{v}_{2\mathrm{nd},-}} \\ &= \mathrm{LLR}_{\mathrm{mth}}(b_k) + \Delta_{2\mathrm{nd}}(b_k), \end{aligned} \qquad (9.22)$$

where

$$\Delta_{2\mathrm{nd}}(b_k) = \log \frac{1 + (\phi^+/N_0)}{1 + (\phi^-/N_0)}. \qquad (9.23)$$

Here,

$$\phi^- = \sum_p E[|\tilde{a}_p|^2]\left(\frac{2}{N_0}|\bar{u}_p^-|^2 - [\mathbf{S}^{\mathsf{T}}\mathbf{S}]_{p,p}\right),$$

where $\bar{u}_p^-$ stands for the pth element of $\bar{\mathbf{u}}^-$ given by

$$\bar{\mathbf{u}}^- = \mathbf{S}^{\mathsf{T}}(\mathbf{S}\bar{\mathbf{a}} - \mathbf{x}^-).$$

As shown in Eq. (9.22), the second-order approximation of the LLR requires a minor additional increase in complexity to find $\Delta_{2\text{nd}}(b_k)$ from the MF-SC detector. This detector will be referred to as the MF-SC2 detector.

9.4 MMSE-SC detection within iterative receivers

In Chapter 6, we introduced the MMSE-SC detector[†] for channel equalization within the iterative receiver. The same approach can be used for the multiuser detection to mitigate the MAI effectively. In this section, we study the MMSE-SC detector as a multiuser detector for the iterative receiver.

9.4.1 MMSE-SC and Gaussian approximation for the LLR

In Eq. (9.13), there exists the residual interference after SC. Since this residual interference can degrade the performance, it would be desirable to mitigate further if possible. In Wang and Poor (1999), the use of the MMSE filter is proposed to reduce the residual interference. The resulting detector is called the MMSE detector with SC or simply the MMSE-SC detector. After applying MMSE filtering to mitigate the residual interference, the output of the MMSE filter is assumed to be a Gaussian random variable in order to find the LLR of b_k.

In summary, the operation of the MMSE-SC detector that provides an approximate LLR is given in the following:

(F1) The MMSE filtering is applied after SC. The output of the (instantaneous) MMSE filter used to reduce the residual interference is given by

$$\hat{b}_{\text{mmse},k} = \mathbf{g}_k^{\text{T}} \left(\mathbf{y} - \mathbf{C}_k \mathbf{A}_k \bar{\mathbf{b}}_k \right), \tag{9.24}$$

where $\mathbf{g}_k$ is the MMSE filtering vector, which will be found later.

(F2) This output, $\hat{b}_{\text{mmse},k}$, is assumed to be an output of the AWGN channel as follows:

$$\hat{b}_{\text{mmse},k} = \mu_k b_k + \eta_k, \tag{9.25}$$

where μ_k is the channel gain and η_k is the AWGN. This assumption is essential when we obtain the LLR, which becomes the input to the following channel decoder. From Eq. (9.25), the LLR of b_k from $\hat{b}_{\text{mmse},k}$ is given by

$$\begin{aligned}
\text{LLR}(b_k; \hat{b}_{\text{mmse},k}) &= \log \frac{\Pr(b_k = +1 | \hat{b}_{\text{mmse},k})}{\Pr(b_k = -1 | \hat{b}_{\text{mmse},k})} - L_{\text{pi}}(b_k) \\
&= \log \frac{f(\hat{b}_{\text{mmse},k} | b_k = +1)}{f(\hat{b}_{\text{mmse},k} | b_k = -1)} \\
&= \frac{2\mu_k}{\sigma_{\eta,k}^2} \hat{b}_{\text{mmse},k}, \tag{9.26}
\end{aligned}$$

where $\sigma_{\eta,k}^2$ is the variance of η_k.

[†] The MMSE-SC detector was originally proposed for multiuser detection within the iterative receiver in CDMA systems (Wang and Poor, 1999).

To find the MMSE filter, define

$$\mathbf{y}_k = \mathbf{y} - \mathbf{C}_k \mathbf{A}_k \bar{\mathbf{b}}_k. \tag{9.27}$$

The MMSE problem is as follows:

$$\mathbf{g}_k = \arg \min_{\mathbf{g}} E[|b_k - \mathbf{g}^{\mathrm{T}} \mathbf{y}_k|^2]. \tag{9.28}$$

Note that

$$\mathbf{y}_k = \mathbf{c}_k A_k b_k + \mathbf{C}_k \mathbf{A}_k \tilde{\mathbf{b}}_k + \mathbf{n},$$

where $\tilde{\mathbf{b}}_k = \mathbf{b}_k - \bar{\mathbf{b}}_k$. Hence, we can easily show that

$$\begin{aligned}
\mathbf{Q}_k &= E\left[\mathbf{y}_k \mathbf{y}_k^{\mathrm{T}}\right] \\
&= A_k^2 \mathbf{c}_k \mathbf{c}_k^{\mathrm{T}} + \mathbf{C}_k \mathbf{A}_k E[\tilde{\mathbf{b}}_k \tilde{\mathbf{b}}_k^{\mathrm{T}}] \mathbf{A}_k^{\mathrm{T}} \mathbf{C}_k^{\mathrm{T}} + \frac{N_0}{2} \mathbf{I}
\end{aligned}$$

when we assume that b_k and $\tilde{\mathbf{b}}_k$ are uncorrelated, i.e.

$$E[b_k \tilde{\mathbf{b}}_k] = E[b_k] E[\mathbf{b}_k - \bar{\mathbf{b}}_k] = \mathbf{0}.$$

Under the same assumption, we have

$$E[\mathbf{y}_k b_k] = \mathbf{c}_k A_k.$$

These lead to the MMSE filter:

$$\mathbf{g}_k = A_k \mathbf{Q}_k^{-1} \mathbf{c}_k. \tag{9.29}$$

From Eq. (9.24), $\hat{b}_{\mathrm{mmse},k}$ becomes

$$\hat{b}_{\mathrm{mmse},k} = A_k \mathbf{c}_k^{\mathrm{T}} \mathbf{Q}_k^{-1} \mathbf{y}_k. \tag{9.30}$$

To find the LLR of b_k from $\hat{b}_{\mathrm{mmse},k}$ under the Gaussian assumption in (F2), the mean and variance, μ_k and $\sigma_{\eta,k}^2$, respectively, are required. From Eq. (9.30), it follows that

$$\begin{aligned}
\mu_k &= E[\hat{b}_{\mathrm{mmse},k} b_k] \\
&= A_k \mathbf{c}_k^{\mathrm{T}} \mathbf{Q}_k^{-1} E[\mathbf{y}_k b_k] \\
&= A_k^2 \mathbf{c}_k^{\mathrm{T}} \mathbf{Q}_k^{-1} \mathbf{c}_k
\end{aligned} \tag{9.31}$$

and

$$\begin{aligned}
\sigma_{\eta,k}^2 &= E[\hat{b}_{\mathrm{mmse},k}^2] - \mu_k^2 \\
&= A_k^2 \mathbf{c}_k^{\mathrm{T}} \mathbf{Q}_k^{-1} \mathbf{c}_k - \mu_k^2 \\
&= \mu_k - \mu_k^2.
\end{aligned} \tag{9.32}$$

Hence, from Eq. (9.26), the LLR from $\hat{b}_{\mathrm{mmse},k}$ is found as follows:

$$\mathrm{LLR}(b_k; \hat{b}_{\mathrm{mmse},k}) = \frac{2\hat{b}_{\mathrm{mmse},k}}{1 - \mu_k}. \tag{9.33}$$

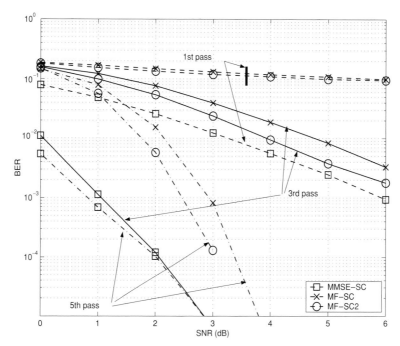

Figure 9.3. Bit error rate performance of the iterative receivers with the MF-SC, MF-SC2, and MMSE-SC detectors. Full load: $N = 10$, $K = 10$.

It is important to note that this LLR is not the same as the LLR in Eq. (9.8), as there would be a loss of information through linear filtering and SC. Hence, the LLR from $\hat{b}_{\mathrm{mmse},k}$ would not provide a better performance than the LLR in Eq. (9.8).

For performance comparison, three iterative receivers are considered with the MF-SC, MF-SC2, and MMSE-SC detectors. Simulations are carried out with $N = 10$ and $K = 10$. Random spreading codes are assumed with the same amplitude, $A_1 = \cdots = A_K = 1$. A rate-half convolutional code with generator polynomial (7, 5) in octal is used for all users. The length of coded sequences is $L = 2^{11}$. The simulation results are shown in Fig. 9.3, and it is shown that the iterative receiver with the MMSE-SC detector outperforms the other receivers.

Through EXIT charts, we can also see the performance and the number of iterations required for convergence. Figure 9.4 shows the EXIT charts when the SNR is 2 dB. According to Fig. 9.3, the iterative receiver with the MMSE-SC detector can converge after three iterations at SNR of 2 dB. It can also be confirmed by the EXIT charts in Fig. 9.4. When the MF-SC detector is used, more iterations are required before convergence as the tunnel (the gap between the EXIT charts of the decoder and MF-SC detector) is narrower.

It is also shown that the MF-SC2 detector has a slightly higher mutual information than the MF-SC detector. Thus, the performance of the iterative receiver with the MF-SC2 detector is marginally better than that with the MF-SC detector. In addition, the iterative receiver with the MF-SC2 detector would have a (slightly) smaller number of iterations for convergence than that with the MF-SC detector.

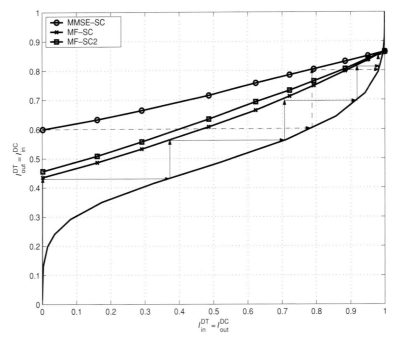

Figure 9.4. EXIT charts for three iterative receivers; I_{in}^{DT} and I_{in}^{DC} = input extrinsic information to the detector and decoder, respectively; I_{out}^{DT} and I_{out}^{DC} = output extrinsic information to the detector and decoder, respectively. (From Choi, J. (2006). "Low complexity MAP detection using approximate LLR for CDMA iterative receivers," *IEEE Commun. Lett.*, **10**, 321–323.)

To see the number of iterations for convergence clearly, simulations are carried out with a fixed SNR of 2 dB, and the results are shown in Fig. 9.5. It is shown that the iterative receiver with the MMSE-SC detector needs three iterations to converge, while that with the MF-SC2 detector needs seven or eight iterations. When the MF-SC detector is used, more than eight iterations are required.

There is a trade-off between the complexity and performance. The MF-SC and MF-SC2 detectors have lower complexity than the MMSE-SC detector, while the MMSE-SC detector outperforms the others. If the complexity of the iterative receiver is limited (such as matrix inversion operations cannot be accommodated), the MF-SC or MF-SC2 detector is preferable at the expense of more iterations and/or worse performance.

9.4.2 Approximation of the original LLR using Gaussian approximation

We can use the Gaussian approximation for interfering signals to find the LLR of the desired signal based on the central limit theorem (CLT). This approach is not new in CDMA systems. In conventional single-user detection, the correlator detector is often assumed to be an "asymptotically" optimal detector (under the ML detection criterion) if the sum of interfering signals can be approximated by a Gaussian random process for a large number of users (Viterbi, 1995). Interestingly, the LLR obtained on using the CLT for interfering signals is identical to that obtained by the MMSE-SC detector as shown below.

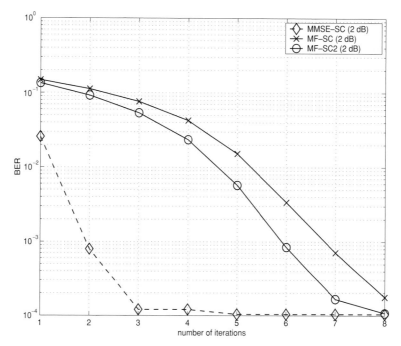

Figure 9.5. Coded BER versus number of iterations with SNR $(1/N_0) = 2$ dB ($K = N = 10$, $A_k = 1$ for all k). (From Choi, J. (2006). "Low complexity MAP detection using approximate LLR for CDMA iterative receivers," *IEEE Commun. Lett.*, **10**, 321–323.)

The received signal vector can be rewritten as follows:

$$
\begin{aligned}
\mathbf{y} &= A_k \mathbf{c}_k b_k + \mathbf{C}_k \mathbf{A}_k \mathbf{b}_k + \mathbf{n} \\
&= A_k \mathbf{c}_k b_k + \mathbf{C}_k \mathbf{A}_k \bar{\mathbf{b}}_k + \mathbf{e}_k,
\end{aligned}
\tag{9.34}
$$

where $\mathbf{e}_k = \mathbf{C}_k \mathbf{A}_k \tilde{\mathbf{b}}_k + \mathbf{n}$. From $\mathbf{y}$, we need to find the LLR of b_k. Since $\bar{\mathbf{b}}_k$ is a known constant vector, it does not affect the LLR. Thus, the LLR of b_k can also be found from $\mathbf{y}_k$ in Eq. (9.27). From Eq. (9.34), $\mathbf{y}_k$ is rewritten as follows:

$$
\mathbf{y}_k = A_k \mathbf{c}_k b_k + \mathbf{e}_k.
\tag{9.35}
$$

If K is large, $\mathbf{e}_k$ can be assumed to be a Gaussian random vector based on the CLT. The mean of $\mathbf{e}_k$ is zero and the covariance is given by

$$
\begin{aligned}
\mathbf{R}_k &= E\left[\mathbf{e}_k \mathbf{e}_k^{\mathsf{T}}\right] \\
&= \mathbf{C}_k \mathbf{A}_k E\left[\tilde{\mathbf{b}}_k \tilde{\mathbf{b}}_k^{\mathsf{T}}\right] \mathbf{A}_k^{\mathsf{T}} \mathbf{C}_k^{\mathsf{T}} + \frac{N_0}{2}\mathbf{I}.
\end{aligned}
\tag{9.36}
$$

Then, the LLR of b_k can be found as follows:

$$
\text{LLR}_{\text{clt}}(b_k) = \log \frac{f_{\text{clt}}(\mathbf{y}_k | b_k = +1)}{f_{\text{clt}}(\mathbf{y}_k | b_k = -1)},
\tag{9.37}
$$

where $f_{\text{clt}}(\mathbf{y}_k|b_k)$ denotes the likelihood function under the assumption that $\mathbf{e}_k$ is Gaussian. Precisely, $f_{\text{clt}}(\mathbf{y}_k|b_k)$ is given by

$$f_{\text{clt}}(\mathbf{y}_k|b_k) = C \exp\left(-\frac{1}{2}(\mathbf{y}_k - A_k\mathbf{c}_kb_k)^{\mathsf{T}}\mathbf{R}_k^{-1}(\mathbf{y}_k - A_k\mathbf{c}_kb_k)\right), \qquad (9.38)$$

where C is a normalizing constant. Substituting Eq. (9.38) into Eq. (9.37), the LLR of b_k becomes

$$\text{LLR}_{\text{clt}}(b_k) = 2A_k\mathbf{c}_k^{\mathsf{T}}\mathbf{R}_k^{-1}\mathbf{y}_k. \qquad (9.39)$$

Using the matrix inversion lemma, we can further show that

$$\text{LLR}_{\text{clt}}(b_k) = \text{LLR}(b_k; \hat{b}_{\text{mmse},k}). \qquad (9.40)$$

That is, the LLR based on the CLT is identical to the LLR from $\hat{b}_{\text{mmse},k}$ in Eq. (9.33). This observation shows that the MMSE-SC detector can provide an asymptotically optimal performance as $K \to \infty$.

The result in Eq. (9.40) can be explained using the notion of sufficient statistics. The following theorem is from Porat (1994).

Theorem 9.4.1 *The statistic $\mathcal{T}(\mathbf{y})$ is sufficient for the parameter θ if and only if there exists a factorization of $f(\mathbf{y}|\theta)$ of the following form:*

$$f(\mathbf{y}|\theta) = g_\theta(\mathcal{T}(\mathbf{y}))h(\mathbf{y}). \qquad (9.41)$$

Let $\mathcal{T}(\mathbf{y}_k)$ be a sufficient statistic for b_k. Then, from Eq. (9.41) we can show that

$$\text{LLR}(b_k; \mathbf{y}_k) = \text{LLR}(b_k; \mathcal{T}(\mathbf{y}_k)), \qquad (9.42)$$

where $\text{LLR}(b_k; \mathbf{x})$ stands for the LLR of b_k from statistic $\mathbf{x}$. Indeed, $\hat{b}_{\text{mmse},k}$ in Eq. (9.33) is a sufficient statistic for b_k (under the Gaussian assumption). This can be shown by factorizing the conditional pdf of $\mathbf{y}_k$ (given b_k). From Eq. (9.38), we can show that

$$f_{\text{clt}}(\mathbf{y}_k|b_k) = C \exp\left(-\frac{1}{2}\left(\mathbf{y}_k^{\mathsf{T}}\mathbf{R}_k^{-1}\mathbf{y}_k + A_k^2\mathbf{c}_k^{\mathsf{T}}\mathbf{R}_k^{-1}\mathbf{c}_k\right)\right) \exp\left(A_k\mathbf{c}_k^{\mathsf{T}}\mathbf{R}_k^{-1}\mathbf{y}_kb_k\right).$$

Hence, $\mathcal{T}(\mathbf{y}_k) = A_k\mathbf{c}_k^{\mathsf{T}}\mathbf{R}_k^{-1}\mathbf{y}_k$ becomes a sufficient statistic according to Eq. (9.41). Furthermore, we have

$$\mathcal{T}(\mathbf{y}_k) = A_k\mathbf{c}_k^{\mathsf{T}}\mathbf{R}_k^{-1}\mathbf{y}_k$$
$$\propto A_k\mathbf{c}_k^{\mathsf{T}}\left(\mathbf{R}_k + A_k^2\mathbf{c}_k\mathbf{c}_k^{\mathsf{T}}\right)^{-1}\mathbf{y}_k$$
$$= \hat{b}_{\text{mmse},k}$$

from the matrix inversion lemma. Consequently, we can show that $\hat{b}_{\text{mmse},k}$ is a sufficient statistic for b_k.

9.5   Iterative receiver with unknown interferers

In order to use the iterative receiver in CDMA systems, there is a fundamental assumption that spreading codes and channel codes of users are known. However, as far as cellular systems are concerned, this fundamental assumption may not be valid due to the intercell interference. For example, suppose that a mobile station in a cell is to use the iterative receiver. The mobile station may know the spreading codes and channel codes of the other users within the cell, and can run the iterative receiver. However, in practice it may not be straightforward to run the iterative receiver including the users in adjacent cells due to the high complexity; thus the users in adjacent cells may be assumed to be unknown interferers. This problem was addressed with a blind detector in Reynolds and Wang (2002) for unknown interferers for the iterative receiver.

In this section, we discuss the iterative receiver with unknown interferers as in Reynolds and Wang (2002). The MMSE-SC detector will be considered. The EM algorithm is used to estimate the covariance matrix of unknown interferers.

9.5.1 System model with unknown interferers

Assume that the spreading codes and channel encoders of the first $\bar{K}(< K)$ users are known, but that there is no information on the other users. This case can happen in a downlink channel of cellular systems. Suppose that a mobile user is close to a cell edge. There are $\bar{K}$ signals from its basestation and $K_{\text{un}} = K - \bar{K}$ signals from adjacent cells, which can be equally as strong as the signals from the desired basestation at the cell edge. In this case, the received signal is given by

$$
\begin{aligned}
\mathbf{y}_l &= \sum_{k=1}^{\bar{K}} A_k \mathbf{c}_k b_{k,l} + \mathbf{u}_l \\
&= \mathbf{CAb}_l + \mathbf{u}_l,
\end{aligned}
\tag{9.43}
$$

where the following quantities are redefined:

$$
\begin{aligned}
\mathbf{A} &= \text{Diag}(A_1, A_2, \ldots, A_{\bar{K}}), \\
\mathbf{C} &= [\mathbf{c}_1 \; \mathbf{c}_2 \; \ldots \; \mathbf{c}_{\bar{K}}], \\
\mathbf{b}_l &= [b_{1,l} \; b_{2,l} \; \ldots \; b_{\bar{K},l}]^{\text{T}}, \\
\mathbf{u}_l &= \sum_{k=\bar{K}+1}^{K} A_k \mathbf{c}_k b_{k,l} + \mathbf{n}_l.
\end{aligned}
$$

Clearly, the statistical properties of $\mathbf{u}_l$ are unknown as the spreading codes and amplitudes of the last K_{un} users are unknown.

9.5.2 Iterative receiver and EM algorithm to estimate interferer covariance matrix

In this subsection, we consider first an iterative receiver under the assumption that the covariance matrix of $\mathbf{u}_l$ is known. Next, we consider the ML estimation of the covariance matrix of $\mathbf{u}_l$ with the EM algorithm. This results in an iterative receiver with the estimation of the covariance matrix of interferers.

Iterative receiver with known interferer covariance matrix

To detect/decode the signals of interest (i.e. $b_{k,l}$ for $k = 1, 2, \ldots, \bar{K}$), an iterative receiver is considered under the assumption that the covariance matrix of $\mathbf{u}_l$ is known.

Consider the iterative receiver consisting of the MMSE-SC detectors and channel decoders. For convenience, we omit the time index l. Suppose that the *a priori* probabilities of b_k, $k = 1, 2, \ldots, \bar{K}$, are available. In the MMSE-SC detector for the kth signal, we first perform the SC and the resulting signal vector is given by

$$\begin{aligned}
\mathbf{r}_k &= \mathbf{y} - \mathbf{C}_k \mathbf{A}_k \bar{\mathbf{b}}_k \\
&= A_k \mathbf{c}_k b_k + \mathbf{C}_k \mathbf{A}_k \tilde{\mathbf{b}}_k + \mathbf{u},
\end{aligned} \tag{9.44}$$

where

$$\begin{aligned}
\mathbf{A}_k &= \mathrm{Diag}(A_1, \ldots, A_{k-1}, A_{k+1}, \ldots, A_{\bar{K}}), \\
\mathbf{C}_k &= [\mathbf{c}_1 \cdots \mathbf{c}_{k-1} \mathbf{c}_{k+1} \cdots \mathbf{c}_{\bar{K}}], \\
\bar{\mathbf{b}}_k &= [\bar{b}_1 \cdots \bar{b}_{k-1} \bar{b}_{k+1} \cdots \bar{b}_{\bar{K}}], \\
\tilde{\mathbf{b}}_k &= [\tilde{b}_1 \cdots \tilde{b}_{k-1} \tilde{b}_{k+1} \cdots \tilde{b}_{\bar{K}}].
\end{aligned}$$

Here, $\bar{b}_k$ stands for the mean value of b_k from the *a priori* probability, i.e. $\bar{b}_k = \Pr(b_k = +1) - \Pr(b_k = -1)$, and $\tilde{b}_k = b_k - \bar{b}_k$. To mitigate the residual interference, the MMSE filtering is applied after SC. The MMSE filtering vector can be obtained as follows:

$$\begin{aligned}
\mathbf{w}_k &= \arg \min_{\mathbf{w}} E[|\mathbf{w}^\mathrm{T} \mathbf{r}_k - b_k|^2] \\
&= A_k \left(A_k^2 \mathbf{c}_k \mathbf{c}_k^\mathrm{T} + \mathbf{C}_k \mathbf{A}_k \bar{\mathbf{Q}}_k \mathbf{A}_k^\mathrm{T} \mathbf{C}_k^\mathrm{T} + \mathbf{R_u} \right)^{-1} \mathbf{c}_k,
\end{aligned} \tag{9.45}$$

where $\mathbf{R_u} = E[\mathbf{u}\mathbf{u}^\mathrm{T}]$ and

$$\begin{aligned}
\bar{\mathbf{Q}}_k &= E[\tilde{\mathbf{b}}_k \tilde{\mathbf{b}}_k^\mathrm{T}] \\
&= \mathrm{Diag}(1 - \bar{b}_1^2, \ldots, 1 - \bar{b}_{k-1}^2, 1 - \bar{b}_{k+1}^2, \ldots, 1 - \bar{b}_{\bar{K}}^2).
\end{aligned}$$

The output of the MMSE-SC detector is given by

$$z_k = \mathbf{w}_k^\mathrm{T} \mathbf{r}_k. \tag{9.46}$$

Using a forced Gaussian assumption, the LLR can be obtained from the output of the MMSE filter. The LLR is given by

$$\mathrm{LLR}(b_k; z_k) = \frac{2z_k}{1 - \mu_k}, \tag{9.47}$$

where $\mu_k = A_k \mathbf{w}_k^\mathrm{T} \mathbf{c}_k$. The difference from the standard MMSE-SC detector is that $\mathbf{u}$ is a colored noise. Since the covariance matrix of $\mathbf{u}$ is not known, it has to be estimated.

EM algorithm for the ML approach to estimate $\mathbf{R_u}$

The ML estimation of $\mathbf{R_u}$ is formulated as follows:

$$\begin{aligned}
\hat{\mathbf{R}}_\mathbf{u} &= \arg \max_{\mathbf{R_u}} f(\mathbf{Y}|\mathbf{R_u}) \\
&= \arg \max_{\mathbf{R_u}} \prod_{l=0}^{L-1} f(\mathbf{y}_l|\mathbf{R_u}),
\end{aligned} \tag{9.48}$$

where $\mathbf{Y} = [\mathbf{y}_0 \ \mathbf{y}_1 \ \cdots \ \mathbf{y}_{L-1}]$ and $f(\mathbf{y}_l|\mathbf{R_u})$ is the likelihood function of $\mathbf{R_u}$ given $\mathbf{y}_l$. In general, it is not easy to find the ML estimate of $\mathbf{R_u}$ as the likelihood function $f(\mathbf{y}_l|\mathbf{R_u})$ is highly nonlinear. However, the EM algorithm, which is an iterative algorithm, can be used to find the ML estimate numerically.

Suppose that $\mathbf{u}$ is a Gaussian random vector. This would be a proper assumption if K_{un} is large. For the EM algorithm to estimate $\mathbf{R_u}$, we define the complete data as $\{\mathbf{Y}, \mathbf{B}\}$, where $\mathbf{B} = [\mathbf{b}_0 \ \mathbf{b}_1 \ \cdots \ \mathbf{b}_{L-1}]$ is the missing data. In the EM algorithm, each iteration consists of the E-step and the M-step. Denote by $\hat{\mathbf{R}}_{\mathbf{u}}^{(q)}$ the estimate of $\mathbf{R_u}$ at the qth iteration. Suppose that the initial estimate, $\mathbf{R}_{\mathbf{u}}^{(0)}$, is available. The E-step yields the following function:

$$Q\left(\mathbf{R_u}|\hat{\mathbf{R}}_{\mathbf{u}}^{(q)}\right) = E\left[\log f(\mathbf{B}, \mathbf{Y}|\mathbf{R_u})|\mathbf{Y}, \hat{\mathbf{R}}_{\mathbf{u}}^{(q)}\right], \tag{9.49}$$

where the expectation takes place with respect to $\mathbf{B}$. It can be shown that

$$\begin{aligned} \log f(\mathbf{B}, \mathbf{Y}|\mathbf{R_u}) &= \log\left(f(\mathbf{Y}|\mathbf{B}, \mathbf{R_u})\Pr(\mathbf{B}|\mathbf{R_u})\right) \\ &= \log\left(f(\mathbf{Y}|\mathbf{B}, \mathbf{R_u})\Pr(\mathbf{B})\right) \\ &= \log f(\mathbf{Y}|\mathbf{B}, \mathbf{R_u}) + C, \end{aligned} \tag{9.50}$$

where C is a constant. In Eq. (9.50), the first equality is due to the Bayes rule, the second equality is due to the fact that $\mathbf{B}$ and $\mathbf{R_u}$ are independent, and the third equality is due to the fact that $\mathbf{B}$ is equally likely. Since the likelihood function with the missing data is given by

$$f(\mathbf{y}_l|\mathbf{b}_l, \mathbf{R_u}) = C'\,(|\mathbf{R_u}|)^{-1/2} \exp\left(-\frac{1}{2}(\mathbf{y}_l - \mathbf{CAb}_l)^{\mathsf{T}}\mathbf{R_u}^{-1}(\mathbf{y}_l - \mathbf{CAb}_l)\right), \tag{9.51}$$

where C' is a constant and $|\mathbf{R_u}|$ stands for the determinant of $\mathbf{R_u}$, and $f(\mathbf{Y}|\mathbf{B}, \mathbf{R_u}) = \prod_l f(\mathbf{y}_l|\mathbf{b}_l, \mathbf{R_u})$, we have

$$\log f(\mathbf{Y}|\mathbf{B}, \mathbf{R_u}) = -\frac{L}{2}\log|\mathbf{R_u}| - \frac{1}{2}\sum_{l=0}^{L-1}(\mathbf{y}_l - \mathbf{CAb}_l)^{\mathsf{T}}\mathbf{R_u}^{-1}(\mathbf{y}_l - \mathbf{CAb}_l) + C'', \tag{9.52}$$

where C'' is a constant. After ignoring constants, we can show that

$$Q\left(\mathbf{R_u}|\hat{\mathbf{R}}_{\mathbf{u}}^{(q)}\right) = -\frac{L}{2}\left(\log|\mathbf{R_u}| + \mathrm{Tr}\left(\mathbf{R_u}^{-1}E\left[\mathbf{S}|\mathbf{Y}, \hat{\mathbf{R}}_{\mathbf{u}}^{(q)}\right]\right)\right), \tag{9.53}$$

where

$$\mathbf{S} = \frac{1}{L}\sum_{l=0}^{L-1}(\mathbf{y}_l - \mathbf{CAb}_l)(\mathbf{y}_l - \mathbf{CAb}_l)^{\mathsf{T}}.$$

The M-step yields the next estimate that maximizes $Q(\mathbf{R_u}|\hat{\mathbf{R}}_{\mathbf{u}}^{(q)})$; i.e.

$$\begin{aligned} \hat{\mathbf{R}}_{\mathbf{u}}^{(q+1)} &= \arg\max_{\mathbf{R_u}} Q\left(\mathbf{R_u}|\hat{\mathbf{R}}_{\mathbf{u}}^{(q)}\right) \\ &= \arg\min_{\mathbf{R_u}}\left\{\log|\mathbf{R_u}| + \mathrm{Tr}\left(\mathbf{R_u}^{-1}E\left[\mathbf{S}|\mathbf{Y}, \hat{\mathbf{R}}_{\mathbf{u}}^{(q)}\right]\right)\right\}. \end{aligned} \tag{9.54}$$

From Burg, Luenberger, and Wenger (1982), $\hat{\mathbf{R}}_{\mathbf{u}}^{(q+1)}$ can be readily found as follows:

$$\hat{\mathbf{R}}_{\mathbf{u}}^{(q+1)} = E\left[\mathbf{S}|\mathbf{Y}, \hat{\mathbf{R}}_{\mathbf{u}}^{(q)}\right]. \tag{9.55}$$

The EM algorithm requires knowledge of $\Pr(\mathbf{B}|\mathbf{Y}, \hat{\mathbf{R}}_{\mathbf{u}}^{(q)})$ to find the next estimate.

Approximations to iterative receiver

Even though it is possible to find $\Pr(\mathbf{B}|\mathbf{Y}, \hat{\mathbf{R}}_{\mathbf{u}}^{(q)})$ by a straightforward approach, its complexity is prohibitively high. Thus, we will consider a couple of approximations. Firstly, we assume that the $b_{k,l}$'s, $k = 1, 2, \ldots, \bar{K}$, conditioned on $\mathbf{Y}$ and $\hat{\mathbf{R}}_{\mathbf{u}}^{(q)}$ are independent. That is, we assume (or approximate) that

$$\Pr\left(\mathbf{B}|\mathbf{Y}, \hat{\mathbf{R}}_{\mathbf{u}}^{(q)}\right) = \prod_{k=1}^{\bar{K}} \prod_{l=0}^{L-1} \Pr\left(b_{k,l}|\mathbf{Y}, \hat{\mathbf{R}}_{\mathbf{u}}^{(q)}\right). \tag{9.56}$$

As in Chapter 7, we can apply the iterative detector and decoder (IDD) to find the *a posteriori* probability. Generally, several iterations are required for convergence when the IDD is employed. This results in a doubly iterative receiver in which the inner iteration is for the detection and decoding and the outer iteration is for the EM algorithm (EM iteration) to update the estimate of $\mathbf{R}_{\mathbf{u}}$. However, this structure requires a number of iterations.

A singly iterative receiver is available. For each detection and decoding iteration, the estimate of the covariance matrix can be updated. We summarize the operation of the singly iterative receiver as follows.

(S0) Let $q = 0$ (here, q is the index for iteration) and find $\hat{\mathbf{R}}_{\mathbf{u}}^{(0)}$. Let the LLR from the MAP decoder be zero (no prior information).

(S1) Perform the MMSE-SC detection with $\hat{\mathbf{R}}_{\mathbf{u}}^{(q)}$ and prior information (i.e. the LLR from the MAP decoder after bit interleaving) and obtain the LLR using a forced Gaussian approximation.

(S2) After bit deinterleaving, perform the MAP decoding and obtain the LLR and *a posteriori* probability of $b_{k,l}$.

(S3) With the *a posteriori* probability, perform the E-step and the M-step to find $\hat{\mathbf{R}}_{\mathbf{u}}^{(q+1)}$.

(S4) If $\hat{\mathbf{R}}_{\mathbf{u}}^{(q+1)}$ converges, stop. Otherwise, $q = q + 1$ and go to (S1).

Note that according to Eq. (9.55) the initial covariance matrix $\mathbf{R}_{\mathbf{u}}^{(0)}$ is given by

$$\mathbf{R}_{\mathbf{u}}^{(0)} = \left(\frac{1}{L} \sum_{l=0}^{L-1} \mathbf{y}_l \mathbf{y}_l^{\mathrm{T}}\right) + \mathbf{CAA}^{\mathrm{T}}\mathbf{C}^{\mathrm{T}}.$$

since $E[\mathbf{b}_l] = 0$ and $E[\mathbf{b}_l\mathbf{b}_l^{\mathrm{T}}] = \mathbf{I}$ for $\mathbf{b}_l$, in which the data symbols are independent and equally likely.

For simulations, assume that there are $K = 16$ users and that the channel codes and spreading codes of the first $\bar{K} = 5$ users are known. It is also assumed that the processing gain, N, is 16 and the signal amplitudes of all users are set to unity (i.e. $A_k = 1$, $k = 1, 2, \ldots, K$). Thus, the interferers are equally as strong as the signals of interest. A signal block from each user consists of $L = 2^{11}$ coded bits. A rate-half convolutional code with generator polynomial $(7, 5)$ in octal is used for all users, while the spreading codes with $N = 16$ are randomly generated. A different random bit interleaver is assumed for each user.

Firstly, we consider the case that $\mathbf{R}_{\mathbf{u}}$ is known. Simulation results are presented in Fig. 9.6. It is shown that two or three iterations are sufficient to achieve a good performance.

Figure 9.7 shows the performance of the EM-based iterative receiver. The estimated covariance matrix of unknown interferers is updated for each iteration. Comparing with the

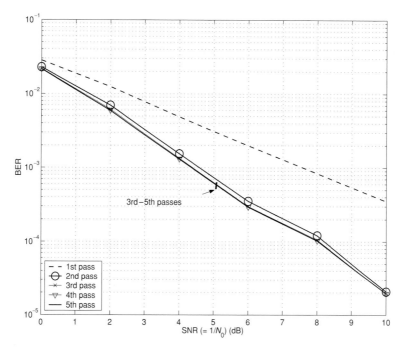

Figure 9.6. Bit error rate performance of the iterative receiver with known $\mathbf{R_u}$. Full load: $N = 16$, $K = 16$.

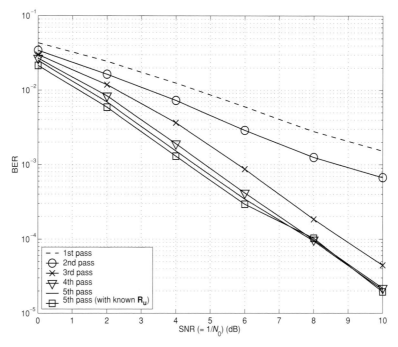

Figure 9.7. Bit error rate performance of the EM-based iterative receiver with estimated $\mathbf{R_u}$. Full load: $N = 16$, $K = 16$.

case that $\mathbf{R_u}$ is known, the performance improvement through iterations is slower. However, after four or five iterations, the performance approaches that of the case with known $\mathbf{R_u}$.

9.6   Summary and notes

In this chapter, we discussed the iterative receiver for coded CDMA systems. Suboptimal multiuser detectors were studied to reduce the complexity. It was shown that the MMSE-SC detector provides an approximate LLR under the assumption that a sum of interfering signals is Gaussian. For unknown interferers, we showed that the iterative receiver can be extended with the EM algorithm to estimate the covariance matrix of unknown interferers.

The MMSE-SC detector for the iterative receiver was first proposed in Wang and Poor (1999) for CDMA systems. An EXIT chart analysis was presented in Li and Wang (2005). An in-depth performance analysis of the iterative receiver with the MMSE-SC detector can be found in Boutros and Caire (2002).

In this chapter, we assume the AWGN channel and ignore multipaths. If there are multipaths, the spreading code can be modified as a superposition of delayed spreading codes with corresponding attenuation factors. An extension of the MMSE-SC detector to multipath channels is also discussed in Wang and Poor (1999). The iterative receiver can also be extended to include the channel estimation. This is addressed in Sellami, Lasaulce, and Fijalkow (2003) using the EM algorithm.

For uncoded signals, the EM algorithm is applied to derive iterative receivers (in this case, the data symbols are generally considered as the parameters to be estimated). Multiuser detection is considered using the EM and SAGE algorithms in Nelson and Poor (1996) (the SAGE algorithm is a generalization of the EM algorithm). For joint channel estimation and data detection of CDMA signals, the EM algorithm is applied to derive iterative receivers in Kocian and Fleury (2003).

10 Iterative receivers for multiple antenna systems

Multiple antennas have been widely used in wireless communications to introduce diversity gain, because wireless communications suffer from fading. There have been various diversity techniques employed at both transmitters and receivers using multiple antennas (Lee, 1982).

The channel capacity of multiple antenna systems at both transmitter and receiver was investigated in Foschini and Gans (1998) and Telatar (1999), which revealed that the channel capacity can be significantly improved by using multiple antennas. Following this discovery, the fundamental issues underlying the practical implementation of multiple input multiple output (MIMO) channels have been studied extensively.

In this chapter, we study several detection techniques and iterative receivers for MIMO channels. An iterative receiver with channel estimation is also discussed. Again, the EM algorithm plays a key role in deriving the doubly iterative receiver that can detect and decode signals as well as estimate channel matrices. The reader is referred to Paulraj, Nabar, and Gore (2003) and Tse and Viswanath (2005) for detailed accounts of the channel capacity of MIMO channels and space-time processing techniques, subjects that are not addressed in this chapter.

10.1 Space diversity

In wireless communications, diversity techniques are important in combatting fading. To introduce diversity, various approaches can be taken. In this section, we focus on space diversity introduced by multiple antennas.

10.1.1 Diversity combining with multiple receive antennas

Suppose that there are N receive antennas for a flat fading channel. The received signal at the mth antenna is given by

$$y_{m,l} = h_{m,l}b_l + n_{m,l}, \tag{10.1}$$

where $h_{m,l}$ and $n_{m,l}$ are the channel coefficient and the noise at the mth antenna at time l, respectively, and $b_l \in \{-1, +1\}$ is the data symbol. Figure 10.1 shows a block diagram for multiple-antenna reception. The receiver has N received signals to detect a data symbol.

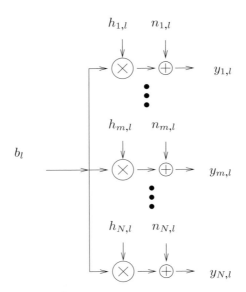

Figure 10.1. Multiple-antenna reception.

Throughout this chapter, we will assume that $h_{m,l}$ and $n_{m,l}$ are complex random variables as $y_{m,l}$ has both in-phase and quadrature-phase components.

For Rayleigh fading, the channel coefficient is assumed to be a zero-mean circular complex Gaussian random variable. If X is a zero-mean circular complex Gaussian random variable, the real and imaginary components of X are independent and

$$\frac{1}{2} E[|X|^2] = E[(\Re(X))^2] = E[(\Im(X))^2] \text{ and } E[X^2] = 0.$$

In Eq. (10.1), we also assume that the background noise, $n_{m,l}$, is a zero-mean circular complex Gaussian random variable with $E[|n_{m,l}|^2] = N_0$.

For convenience, we will omit the time index l if there is no confusion. Define the instantaneous SNR of the mth branch as $\gamma_m = |h_m|^2/N_0$. If the antennas are spaced sufficiently far apart, the signals received at each antenna can experience independent fading. In other words, the $\{h_m, \forall m\}$ become independent. Thus, there would be some antennas receiving strong signals without severe fading, while others receive faded signals.

There are several combining techniques for the space diversity. We will discuss three major techniques: (1) selective diversity, (2) maximal ratio combining, and (3) equal gain combining.

Selection diversity

The idea of selection diversity is simple. Among N received signals, the signal that has the highest instantaneous SNR is chosen at each time instant. The resulting SNR of the selection diversity is given by

$$\gamma_{\text{sd}} = \max\{\gamma_1, \gamma_2, \ldots, \gamma_N\}. \tag{10.2}$$

Since each SNR is a random variable, the SNR of the selection diversity is also a random variable. Each SNR, γ_m, has a chi-square pdf with two degrees of freedom:

$$f_m(\gamma_m) = \frac{1}{\bar{\gamma}_m} e^{-\gamma_m/\bar{\gamma}_m},$$

where $\bar{\gamma}_m = E[\gamma_m]$. The cdf is given by

$$F_m(\gamma) = \Pr(\gamma_m \leq \gamma) = 1 - e^{-\gamma/\bar{\gamma}_m}, \quad \gamma \geq 0.$$

From the cdf of γ_m, the cdf of γ_{sd} can be found as follows:

$$\begin{aligned}
F_{sd}(\gamma) &= \Pr(\gamma_{sd} \leq \gamma) \\
&= \Pr(\gamma_1 \leq \gamma, \gamma_2 \leq \gamma, \ldots, \gamma_N \leq \gamma) \\
&= \prod_{m=1}^{N} \Pr(\gamma_m \leq \gamma) \\
&= \prod_{m=1}^{N} \left(1 - e^{-\gamma/\bar{\gamma}_m}\right).
\end{aligned} \tag{10.3}$$

We note that $F_{sd}(\gamma)$ is the probability that γ_{sd} is smaller than or equal to a threshold SNR γ. This probability is called the outage probability, which is the probability of not getting a reliable communication as the SNR falls below the threshold, γ. It is clear from Eq. (10.3) that this probability decreases with N, because $(1 - e^{-\gamma/\bar{\gamma}_m}) \leq 1$. The chance of a reliable communication therefore becomes higher as the number of receive antennas increases.

Example 10.1.1 If $\bar{\gamma}_m = \bar{\gamma}$ for all m, the pdf of γ_{sd} becomes

$$f_{sd}(\gamma) = \frac{N}{\bar{\gamma}} e^{-\gamma/\bar{\gamma}} \left(1 - e^{-\gamma/\bar{\gamma}}\right)^{N-1}, \quad \gamma \geq 0.$$

Given an instantaneous SNR, γ_{sd}, the (conditional) BER is given by

$$P_b(\gamma_{sd}) = Q(\sqrt{2\gamma_{sd}}).$$

Since $Q(\sqrt{2\gamma_{sd}}) \leq \exp(-\gamma_{sd})$, the average BER is bounded as follows:

$$\begin{aligned}
P_b &= E[Q(\sqrt{2\gamma_{sd}})] \\
&\leq E[\exp(-\gamma_{sd})] \\
&= \int_0^{\infty} e^{-\gamma} f_{sd}(\gamma) \, d\gamma \\
&= \frac{N}{\bar{\gamma}} \int_0^{\infty} e^{-(1+\frac{1}{\bar{\gamma}})\gamma} \left(1 - e^{-\gamma/\bar{\gamma}}\right)^{N-1} d\gamma.
\end{aligned}$$

Using a binomial expansion, we can show that

$$\left(1 - e^{-\gamma/\bar{\gamma}}\right)^{N-1} = \sum_{k=0}^{N-1} \binom{N-1}{k} (-1)^k e^{-\frac{k\gamma}{\bar{\gamma}}}.$$

From this, it follows that

$$\int_0^{\infty} e^{-(1+\frac{1}{\bar{\gamma}})\gamma} \left(1 - e^{-\gamma/\bar{\gamma}}\right)^{N-1} d\gamma = \sum_{k=0}^{N-1} \binom{N-1}{k} (-1)^k \frac{\bar{\gamma}}{\bar{\gamma}+1+k}.$$

Finally, an upper bound of the BER of selection diversity is given by

$$P_b \leq N \sum_{k=0}^{N-1} \binom{N-1}{k} (-1)^k \frac{1}{\bar{\gamma} + 1 + k}. \tag{10.4}$$

An exact expression for the BER with the Q-function is also available.

Maximal ratio combining

In contrast to selection diversity, maximal ratio combining uses all the received signals. The combined signal is given by

$$x = \sum_{m=1}^{N} g_m y_m$$

$$= \sum_{m=1}^{N} g_m h_m b + \sum_{m=1}^{N} g_m n_m, \tag{10.5}$$

where g_m is the combining weight for the mth signal. The combining weights can be determined to maximize the SNR for x, which is given by

$$\gamma(\mathbf{g}) = \frac{\left| \sum_{m=1}^{N} g_m h_m \right|^2}{N_0 \sum_{m=1}^{N} |g_m|^2},$$

where $\mathbf{g} = [g_1 \ g_2 \ \cdots \ g_N]^{\mathrm{T}}$. From the Cauchy–Schwarz inequality, we can show that

$$\frac{\left| \sum_{m=1}^{N} g_m h_m \right|^2}{N_0 \sum_{m=1}^{N} |g_m|^2} \leq \frac{\sum_{m=1}^{N} |g_m|^2 \sum_{m=1}^{N} |h_m|^2}{N_0 \sum_{m=1}^{N} |g_m|^2} = \frac{\sum_{m=1}^{N} |h_m|^2}{N_0},$$

where the equality is achieved when $g_m \propto h_m^*$ or simply $g_m = h_m^*$. This results in the maximal ratio combining (MRC). The instantaneous SNR of the MRC is given by

$$\gamma_{\mathrm{mrc}} = \frac{\sum_{m=1}^{N} |h_m|^2}{N_0} = \sum_{m=1}^{N} \gamma_m. \tag{10.6}$$

Example 10.1.2 If $\bar{\gamma}_m = \bar{\gamma}$ for all m, γ_{mrc} has a chi-square pdf with $2N$ degrees of freedom, which is given by (Proakis, 1995)

$$f_{\mathrm{mrc}}(\gamma) = \frac{1}{(N-1)! \bar{\gamma}^N} \gamma^{N-1} e^{-\gamma/\bar{\gamma}}, \quad \gamma \geq 0.$$

The average BER of the MRC is upper bounded as

$$P_b = E[Q(\sqrt{2\gamma_{\mathrm{mrc}}})]$$

$$\leq E[\exp(-\gamma_{\mathrm{mrc}})]$$

$$= \int_0^\infty e^{-\gamma} f_{\mathrm{mrc}}(\gamma) \, d\gamma$$

$$= \frac{1}{(N-1)! \bar{\gamma}^N} \int_0^\infty e^{-(1+\frac{1}{\bar{\gamma}})\gamma} \gamma^{N-1} \, d\gamma.$$

Because

$$\int_0^\infty e^{-ax} x^n \, dx = \Gamma(n+1)/a^{n+1},$$

where $\Gamma(x)$ is a Gamma function, we have

$$P_b \le (1+\bar{\gamma})^{-N}$$
$$\simeq \frac{1}{\bar{\gamma}^N}, \quad \bar{\gamma} \gg 1, \tag{10.7}$$

where $\Gamma(n) = (n-1)!$ for a positive integer n. Now we can see that the BER decreases rapidly with the average SNR, $\bar{\gamma}$, for large N. For Rayleigh fading, if the BER of a diversity combining technique is given by

$$P_b \simeq \kappa \bar{\gamma}^{-n}, \quad \bar{\gamma} \gg 1, \tag{10.8}$$

where κ is a constant independent of $\bar{\gamma}$, then the exponent n in Eq. (10.8) is referred to as the *diversity order*. As shown in Eq. (10.8), it is clear that a larger n is required for a lower BER. In this way, we can see that MRC provides a diversity order of N. An exact expression for the BER is also available (Proakis, 1995) .

It is interesting to compare the BER of selection diversity with that of MRC. It is not easy to see the diversity order of the selection diversity from Eq. (10.4). Noting that

$$1 - e^{-\gamma/\bar{\gamma}} \simeq \frac{\gamma}{\bar{\gamma}}, \quad \bar{\gamma} \gg 1,$$

from a Taylor series, we can show that

$$E[\exp(-\gamma_{sd})] = \frac{N}{\bar{\gamma}} \int_0^\infty e^{-(1+\frac{1}{\bar{\gamma}})\gamma} \left(1 - e^{-\gamma/\bar{\gamma}}\right)^{N-1} d\gamma$$
$$\simeq \frac{N}{\bar{\gamma}} \int_0^\infty e^{-(1+\frac{1}{\bar{\gamma}})\gamma} \frac{\gamma^{N-1}}{\bar{\gamma}^{N-1}} d\gamma$$
$$= N! \left(\frac{1}{1+\bar{\gamma}}\right)^N$$
$$\simeq N! \bar{\gamma}^{-N}.$$

This shows that the diversity order of the selection diversity is also N. However, the performance is worse than that of MRC due to a multiplication factor, $N!$.

Equal gain combining

In Eq. (10.5), if $g_m = h_m^*/|h_m|$, the resulting combining is called the equal gain combining (EGC). We note that the amplitude of g_m is the same for all m.

We make the following remarks about diversity combining.

- To employ diversity combining techniques, the channel state information (CSI) is required. For selection diversity, we need to know the amplitudes of the h_m's so we can choose the signal of the highest (instantaneous) SNR. On the other hand, the phases of the h_m's are required for combining signals in EGC. The MRC requires both amplitudes and phases of the h_m's for combining.

- In practice, h_m has to be estimated. Due to the estimation error, the combining gain would be degraded.
- Selection diversity can be generalized using the idea of MRC. Among N signals, the $\bar{N}(< N)$ signals that have the highest SNRs can be chosen and combined by MRC. This technique is called generalized selection diversity (GSD).
- MMSE combining can be considered as

$$\mathbf{g}_{\text{mmse}} = E[|b - \mathbf{g}^T \mathbf{y}|^2],$$

where $\mathbf{y} = [y_1 \ y_2 \ \cdots \ y_N]^T$; MMSE combining becomes MRC since $\mathbf{g}_{\text{mmse}}$ is proportional to $\mathbf{h}^* = [h_1^* \ h_2^* \ \cdots \ h_N^*]^T$.

10.1.2 Transmit diversity with multiple transmit antennas

The diversity can be introduced by the transmitter. Generally, there are three different types of transmit diversity, depending on the use of CSI and signal mapping (or coding). Various approaches have been proposed for each type. Some approaches for transmit diversity are summarized in the following.

- Type I (coding only). For transmit diversity, coding is used without CSI. Space-time codes (Alamouti, 1998; Tarokh, Jafarkhani, and Calderbank, 1999; Tarokh, Seshadri, and Calderbank, 1998) belong to this group.
- Type II (CSI and coding). Both CSI and mapping can be used for transmit diversity.
- Type III (CSI only). Beamforming can be used to transmit signals with CSI.

A simple example of transmit diversity without CSI is the transmit delay diversity. Suppose there are two transmit antennas. The second antenna transmits the same signal as the first antenna, but with one symbol delay. Then, the received signal over flat fading channels is given by

$$y_l = h_{1,l} b_l + h_{2,l} b_{l-1} + n_l, \tag{10.9}$$

where $h_{k,l}$ denotes the channel coefficient from transmit antenna k to the receive antenna at time l (here, we assume one receive antenna for convenience). As shown in Eq. (10.9), the received signal behaves as if it comes through an ISI channel with two multipath components. Then, the MLSD is applicable. When the two transmit antennas are spaced far apart, the two channel coefficients can be independent. Thus, when one channel is not reliable, the other channel can be reliable with a high probability. This provides a diversity gain.

Transmit diversity has been studied extensively due to several important features for wireless communications. Since the diversity is introduced by the transmitter, the receiver needs only a single multiple receive antenna, which becomes very attractive because the mobile unit can offer the diversity gain with one receive antenna to improve the downlink's reliability. For a detailed account of transmit antenna diversity and space-time codes, the reader is referred to Jafarkhani (2005) and Larsson and Stoica (2003) .

10.2 MIMO detection

In this section, we will introduce several MIMO detection methods. As CDMA channels are MIMO channels, most multiuser detection methods for CDMA systems are applicable to MIMO detection and vice versa.

10.2.1 Multiple transmit and receive antennas

Suppose there are K transmit and N receive antennas. The channel is assumed to be flat fading. Let $\mathbf{H}_l$ represent the $N \times K$ channel matrix at symbol time l, where $[\mathbf{H}_l]_{n,k} = h_{n,k,l}$ denotes the channel coefficient from the kth transmit antenna to the nth receive antenna. Denote by $b_{k,l}$ and $y_{n,l}$ the symbol transmitted by the kth transmit antenna and the received signal from the nth receive antenna during the lth symbol interval, respectively. Then, the received signal vector is given by

$$\mathbf{y}_l = [y_{1,l} \ y_{2,l} \ \cdots \ y_{N,l}]^{\mathrm{T}}$$
$$= \mathbf{H}_l \mathbf{b}_l + \mathbf{n}_l, \quad l = 0, 1, \ldots, L - 1, \tag{10.10}$$

where L is the length of a data packet, $\mathbf{b}_l = [b_{1,l} \ b_{2,l} \ \cdots \ b_{K,l}]^{\mathrm{T}}$, and $\mathbf{n}_l = [n_{1,l} \ n_{2,l} \ \cdots \ n_{N,l}]^{\mathrm{T}}$ is the noise vector generally assumed to be a complex Gaussian random vector with $E[\mathbf{n}_l] = \mathbf{0}$ and $E[\mathbf{n}_l \mathbf{n}_{l'}^H] = N_0 \mathbf{I} \delta_{l,l'}$. Note that the indexes n, k, and l are used for the receive antenna, transmit antenna, and the symbol, respectively.

It is shown in Foschini and Gans (1998) and Telatar (1999) that the capacity of an MIMO Rayleigh channel grows *linearly* with the minimum of the numbers of transmit and receive antennas. That is, the capacity is linearly proportional to $\min(N, K)$. It is also known that there is a fundamental trade-off between diversity gain and spatial multiplexing gain (Zheng and Tse, 2003). The reader is referred to Paulraj *et al.* (2003) and Tse and Viswanath (2005) for a detailed account of MIMO capacity and related issues.

10.2.2 Optimal detectors

For convenience, we omit the symbol-duration index l in Eq. (10.10). The resulting received signal at each symbol interval is given by

$$\mathbf{y} = \mathbf{Hb} + \mathbf{n}. \tag{10.11}$$

In addition, assume that b_k is binary, i.e. $b_k \in \{-1, +1\}$.

There are several performance criteria for MIMO detection. Firstly, we consider the ML detection using Eq. (10.11). The ML detection finds the data symbol vector that maximizes the likelihood function as follows:

$$\mathbf{b}_{\mathrm{ml}} = \arg \max_{\mathbf{b}} f(\mathbf{y}|\mathbf{b})$$
$$= \arg \min_{\mathbf{b}} ||\mathbf{y} - \mathbf{Hb}||^2. \tag{10.12}$$

To identify the ML vector, an exhaustive search is required. Because the number of candidate vectors for $\mathbf{b}$ is 2^K, the complexity grows exponentially with K.

If the *a priori* probability of **b** is available, the MAP sequence detection can be formulated. The MAP vector becomes

$$
\begin{aligned}
\mathbf{b}_{\mathrm{map}} &= \arg\max_{\mathbf{b}} \Pr(\mathbf{b}|\mathbf{y}) \\
&= \arg\max_{\mathbf{b}} f(\mathbf{y}|\mathbf{b})\Pr(\mathbf{b}),
\end{aligned}
\tag{10.13}
$$

where $\Pr(\mathbf{b})$ denotes the *a priori* probability of **b**. In addition, for the MAP symbol detection using Eq. (10.11), the *a posteriori* probability of each data symbol can be found by marginalization as follows:

$$
\begin{aligned}
\Pr(b_k = +1|\mathbf{y}) &= \sum_{\mathbf{b}\in\mathcal{B}_k^+} \Pr(\mathbf{b}|\mathbf{y}), \\
\Pr(b_k = -1|\mathbf{y}) &= \sum_{\mathbf{b}\in\mathcal{B}_k^-} \Pr(\mathbf{b}|\mathbf{y}),
\end{aligned}
\tag{10.14}
$$

where $\mathcal{B}_k^+ = \{[b_1 \ b_2 \ \cdots \ b_K]^{\mathrm{T}} \mid b_k = 1, b_m \in \{+1, -1\}, \ \forall m \neq k\}$ and $\mathcal{B}_k^- = \{[b_1 \ b_2 \ \cdots \ b_K]^{\mathrm{T}} \mid b_k = -1, b_m \in \{+1, -1\}, \ \forall m \neq k\}$. For both MAP sequence detection and MAP symbol detection, the complexity grows exponentially with K.

It is easy to perform the (linear) MMSE detection if the constraint on the symbol vector, $b_k \in \{-1, +1\}, \forall k$, is not imposed. Using the orthogonality principle, the MMSE estimator for **b** can be found as

$$
\begin{aligned}
\mathbf{W}_{\mathrm{mmse}} &= \arg\min_{\mathbf{W}} E[||\mathbf{b} - \mathbf{W}^{\mathrm{H}}\mathbf{y}||^2] \\
&= \left(E[\mathbf{y}\mathbf{y}^{\mathrm{H}}]\right)^{-1} E[\mathbf{y}\mathbf{b}^{\mathrm{H}}].
\end{aligned}
\tag{10.15}
$$

We can show that

$$
\begin{aligned}
E[\mathbf{y}\mathbf{y}^{\mathrm{H}}] &= \mathbf{H}\mathbf{H}^{\mathrm{H}} + N_0\mathbf{I}, \\
E[\mathbf{y}\mathbf{b}^{\mathrm{H}}] &= \mathbf{H}.
\end{aligned}
$$

It follows that

$$
\mathbf{W}_{\mathrm{mmse}} = (\mathbf{H}\mathbf{H}^{\mathrm{H}} + N_0\mathbf{I})^{-1}\mathbf{H}
$$

and

$$
\begin{aligned}
\hat{\mathbf{b}}_{\mathrm{mmse}} &= \mathbf{W}_{\mathrm{mmse}}^{\mathrm{H}}\mathbf{y} \\
&= \mathbf{H}^{\mathrm{H}}(\mathbf{H}\mathbf{H}^{\mathrm{H}} + N_0\mathbf{I})^{-1}\mathbf{y}.
\end{aligned}
\tag{10.16}
$$

Using the matrix inversion lemma, it follows that

$$
\begin{aligned}
\mathbf{R}_{\mathrm{mmse}} &= E\left[\left(\mathbf{b} - \mathbf{W}_{\mathrm{mmse}}^{\mathrm{H}}\mathbf{y}\right)\left(\mathbf{b} - \mathbf{W}_{\mathrm{mmse}}^{\mathrm{H}}\mathbf{y}\right)^{\mathrm{H}}\right] \\
&= \mathbf{I} - \mathbf{H}^{\mathrm{H}}(\mathbf{H}\mathbf{H}^{\mathrm{H}} + N_0\mathbf{I})^{-1}\mathbf{H} \\
&= \left(\mathbf{I} + \frac{1}{N_0}\mathbf{H}^{\mathrm{H}}\mathbf{H}\right)^{-1}.
\end{aligned}
\tag{10.17}
$$

The MSE for each symbol estimate of the MMSE detection can be obtained from the corresponding diagonal element of $\mathbf{R}_{\mathrm{mmse}}$.

Example 10.2.1   From Eq. (10.16), the MMSE estimate of the single symbol b_k is given by

$$\hat{b}_{\mathrm{mmse},k} = \mathbf{h}_k^{\mathrm{H}} \mathbf{A}^{-1} \mathbf{y}, \tag{10.18}$$

where $\mathbf{A} = \mathbf{H}\mathbf{H}^{\mathrm{H}} + N_0 \mathbf{I}$. We will next show that the SINR of $\hat{b}_{\mathrm{mmse},k}$ is a function of the MSE. From Eq. (10.11), Eq. (10.18) is rewritten as follows:

$$\hat{b}_{\mathrm{mmse},k} = \mathbf{h}_k^{\mathrm{H}} \mathbf{A}^{-1} \mathbf{h}_k b_k + \mathbf{h}_k^{\mathrm{H}} \mathbf{A}^{-1} \left(\sum_{q \neq k} \mathbf{h}_q b_q + \mathbf{n} \right). \tag{10.19}$$

The SINR becomes

$$\mathrm{SINR}_k = \frac{\left| \mathbf{h}_k^{\mathrm{H}} \mathbf{A}^{-1} \mathbf{h}_k \right|^2}{\mathbf{h}_k^{\mathrm{H}} \mathbf{A}^{-1} \left(\mathbf{A} - \mathbf{h}_k \mathbf{h}_k^{\mathrm{H}} \right) \mathbf{A}^{-1} \mathbf{h}_k}$$

$$= \frac{\mathbf{h}_k^{\mathrm{H}} \mathbf{A}^{-1} \mathbf{h}_k}{1 - \mathbf{h}_k^{\mathrm{H}} \mathbf{A}^{-1} \mathbf{h}_k}.$$

From Eq. (10.17), the MSE of $\hat{b}_{\mathrm{mmse},k}$ is upper-bounded as

$$\mathrm{MSE}_k = 1 - \mathbf{h}_k^{\mathrm{H}} \mathbf{A}^{-1} \mathbf{h}_k \leq 1.$$

Consequently,

$$\mathrm{SINR}_k = \frac{1 - \mathrm{MSE}_k}{\mathrm{MSE}_k} \tag{10.20}$$

or

$$\mathrm{MSE}_k = \frac{1}{1 + \mathrm{SINR}_k}. \tag{10.21}$$

This relationship shows that minimizing the MSE is equivalent to maximizing the SINR.

10.2.3   Nulling and cancelation detectors

A category of transmission schemes over MIMO channels has been proposed by Foschini (1996) and analyzed in Foschini *et al.* (2003). A nulling and cancelation (N/C) detector is also proposed.

A simple N/C detector is based on the QR factorization of the channel matrix $\mathbf{H}$. For convenience, assume that $N = K$. The channel matrix $K \times K$ $\mathbf{H}$ can be factorized as

$$\mathbf{H} = \mathbf{Q}\mathbf{R}, \tag{10.22}$$

where $\mathbf{Q}$ is unitary and $\mathbf{R}$ is upper triangular. Multiplying $\mathbf{Q}^{\mathrm{H}}$ to $\mathbf{y}$ in Eq. (10.11) results in

$$\mathbf{x} = \mathbf{Q}^{\mathrm{H}} \mathbf{y}$$

$$= \mathbf{R}\mathbf{b} + \mathbf{Q}^{\mathrm{H}} \mathbf{n}. \tag{10.23}$$

We note that $\mathbf{Q}^{\mathrm{H}} \mathbf{n}$ has the same statistical properties as $\mathbf{n}$: both $\mathbf{Q}^{\mathrm{H}} \mathbf{n}$ and $\mathbf{n}$ are zero-mean complex Gaussian random vectors. In addition, we can show that

$$E[\mathbf{Q}^{\mathrm{H}} \mathbf{n}\mathbf{n}^{\mathrm{H}} \mathbf{Q}] = \mathbf{Q}^{\mathrm{H}} \underbrace{E[\mathbf{n}\mathbf{n}^{\mathrm{H}}]}_{=N_0\mathbf{I}} \mathbf{Q} = N_0 \mathbf{I}.$$

As they have the same statistical properties, we simply use $\mathbf{n}$ to denote $\mathbf{Q}^H\mathbf{n}$. The resulting signal is given by

$$\mathbf{x} = \mathbf{R}\mathbf{b} + \mathbf{n}. \tag{10.24}$$

From Eq. (10.24), we can show that

$$x_K = r_{K,K}b_K + n_K,$$
$$x_{K-1} = r_{K-1,K}b_K + r_{K-1,K-1}b_{K-1} + n_{K-1},$$
$$\vdots \tag{10.25}$$

This suggests a sequential detection procedure.

Firstly, b_K can be detected from x_K. Note that there is no interference. Secondly, the hard-decision of b_K is to be canceled in detecting b_{K-1} from x_{K-1}. The kth data symbol, b_k, can be detected, after canceling $K - k$ data symbols, as follows:

$$u_k = x_k - \sum_{q=k+1}^{K} r_{k,q}\hat{b}_q, \quad k \in \{1, 2, \ldots, K - 1\},$$

where $\hat{b}_q$ denotes the hard-decision estimate of b_q from u_q. This is the QR factorization-based N/C detector.

This approach has some drawbacks. For example, the performance is degraded by error propagation. If incorrect decision(s) are made in the early stages, the subsequent cancelation becomes erroneous and more incorrect decisions can follow. This implies that the overall performance is strongly dependent on the first detection for b_K. A detailed performance analysis can be found in Choi (2006).

In the QR factorization-based N/C detector, the transformation, $\mathbf{Q}^H$, performs progressing nulling. More signals are nullified as k increases in Eqs (10.25). We can consider this N/C detector as analogous to the ZF-DFE over ISI channels. Accounting for the background noise in the nulling operation, the N/C detector would be improved. For example, an MMSE filtering approach can be employed as in the MMSE-DFE.

From $\mathbf{y}$ in Eq. (10.11), symbol b_1 can be detected using the MMSE filter:

$$\mathbf{w}_{\text{mmse},1} = \arg\min_{\mathbf{w}} E[|b_1 - \mathbf{w}^H\mathbf{y}|^2]$$
$$= (\mathbf{H}\mathbf{H}^H + N_0\mathbf{I})^{-1}\mathbf{h}_1, \tag{10.26}$$

where $\mathbf{h}_k$ stands for the kth column of $\mathbf{H}$. Once b_1 is estimated, a hard-decision is used to cancel its component to detect the other symbols. Let $\hat{b}_1$ denote a hard-decision of b_1. Then, the signal after cancelation is given by

$$\mathbf{y}_1 = \mathbf{y} - \mathbf{h}_1\hat{b}_1. \tag{10.27}$$

Assuming that $\hat{b}_1 = b_1$,

$$\mathbf{y}_1 = \sum_{k=2}^{K} \mathbf{h}_k b_k + \mathbf{n}. \tag{10.28}$$

With $\mathbf{y}_1$, the MMSE filtering to detect b_2 can be obtained. Repeating cancelation and MMSE filtering, the detection of the b_k's can be performed. This N/C detector also suffers from error propagation. Hence, the overall performance is limited by the first detection. To overcome this problem, an ordering can be considered.

In an ordered N/C detector, the first symbol to be detected is the symbol that has the smallest MSE or (equivalently) highest SINR:

$$\{\alpha(1), \mathbf{w}_{\alpha(1)}\} = \arg \min_k \min_{\mathbf{w}} E[|b_k - \mathbf{w}^H \mathbf{y}|^2],$$

where $\alpha(1)$ denotes the index of the data symbol that has the smallest MSE and $\mathbf{w}_{\alpha(1)}$ denotes the corresponding MMSE filtering vector. Then, the cancelation is carried out as follows:

$$\mathbf{y}_1 = \mathbf{y}_0 - \mathbf{h}_{\alpha(1)} \hat{b}_{\alpha(1)}, \qquad (10.29)$$

where $\mathbf{y}_0 = \mathbf{y}$ and $\hat{b}_{\alpha(1)}$ denotes a hard-decision of $b_{\alpha(1)}$ from $\mathbf{w}_{\alpha(1)}^H \mathbf{y}_0$. With $\mathbf{y}_1$, the next symbol to be detected is found as

$$\{\alpha(2), \mathbf{w}_{\alpha(2)}\} = \arg \min_{k \in I_1} \min_{\mathbf{w}} E[|b_k - \mathbf{w}^H \mathbf{y}_1|^2],$$

where $I_1 = I_0 \setminus \alpha(1)$ and $I_0 = \{1, 2, \ldots, K\}$. The cancelation and MMSE filtering can be repeated until all symbols are detected. This algorithm is summarized as follows.

(S0) Let $\mathbf{y}_0 = \mathbf{y}$ and $I_0 = \{1, 2, \ldots, K\}$. Set $k = 1$.
(S1) Solve the following optimization problem:

$$\{\alpha(k), \mathbf{w}_{\alpha(k)}\} = \arg \min_{q \in I_{k-1}} \min_{\mathbf{w}} E[|b_q - \mathbf{w}^H \mathbf{y}_{k-1}|^2] \qquad (10.30)$$

(S2) Cancel the detected signal:

$$\mathbf{y}_k = \mathbf{y}_{k-1} - \mathbf{h}_{\alpha(k)} \hat{b}_{\alpha(k)} \qquad (10.31)$$

and update $I_k = I_{k-1} \setminus \alpha(k)$.
(S3) If $k = K$, stop. Otherwise, $k = k + 1$ and go to (S1).

The ordered MMSE N/C detector discussed above was proposed in Wolniansky *et al.* (1998). A discussion on the N/C detector and its application can be found in Foschini *et al.* (2003).

Example 10.2.2 Consider a 3×3 MIMO channel as follows:

$$\mathbf{H} = \begin{bmatrix} 1 & \frac{1}{2} & 2 \\ 2 & -1 & 1 \\ 0 & 1 & 3 \end{bmatrix}.$$

With $N_0 = 1$, the MMSE filtering vectors can be found as follows:

$$\begin{aligned}
\mathbf{W}_{\mathrm{mmse}} &= [\mathbf{w}_{\mathrm{mmse},1} \ \mathbf{w}_{\mathrm{mmse},2} \ \mathbf{w}_{\mathrm{mmse},3}] \\
&= (\mathbf{H}\mathbf{H}^H + N_0\mathbf{I})^{-1}\mathbf{H} \\
&= \begin{bmatrix} 0.1884 & 0.2013 & 0.0428 \\ 0.2355 & -0.2484 & 0.0535 \\ -0.1542 & 0.0171 & 0.2377 \end{bmatrix}.
\end{aligned}$$

With the MMSE filters, the MSEs are given by $\{0.3405, 0.6338, 0.1478\}$. Thus, $\alpha(1) = 3$, i.e. the third signal is chosen for the first detection and cancelation, because its MSE is the smallest – the MSE of the third signal is 0.1478. Then, after cancelling the third signal, the MMSE filtering vectors for the first and second signals are given by

$$
\begin{aligned}
\mathbf{W}_{\mathrm{mmse}} &= [\mathbf{w}_{\mathrm{mmse},1} \ \ \mathbf{w}_{\mathrm{mmse},2}] \\
&= ([\mathbf{h}_1 \ \ \mathbf{h}_2][\mathbf{h}_1 \ \ \mathbf{h}_2]^{\mathrm{H}} + N_0\mathbf{I})^{-1}[\mathbf{h}_1 \ \ \mathbf{h}_2] \\
&= \begin{bmatrix} 0.2319 & 0.2609 \\ 0.2899 & -0.1739 \\ 0.0870 & 0.3478 \end{bmatrix}.
\end{aligned}
$$

The MSEs are given by $\{0.1884, 0.3478\}$. Then, the first signal is chosen for the next (i.e. second) detection and cancelation, i.e. $\alpha(2) = 1$. The MSE of the first signal is 0.1884. As $\alpha(3) = 2$ (the second signal is the last signal), the MMSE filtering vector for the second signal is now found as follows:

$$
\mathbf{w}_{\mathrm{mmse},2} = \left(\mathbf{h}_2\mathbf{h}_2^{\mathrm{H}} + N_0\mathbf{I}\right)^{-1}\mathbf{h}_2 = [0.1538 \ -0.3077 \ 0.3077]^{\mathrm{T}}.
$$

The MSE becomes 0.3077.

We note that, without cancelation, the MSEs are $\{0.3405, 0.6338, 0.1478\}$, which are given in the first stage. If the cancelation was perfect, the MSEs of the MMSE N/C detector would be $\{0.3077, 0.1884, 0.1478\}$.

10.2.4 Probabilistic approaches

Applied probabilistic data association (PDA) was developed in Luo *et al.* (2001) for signal tracking (Bar-Shalom and Li, 1993) to multiuser detection. This approach can also be applied to the MIMO detection as in Pham *et al.* (2004). The PDA approach is closely related to the MMSE-SC detector (see Chapter 9). We will first derive the PDA algorithm for MIMO detection based on Luo *et al.* (2001), and then we discuss its relation to the MMSE-SC detector.

Assume that $N \geq K$ and the rank of $\mathbf{H}$ is K. Then, there exists a pseudo-inverse of $\mathbf{H}$, denoted by $\mathbf{H}^\dagger$, which satisfies $\mathbf{H}^\dagger\mathbf{H} = \mathbf{I}$. Multiplying both sides of Eq. (10.11) by $\mathbf{H}^\dagger$:

$$
\tilde{\mathbf{y}} = \mathbf{b} + \mathbf{H}^\dagger\mathbf{n}. \tag{10.32}
$$

Define the basis vectors as follows:

$$
\mathbf{u}_k = [0 \ \cdots \ 0 \ \ \underbrace{1}_{\text{the } k\text{th element}} \ \ 0 \cdots 0]^{\mathrm{T}}, \quad k = 1, 2, \ldots, K.
$$

Then, Eq. (10.32) is rewritten as follows:

$$
\tilde{\mathbf{y}} = b_k\mathbf{u}_k + \sum_{q \neq k} b_q\mathbf{u}_q + \tilde{\mathbf{n}}, \tag{10.33}
$$

where $\tilde{\mathbf{n}} = \mathbf{H}^\dagger\mathbf{n}$. Suppose that the *a priori* probabilities of the b_q's are available. The interfering vector $\mathbf{v}_k = \sum_{q \neq k} b_q\mathbf{u}_q + \tilde{\mathbf{n}}$ is now assumed to be a Gaussian vector (this assumption is called the forced Gaussian assumption). The mean and covariance of $\mathbf{v}_k$ are given by

$$E[\mathbf{v}_k] = \sum_{q \neq k} \bar{b}_q \mathbf{u}_q,$$

$$Cov(\mathbf{v}_k) = \sum_{q \neq k} (1 - \bar{b}_q^2) \mathbf{u}_q \mathbf{u}_q^{\mathrm{T}} + N_0 \mathbf{H}^{\dagger} (\mathbf{H}^{\dagger})^{\mathrm{H}}, \tag{10.34}$$

respectively, where $\bar{b}_q = E[b_q] = \Pr(b_q = 1) - \Pr(b_q = -1)$. For convenience, the Gaussian pdf of $\mathbf{v}_k$ is denoted by $\mathcal{N}(E[\mathbf{v}_k], Cov(\mathbf{v}_k))$.

From Eq. (10.33), we have

$$\tilde{\mathbf{y}} = \mathbf{u}_k b_k + \mathbf{v}_k.$$

Then, the (unnormalized) *a posteriori* probability (called associated probability) of b_k can be found as follows:

$$P_{\mathrm{as}}(b_k) = \mathcal{N}(\tilde{\mathbf{y}} - E[\mathbf{v}_k] - \mathbf{u}_k b_k, Cov(\mathbf{v}_k)) \Pr(b_k = 1) \tag{10.35}$$

using the forced Gaussian assumption. Then, from Eq. (10.35), the LLR of b_k becomes

$$\begin{aligned} \mathrm{LLR}(b_k) &= \log \frac{P_{\mathrm{as}}(b_k = +1)}{P_{\mathrm{as}}(b_k = -1)} - \log \frac{\Pr(b_k = +1)}{\Pr(b_k = -1)} \\ &= \log \frac{\mathcal{N}(\tilde{\mathbf{y}} - E[\mathbf{v}_k] - \mathbf{u}_k, Cov(\mathbf{v}_k))}{\mathcal{N}(\tilde{\mathbf{y}} - E[\mathbf{v}_k] + \mathbf{u}_k, Cov(\mathbf{v}_k))} \\ &= 4 \mathbf{u}_k^{\mathrm{T}} (Cov(\mathbf{v}_k))^{-1} E[\mathbf{v}_k]. \end{aligned} \tag{10.36}$$

From this LLR, the *a priori* probability of $\Pr(b_k)$ can be updated for the next iteration to re-compute the associated probability. The PDA algorithm for MIMO detection is summarized as follows.

(S0) Let $q = 0$, where q is the iteration counter. Let $\Pr(b_k; q) = 1/2$ for all k, where $\Pr(b_k; q)$ stands for the *a priori* probability of b_k at iteration q.
(S1) Find the LLR in Eq. (10.36) for all k.
(S2) Let $q = q + 1$. Update the *a priori* probability as $\Pr(b_k = +1; q) = \tanh(\mathrm{LLR}(b_k)/2)$ and $\Pr(b_k = -1; q) = 1 - \Pr(b_k = +1; q)$.
(S3) If $\Pr(b_k; q)$ converges, stop. Otherwise, go to (S1).

This PDA algorithm uses decorrelation, an idea proposed in Luo *et al.* (2001). The MMSE filtering for the MMSE-SC detector described in Chapter 9 is also applicable. This modification increases the complexity because the MMSE filtering vector has to be updated for each iteration. However, the performance can be improved. The PDA algorithm also suffers from error propagation. This implies that the ordering becomes important as in the N/C detector. In Luo *et al.* (2001), an optimal ordering is used to mitigate the error propagation.

10.2.5 Ideal performance

Most suboptimal MIMO detectors cannot fully exploit the diversity gain from multiple receive antennas since the interference from the other antennas, which is called co-antenna interference (CAI), cannot be perfectly canceled.

The received signal vector in Eq. (10.11) can be rewritten as follows:

$$\mathbf{y} = \mathbf{h}_k b_k + \sum_{q \neq k} \mathbf{h}_q b_q + \mathbf{n}.$$

If the *a priori* probability of b_q, $q \neq k$, is available and reliable, we can cancel b_q, $q \neq k$, and the resulting signal vector is given by

$$\mathbf{y}_k = \mathbf{y} - \sum_{q \neq k} \mathbf{h}_q b_q$$
$$= \mathbf{h}_k b_k + \mathbf{n}. \tag{10.37}$$

The LLR of b_k from $\mathbf{y}_k$ is given by

$$\mathrm{LLR}(b_k; \mathbf{y}_k) = \log \frac{\exp\left(-\frac{1}{N_0}||\mathbf{y}_k - \mathbf{h}_k||^2\right)}{\exp\left(-\frac{1}{N_0}||\mathbf{y}_k + \mathbf{h}_k||^2\right)}$$
$$= \frac{4}{N_0}\Re\left(\mathbf{y}_k^{\mathrm{H}}\mathbf{h}_k\right). \tag{10.38}$$

This ideal LLR can be seen as an output of the matched filter or the MRC. Thus, a full diversity gain from multiple receive antennas can be achieved. The performance obtained from Eq. (10.38) is called the matched filter bound (MFB).

For uncoded signals, the MFB is an optimistic (lower) bound. However, for coded signals, the MFB becomes tighter since reliable prior information becomes available from the channel decoder.

10.2.6 Frequency-selective fading

In Eq. (10.10), we assumed a flat fading for MIMO channels. For a frequency-selective fading channel, the received signal vector may be written as follows:

$$\mathbf{y}_l = \sum_{p=0}^{P-1} \mathbf{H}_{p,l} \mathbf{b}_{l-p} + \mathbf{n}_l, \quad l = 0, 1, \ldots, L-1,$$

where $\mathbf{H}_{p,l}$ denotes the pth multipath channel matrix at time l and P denotes the number of multipaths. Assuming that $\mathbf{b}_l = \mathbf{0}$ for $l < 0$ and $L \gg P$, it can be shown that

$$\mathbf{y} = \begin{bmatrix} \mathbf{y}_0^{\mathrm{T}} & \mathbf{y}_1^{\mathrm{T}} & \cdots & \mathbf{y}_{L-1}^{\mathrm{T}} \end{bmatrix}^{\mathrm{T}}$$
$$= \underbrace{\begin{bmatrix} \mathbf{H}_0 & \mathbf{0} & \cdots & \mathbf{0} \\ \mathbf{H}_1 & \mathbf{H}_0 & \cdots & \mathbf{0} \\ \vdots & \vdots & \ddots & \vdots \\ \mathbf{0} & \mathbf{0} & \cdots & \mathbf{H}_0 \end{bmatrix}}_{=\mathbf{H}} \underbrace{\begin{bmatrix} \mathbf{b}_0 \\ \mathbf{b}_1 \\ \vdots \\ \mathbf{b}_{L-1} \end{bmatrix}}_{=\mathbf{b}} + \underbrace{\begin{bmatrix} \mathbf{n}_0 \\ \mathbf{n}_1 \\ \vdots \\ \mathbf{n}_{L-1} \end{bmatrix}}_{=\mathbf{n}}, \tag{10.39}$$

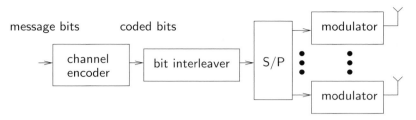

Figure 10.2. Block diagram of MIMO-BICM transmitter.

where the $LN \times LK$ **H** becomes the channel matrix, which is a block Toeplitz matrix. From Eq. (10.39),

$$\mathbf{y} = \mathbf{Hb} + \mathbf{n}.$$

Thus, the previously discussed detection methods (for an MIMO flat fading channel) can also be applied to the MIMO detection for frequency-selective fading. Since a straightforward application results in high computational complexity for large L, it would be neccessary to exploit the trellis structure of an MIMO ISI channel to reduce the complexity.

10.3 Iterative receivers with known channels

The main idea behind the iterative receiver for MIMO channels is basically similar to that for CDMA channels. The structure of the iterative receiver for MIMO channels depends on the transmitter. Several different transmission schemes are discussed in Foschini (1996). In this section, however, we focus on the MIMO-BICM and we study iterative receivers for MIMO-BICM signals.

10.3.1 BICM for MIMO channels

A block diagram of an MIMO-BICM transmitter is shown in Fig. 10.2, in which there is a single channel code. An interleaved coded bit sequence is split into multiple streams to transmit through multiple transmit antennas. Since a single channel code is used, the length of codeword can be long. The interleaver is essential to break the correlation of a coded sequence.

10.3.2 Structure of iterative receiver

The structure of the iterative receiver for MIMO channels is depicted in Fig. 10.3. The MIMO detector and channel decoder exchange the LLR of the data symbols. The principles of the iterative receivers are the same as those for ISI channels discussed in Chapter 6. In ISI channels, the MAP equalizer using the BCJR algorithm can be computationally efficient because of the trellis structure of an ISI channel. However, in MIMO channels, due to the lack of a trellis structure, suboptimal methods for the MAP detection are required to implement a lower complexity implementation.

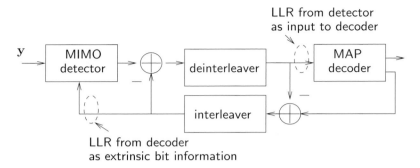

Figure 10.3. Block diagram of the iterative receiver for MIMO-BICM signals.

10.3.3 MIMO detectors for the iterative receiver

There are several requirements for an MIMO detector when used as an iterative receiver as follows.

- Soft input: the input to the MIMO detector takes the form of demodulated signal vectors. Hence, the MIMO detector has to accept soft-inputs so as not to lose any information of demodulated signals.
- Soft output: the output of the MIMO detector becomes the input of channel decoder. Since the performance of channel decoder is better when soft-input is available, the MIMO detector needs to find soft-outputs.
- Extrinsic input: together with demodulated signal vectors, the extrinsic input is available from the channel decoder. Thus, the MIMO detector should make use of the extrinsic input for more reliable outputs. In addition, it is desirable to provide the ideal LLR in Eq. (10.38) if a reliable extrinsic information is available.

The MIMO detectors introduced in Section 10.2 can be used for MIMO detection in the iterative receiver with or without modifications. In addition to these detectors, we introduce some other approaches for MIMO detection with the requirements given above in this section.

Sphere decoding and its modifications

The sphere decoding technique was introduced in Viterbo and Boutros (1999) to perform the ML decoding. The sphere decoding technique can be easily extended to solve the ML detection problem for MIMO channels.

Recall the ML detection in Eq. (10.12). Assume that $\mathbf{H}$ is square (a $K \times K$ matrix) and full-rank. Letting

$$\mathbf{H} = \mathbf{QR}$$

and

$$\hat{\mathbf{b}} = (\mathbf{H}^{\mathrm{H}}\mathbf{H})^{-1}\mathbf{H}^{\mathrm{H}}\mathbf{y},$$

we can show that

$$||\mathbf{y} - \mathbf{H}\mathbf{b}||^2 = ||\mathbf{y}||^2 - ||\mathbf{H}\hat{\mathbf{b}}||^2 + ||\mathbf{H}(\mathbf{b} - \hat{\mathbf{b}})||^2$$
$$= ||\mathbf{y}||^2 - ||\mathbf{H}\hat{\mathbf{b}}||^2 + ||\mathbf{R}(\mathbf{b} - \hat{\mathbf{b}})||^2. \tag{10.40}$$

Here, we assume that the inverse of $\mathbf{H}\mathbf{H}^H$ exists. If it does not, a pseudo-inverse can be used. In Eq. (10.40), the first and second terms are independent of $\mathbf{b}$ and can be deleted.

Because there are a number of candidate vectors for an exhaustive search, we can attempt to find a *subset* of $\mathbf{b}$ in which the ML vector exists to reduce the complexity. If the subset has a small number of candidate vectors, the ML vector can be found with a low computational complexity. To this end, consider a subset $C(r)$ defined as

$$C(r) = \{\mathbf{b} \mid ||\mathbf{R}(\mathbf{b} - \hat{\mathbf{b}})||^2 \le r^2\},$$

where $r > 0$ is a given radius. The ML vector should be found in $C(r)$ as the ML vector has the smallest cost:

$$\mathbf{b}_{\mathrm{ml}} = \arg \min_{\mathbf{b}} ||\mathbf{R}(\mathbf{b} - \hat{\mathbf{b}})||^2.$$

In general, the number of vectors in $C(r)$ increases with r. Thus, it is necessary to find $C(r)$ with a small r. The sphere decoding technique finds $C(r)$ for a given r. Note that if $C(r)$ is empty for some value of r, a larger value of r is needed for a nonempty $C(r)$.

Noting that $\mathbf{R}$ is upper triangular, we can show that

$$||\mathbf{R}(\mathbf{b} - \hat{\mathbf{b}})||^2 = \sum_{k=1}^{K} r_{k,k}^2 \underbrace{\left[(b_k - \hat{b}_k) + \sum_{q=k+1}^{K} \frac{r_{k,q}}{r_{k,k}} (b_q - \hat{b}_q) \right]^2}_{=d_k^2} \le r^2. \tag{10.41}$$

From Eq. (10.41), the following constraints can be derived:

$$r_{K,K}^2 d_K^2 \le r^2,$$
$$r_{K-1,K-1}^2 d_{K-1}^2 \le r^2 - r_{K,K}^2 d_K^2,$$
$$\vdots \tag{10.42}$$
$$r_{1,1}^2 d_1^2 \le r^2 - \sum_{k=2}^{K} r_{k,k}^2 d_k^2.$$

Without loss of generality, we can assume that $r_{k,k} > 0$ for all k. With $k = K$, it follows that

$$r_{K,K}^2 (b_K - \hat{b}_K)^2 \le r^2$$

or

$$\hat{b}_K - \frac{r}{r_{K,K}} \le b_K \le \hat{b}_K + \frac{r}{r_{K,K}}.$$

Define the set of candidate symbols for b_K as follows:

$$S_K(r) = \left\{ b_K | b_K \in \mathcal{B}, \ \hat{b}_K - \frac{r}{r_{K,K}} \le b_K \le \hat{b}_K + \frac{r}{r_{K,K}} \right\},$$

where $\mathcal{B} = \{-1, +1\}$. For a higher order modulation, $\mathcal{B}$ can be generalized.

With the same approach, from Eqs (10.42), we can find the upper and lower bounds for b_{K-1} as follows:

$$b_{K-1} \le \hat{b}_{K-1} + \frac{\sqrt{r^2 - r_{K,K}^2(b_K - \hat{b}_K)}}{r_{K-1,K-1}} - \frac{r_{K-1,K}}{r_{K,K}}(b_K - \hat{b}_K) \qquad (10.43)$$

and

$$b_{K-1} \ge \hat{b}_{K-1} - \frac{\sqrt{r^2 - r_{K,K}^2(b_K - \hat{b}_K)}}{r_{K-1,K-1}} - \frac{r_{K-1,K}}{r_{K,K}}(b_K - \hat{b}_K). \qquad (10.44)$$

With the upper and lower bounds (associated with $b_K \in S_K(r)$) and the constraint $b_{K-1} \in \mathcal{B}$, $S_{K-1}(r)$ can be obtained. The sets of candidates for b_k's, i.e. $S_k(r)$'s, can be found recursively from $k = K$ back to $k = 1$ (backward recursion). If $S_k(r)$ becomes an empty set for some k, r is too small and there is no vector in $C(r)$. Hence, a larger value of r has to be chosen and the procedure has to be repeated.

If $S_1(r)$ is not empty, we can build $C(r)$ from $S_k(r)$, $k = 1, 2, \ldots, K$, i.e.

$$C(r) = S_1(r) \times S_2(r) \times \cdots \times S_K(r),$$

where $\times$ denotes the Cartesian product. Then, an exhaustive search may be performed to find the ML vector with $C(r)$. If there are too many vectors in $C(r)$, there is no benefit in using the sphere decoding technique. On the other hand, if r is too small, the procedure would be repeated as $C(r)$ can be empty.

We note that the sphere decoding technique solves the ML detection and provides a hard-decision, which is the ML vector. Thus, the sphere decoding technique has to be modified for the iterative receiver. MIMO detection algorithms based on modified sphere decoding techniques are proposed in Hochwald and Brink (2003) and Vikalo, Hassibi, and Kailath (2004). One key modification is to include in the iterative receiver the *a priori* probability that comes from the channel decoder. Another essential modification is to provide soft-output from multiple candidate vectors. We will not go into the details of the modifications because we will consider similar modifications to the M-algorithm.

M-algorithm

The M-algorithm is suboptimal for channel decoding (Anderson, 1989; Aulin, 1999). This algorithm is suitable for a tree structure. After the QR factorization, the MAP sequence detection over an MIMO channel can find the optimal path corresponding to the MAP vector in the tree shown in Fig. 10.4.

Suppose that the *a priori* probability of b_k, denoted by $P_{ex}^{(dc)}(b_k)$, is available as the extrinsic bit information from the output of the MAP channel decoder. The superscript "(dc)" is used to represent the quantities which are obtained from the "decoder." Using Eq. (10.40), the *a posteriori* probability of $\mathbf{b}$ can be found as follows:

$$\Pr(\mathbf{b}|\mathbf{y}) = C \exp\left(-\frac{1}{N_0}||\mathbf{R}(\mathbf{b} - \hat{\mathbf{b}})||^2\right) P_{ex}^{(dc)}(\mathbf{b})$$

$$\propto \exp\left(-\frac{1}{N_0}||\mathbf{R}(\mathbf{b} - \hat{\mathbf{b}})||^2 + \log P_{ex}^{(dc)}(\mathbf{b})\right), \qquad (10.45)$$

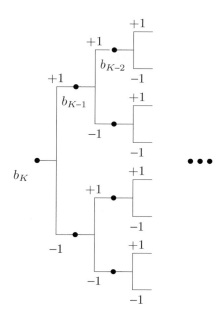

Figure 10.4. Tree diagram for binary vector **b**.

where

$$P_{\text{ex}}^{(\text{dc})}(\mathbf{b}) = \prod_{k=1}^{K} P_{\text{ex}}^{(\text{dc})}(b_k)$$

due to the assumption that the b_k's are independent (this would be reasonable as a bit interleaver is used). From Eq. (10.41), it follows that

$$||\mathbf{R}(\mathbf{b} - \hat{\mathbf{b}})||^2 = \sum_{k=1}^{K} r_{k,k}^2 d_k^2(b_k, b_{k+1}, \ldots, b_K),$$

where

$$d_k(b_k, b_{k+1}, \ldots, b_K) = (b_k - \hat{b}_k) + \sum_{q=k+1}^{K} \frac{r_{k,q}}{r_{k,k}}(b_q - \hat{b}_q).$$

Define a cost function as

$$\Gamma(b_k, \ldots, b_K) = \sum_{q=k}^{K} \left(\frac{r_{q,q}^2}{N_0} d_q^2(b_q, \ldots, b_K) - \log P_{\text{ex}}^{(\text{dc})}(b_q) \right). \tag{10.46}$$

We can see that $\Gamma(b_k, \ldots, b_K)$ is the accumulated cost from K to k along the tree in Fig. 10.4. Thus, any path in conjunction with the accumulated cost can be shown in the tree in Fig. 10.4. We note that $\Gamma(b_1, \ldots, b_K)$ is the total cost and

$$\Pr(\mathbf{b}|\mathbf{y}) \propto \exp(-\Gamma(b_1, \ldots, b_K)).$$

In the M-algorithm, M candidate paths are kept for each stage from $k = K$ to $k = 1$ (backward processing is assumed since **R** is upper triangular). At each stage, the paths corresponding to the M lowest accumulated costs are kept. Since each node has two branches, there are $2M$ paths to the next stage. Among $2M$ paths, the M paths corresponding to the M lowest accumulated costs are again kept. We can readily find the recursion of the accumulated cost as follows:

$$\Gamma(b_k = \pm 1, b_{k+1}, \ldots, b_K) = \Gamma(b_{k+1}, \ldots, b_K)$$
$$+ \frac{r_{k,k}^2}{N_0} d_k^2(b_k = \pm 1, b_{k+1}, \ldots, b_K) - \log P_{\text{ex}}^{(\text{dc})}(b_k).$$

As M paths are kept for all stages, the complexity does not grow exponentially with K.

At the last stage, $k = 1$, there are M candidate vectors of $\mathbf{b}$ for the MAP decision. If $M = 2^K$, all the candidate vectors are found with their costs. In general, however, since $M \ll 2^K$, it is possible that the MAP vector may not be included in the final set of M candidate vectors.

In the iterative receiver, the MIMO detector has to provide a soft-decision. We now need to consider a soft-output from the final M candidate vectors. Thus, it may not be crucial, although the MAP vector would not be in the final set of M candidate vectors. Let $\mathcal{M}$ denote the final set of M candidate vectors from the M-algorithm. We can assume that

$$\Gamma(\mathbf{b}) \gg 0 \text{ if } \mathbf{b} \notin \mathcal{M}. \tag{10.47}$$

That is, a vector not in $\mathcal{M}$ has a large cost and is negligible (in computing the LAPP). From Eqs (10.14) and (10.47), the LAPP as a soft-output is approximated as follows:

$$\log \frac{\Pr(b_k = +1|\mathbf{y})}{\Pr(b_k = -1|\mathbf{y})} = \log \frac{\sum_{\mathbf{b} \in \mathcal{B}_k^+} \Pr(\mathbf{b}|\mathbf{y})}{\sum_{\mathbf{b} \in \mathcal{B}_k^-} \Pr(\mathbf{b}|\mathbf{y})}$$

$$\simeq \log \frac{\sum_{\mathbf{b} \in \mathcal{M}_k^+} \exp(-\Gamma(\mathbf{b}))}{\sum_{\mathbf{b} \in \mathcal{M}_k^-} \exp(-\Gamma(\mathbf{b}))}$$

$$\simeq \log \frac{\max_{\mathbf{b} \in \mathcal{M}_k^+} \exp(-\Gamma(\mathbf{b}))}{\max_{\mathbf{b} \in \mathcal{M}_k^-} \exp(-\Gamma(\mathbf{b}))}$$

$$= \min_{\mathbf{b} \in \mathcal{M}_k^-} \Gamma(\mathbf{b}) - \min_{\mathbf{b} \in \mathcal{M}_k^+} \Gamma(\mathbf{b}), \tag{10.48}$$

where $\mathcal{M}_k^+ = \{\mathbf{b}|\mathbf{b} \in \mathcal{M}, b_k = +1\}$ and $\mathcal{M}_k^- = \{\mathbf{b}|\mathbf{b} \in \mathcal{M}, b_k = -1\}$. The first approximation in Eq. (10.48) is due to Eq. (10.47). Finally, the LLR of b_k is approximated by

$$\text{LLR}(b_k) \simeq \min_{\mathbf{b} \in \mathcal{M}_k^-} \Gamma(\mathbf{b}) - \min_{\mathbf{b} \in \mathcal{M}_k^+} \Gamma(\mathbf{b}) - \log \frac{P_{\text{ex}}^{(\text{dc})}(b_k = +1)}{P_{\text{ex}}^{(\text{dc})}(b_k = -1)}. \tag{10.49}$$

As M is small, there are cases that $\mathcal{M}_k^+ (\mathcal{M}_k^-) = \emptyset$. In this case, a predetermined large number, say $n_\infty \gg 0$, is used to approximate the LLR as $\text{LLR}(b_k) = -n_\infty$ (resp., n_∞). This technique is called the clipping.

For comparison purposes, we consider the iterative receiver with the MAP symbol detector. Figure 10.5 shows simulation results of the iterative receiver with the MAP symbol detector with $K = N = 4$. It is assumed that a single rate-half convolutional code with generator polynomial $(7, 5)$ in octal is used. The length of coded sequence is 2^{11}. The

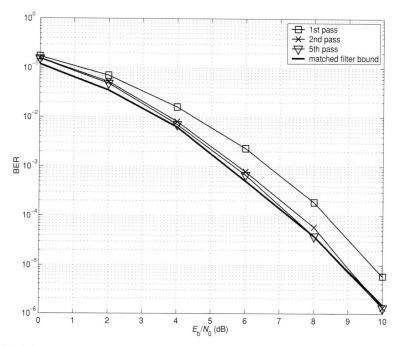

Figure 10.5. Bit error rate performance of the iterative receiver with the MAP detector.

channel coefficients, $h_{n,k}$, are assumed to be independent (circular) complex Gaussian random variables with $E[h_{n,k}] = 0$ and $E[|h_{n,k}|^2] = 1/N$. According to Fig. 10.5, we see that two iterations are enough to approach the ideal performance if the MAP symbol detector is used.

Figure 10.6 shows the same simulation as in Fig. 10.5 except that the M-algorithm with $M = 4$ is used for MIMO detection. It is observed that the performance is worse than that with the MAP symbol detection and that more iterations are required.

MMSE-SC detector

The MMSE-SC detector was introduced in Chapter 9 for CDMA systems. The same approach can be adopted for MIMO systems (Abe and Matsumoto, 2003). Generally, the MMSE-SC detector performs well within the iterative receiver as decoded signals are used effectively for the cancelation. Since a more reliable method is used to estimate the signals used for the cancelation, there would be less error propagation.

Using $P_{\text{ex}}^{(\text{dc})}(b_k)$, the mean and variance of b_k are given by

$$\bar{b}_k^{(\text{dc})} = E[b_k] = P_{\text{ex}}^{(\text{dc})}(b_k = 1) - P_{\text{ex}}^{(\text{dc})}(b_k = -1),$$
$$\left(\bar{\sigma}_k^{(\text{dc})}\right)^2 = 1 - \left(\bar{b}_k^{(\text{dc})}\right)^2,$$

$$(10.50)$$

respectively. The MMSE-SC detector yields the following output:

$$\hat{b}_{\text{mmse},k} = \mathbf{w}_k^{\text{H}}(\mathbf{y} - \mathbf{H}_k \bar{\mathbf{b}}_k),$$

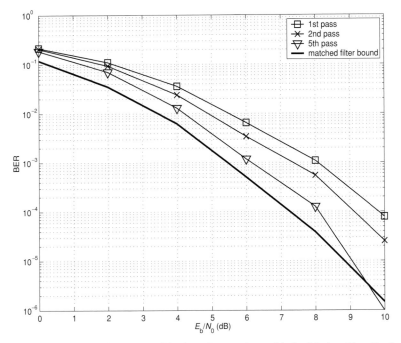

Figure 10.6. Bit error rate performance of the iterative receiver with the M-algorithm for the MIMO detection, where $M = 4$.

where $\mathbf{H}_k$ is a submatrix of $\mathbf{H}$ obtained by deleting the kth column and

$$\bar{\mathbf{b}}_k = \begin{bmatrix} \bar{b}_1^{(\text{dc})} & \cdots & \bar{b}_{k-1}^{(\text{dc})} & \bar{b}_{k+1}^{(\text{dc})} & \cdots & \bar{b}_K^{(\text{dc})} \end{bmatrix}^{\text{T}}.$$

The combining vector $\mathbf{w}_k$ is obtained as follows to minimize the MSE:

$$\mathbf{w}_k = \arg\min_{\mathbf{w}} E[|b_k - \mathbf{w}^{\text{H}}(\mathbf{y} - \mathbf{H}_k \bar{\mathbf{b}}_k)|^2]$$
$$= \mathbf{R}_k^{-1}\mathbf{h}_k, \tag{10.51}$$

where $\mathbf{h}_k$ denotes the kth column of $\mathbf{H}$ and $\mathbf{R}_k = \mathbf{h}_k\mathbf{h}_k^{\text{H}} + \mathbf{H}_k\mathbf{Q}_k\mathbf{H}_k^{\text{H}} + N_0\mathbf{I}$. Here,

$$\mathbf{Q}_k = \text{Diag}\left\{ \left(\bar{\sigma}_1^{(\text{dc})}\right)^2, \ \ldots, \ \left(\bar{\sigma}_{k-1}^{(\text{dc})}\right)^2, \ \left(\bar{\sigma}_{k+1}^{(\text{dc})}\right)^2, \ \ldots, \ \left(\bar{\sigma}_K^{(\text{dc})}\right)^2 \right\}.$$

The complexity in finding the MMSE combining vector $\mathbf{w}_k$ is dominated by the matrix inversion of $\mathbf{R}_k$, which has a complexity of $O(N^3)$. Hence, the total complexity of the detection per iteration becomes $O(N^3 KL)$, since $\mathbf{w}_k$ is different at each time l.

The LLR of b_k from $\hat{b}_{\text{mmse},k}$ is given by

$$\text{LLR}(b_k; \hat{b}_{\text{mmse},k}) = \frac{4\Re(\hat{b}_{\text{mmse},k})}{1 - \mu_k}, \tag{10.52}$$

where $\mu_k = \mathbf{w}_k^{\text{H}}\mathbf{h}_k$. Since b_k is a real number, the real part of $\hat{b}_{\text{mmse},k}$ is taken to find the LLR.

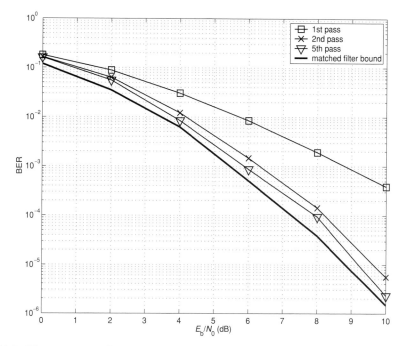

Figure 10.7. Bit error rate performance of the iterative receiver with the MMSE-SC detector.

Figure 10.7 shows the same simulation as in Fig. 10.5 except that the MMSE-SC detector is used for the MIMO detection. It is shown that a good performance is achieved after the second iteration. Comparing with the M-algorithm (see Fig. 10.6), the MMSE-SC can provide a better performance after two iterations. However, if iteration is not allowed (due to limited computing power or delay constraint), it seems the M-algorithm is preferable, as it can provide a better performance without iteration.

10.4 Doubly iterative receivers under uncertain conditions

In this section, we introduce an iterative receiver for MIMO channels with channel estimation. For MIMO channel estimation, the EM algorithm will be employed.

10.4.1 MIMO-BICM transmitter with pilot symbols

Recall that the receive signal vector in (10.10):

$$\mathbf{y}_l = \mathbf{H}_l \mathbf{b}_l + \mathbf{n}_l, \quad l = 0, 1, \dots, L - 1. \tag{10.53}$$

We assume that the length of packet, L, is sufficiently small so that the channel matrix $\mathbf{H}_l = \mathbf{H}$ is invariant during the packet duration (i.e. block fading MIMO channels are considered).

Throughout this section, we assume that a packet or sequence of symbols $b_{k,l}$'s consists of a coded binary sequence and a training sequence. For example, the first L_1 binary vectors,

$\{\mathbf{b}_0, \mathbf{b}_1, \ldots, \mathbf{b}_{L_1-1}\}$ (preamble), become the training sequence, while the last $L_2 = L - L_1$ binary vectors, $\{\mathbf{b}_{L_1}, \mathbf{b}_{L_1+1}, \ldots, \mathbf{b}_{L-1}\}$, become the data sequence.

10.4.2   EM algorithm-based channel estimation

In the EM algorithm, an initial estimate of the channel matrix, $\mathbf{H}$, can be obtained by using a training sequence. Then, a better channel estimate becomes available through EM iterations by utilizing the data sequence.

The ML estimate of the channel matrix $\mathbf{H}$ given a received signal matrix $\mathbf{Y} = [\mathbf{y}_0 \ \mathbf{y}_1 \ \cdots \ \mathbf{y}_{L-1}]$ is given by

$$\hat{\mathbf{H}}_{\mathrm{ml}} = \arg \max_{\mathbf{H}} f(\mathbf{Y}|\mathbf{H}), \tag{10.54}$$

where $f(\mathbf{Y}|\mathbf{H})$ is the likelihood function. In general, it is difficult to find the ML estimate directly since $f(\mathbf{Y}|\mathbf{H})$ is a highly nonlinear function in terms of $\mathbf{H}$. The EM algorithm can be used to find the ML estimate numerically. Let $\{\mathbf{B}, \mathbf{Y}\}$ be the complete data and let $\{\mathbf{Y}\}$ be the incomplete data in the EM algorithm.

Suppose that there exists an initial estimate of $\mathbf{H}$, denoted by $\hat{\mathbf{H}}^{(0)}$ (possibly obtained from the training sequence). In addition, denote by $\hat{\mathbf{H}}^{(q)}$ the qth estimate of $\mathbf{H}$ in the EM iteration. After some manipulations, and ignoring some constants, we have the E-step, which yields the expectation of the (conditional) log-likelihood as follows:

$$Q\big(\mathbf{H}|\hat{\mathbf{H}}^{(q)}\big) = -\sum_{\mathbf{B}} ||\mathbf{Y} - \mathbf{HB}||_{\mathrm{F}}^2 \, \mathrm{Pr}\big(\mathbf{B}|\mathbf{Y}, \hat{\mathbf{H}}^{(q)}\big), \tag{10.55}$$

where the expectation takes place over $\mathbf{B}$. As shown in Eq. (10.55), we need to find the *a posteriori* probability of $\mathbf{B}$ conditioned on $\mathbf{Y}$ and $\mathbf{H}^{(q)}$. If a random bit interleaver is used, we can assume that

$$\mathrm{Pr}\big(\mathbf{B}|\mathbf{Y}, \hat{\mathbf{H}}^{(q)}\big) = \prod_{k=1}^{K} \prod_{l=0}^{L-1} \mathrm{Pr}\big(b_{k,l}|\mathbf{Y}, \hat{\mathbf{H}}^{(q)}\big). \tag{10.56}$$

This assumption simplifies the following M-step.

In the M-step, the matrix $\mathbf{H}$ that maximizes $Q(\mathbf{H}|\hat{\mathbf{H}}^{(q)})$ can be found as follows:

$$\hat{\mathbf{H}}^{(q+1)} = \mathbf{Y} E_{\mathbf{B}}\big[\mathbf{B}^{\mathrm{T}}|\mathbf{Y}, \hat{\mathbf{H}}^{(q)}\big]\big(E_{\mathbf{B}}\big[\mathbf{BB}^{\mathrm{T}}|\mathbf{Y}, \hat{\mathbf{H}}^{(q)}\big]\big)^{-1}. \tag{10.57}$$

Due to the assumption in Eq. (10.56), the M-step can be carried out computationally efficiently. From Eq. (10.56), since

$$\bar{b}_{k,l} = \big(E_{\mathbf{B}}\big[\mathbf{B}|\mathbf{Y}, \hat{\mathbf{H}}^{(q)}\big]\big)_{k,l}$$

$$= \mathrm{Pr}\big(b_{k,l} = 1|\mathbf{Y}, \hat{\mathbf{H}}^{(q)}\big) - \mathrm{Pr}\big(b_{k,l} = -1|\mathbf{Y}, \hat{\mathbf{H}}^{(q)}\big),$$

$$\sigma_{k,k'}^2 = \big(E_{\mathbf{B}}\big[\mathbf{BB}^{\mathrm{T}}|\mathbf{Y}, \hat{\mathbf{H}}^{(q)}\big]\big)_{k,k'}$$

$$= \begin{cases} L, & \text{if } k = k'; \\ \sum_{l=0}^{L-1} \bar{b}_{k,l} \bar{b}_{k',l}, & \text{otherwise}, \end{cases} \tag{10.58}$$

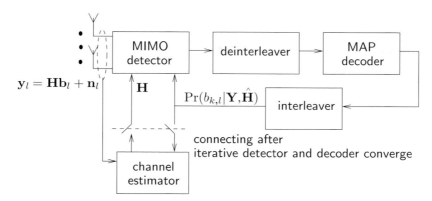

Figure 10.8. Iterative receiver with the iterative detector and decoder and EM-based channel estimation. (From Choi, J. (2006). "MIMO–BICM iterative receiver with the EM based channel estimation and simplified MMSE combining with soft cancellation," *IEEE Trans. Signal Process.*, **54**, 3247–3251.)

the expectation over $\mathbf{B}$ in Eq. (10.57) is not necessary for the M-step. We can readily find the conditional mean and covariance matrices of $\mathbf{B}$ from Eq. (10.58). This greatly reduces the complexity of the M-step.

10.4.3 Doubly iterative receiver

The iterative receiver becomes a doubly iterative receiver if the iterative detector and decoder are used to replace the joint detector and decoder. The inner iteration becomes the detection and decoding (DD) iteration, while the outer iteration becomes the EM iteration. The structure of the doubly iterative receiver is shown in Fig. 10.8. This approach is basically the same as the iterative receiver for ISI channels discussed in Chapter 7.

There are several choices for MIMO detection. For example, the MMSE-SC detector can reduce complexity. Throughout this subsection, we assume that the MMSE-SC detector is used for MIMO detection and that the channel matrix $\mathbf{H}$ is $\hat{\mathbf{H}}^{(q)}$ at the qth iteration. Because the iterative detector has the same operation for every iteration, we simply write $\mathbf{H}$ to denote $\hat{\mathbf{H}}^{(q)}$.

Simplified MMSE-SC detector using the correlation

Although the MMSE-SC detector has a lower complexity, further complexity reduction is needed for use within a doubly iterative receiver.

To reduce the complexity further, it is often desirable to have one MMSE combining vector for $b_{k,l}$ for all $l = 0, 1, \ldots, L - 1$ if possible. To this end, assume that the mean value of $b_{k,l}$ from the channel decoder of the previous iteration is given by

$$\bar{b}_{k,l}^{(dc)} = sb_{k,l}, \quad \text{for all } k \text{ and } l, \tag{10.59}$$

where $s = E[\bar{b}_{k,l}^{(dc)} b_{k,l}] \in [-1, 1]$ stands for the correlation. Note that this is an assumption that will be used to derive the MMSE combining vector, not for the SC. For convenience, let

$$\mathbf{b}_{k,l} = [b_{1,l} \ \ldots \ b_{k-1,l} \ b_{k+1,l} \ \ldots \ b_{K,l}]^{\mathrm{T}},$$

$$\bar{\mathbf{b}}_{k,l} = [\bar{b}_{1,l} \ \ldots \ \bar{b}_{k-1,l} \ \bar{b}_{k+1,l} \ \ldots \ \bar{b}_{K,l}]^{\mathrm{T}}.$$

Then, we have

$$\mathbf{y}_{k,l} = \mathbf{y}_l - \mathbf{H}_k \bar{\mathbf{b}}_{k,l}$$

$$= \mathbf{h}_k b_{k,l} + (1 - s)\mathbf{H}_k \mathbf{b}_{k,l} + \mathbf{n}_l,$$

and the resulting MMSE combining vector becomes

$$\mathbf{w}_{\mathrm{sm},k} = \left(s(2 - s)\mathbf{h}_k \mathbf{h}_k^{\mathrm{H}} + \mathbf{U}(s)\right)^{-1} \mathbf{h}_k, \tag{10.60}$$

where $\mathbf{U}(s) = (1 - s)^2 \mathbf{H}\mathbf{H}^{\mathrm{H}} + N_0 \mathbf{I}$. Applying the matrix inversion lemma to Eq. (10.60) yields

$$\mathbf{w}_{\mathrm{sm},k} = \mathbf{U}^{-1}(s)\mathbf{h}_k - \frac{s(2 - s)}{1 + s(2 - s)\mathbf{h}_k^{\mathrm{H}}\mathbf{U}^{-1}(s)\mathbf{h}_k} \mathbf{h}_k^{\mathrm{H}}\mathbf{U}^{-1}(s)\mathbf{h}_k \mathbf{U}^{-1}(s)\mathbf{h}_k$$

$$= \left(1 + s(2 - s)\mathbf{h}_k^{\mathrm{H}}\mathbf{U}^{-1}(s)\mathbf{h}_k\right)^{-1} \mathbf{U}^{-1}(s)\mathbf{h}_k. \tag{10.61}$$

Thus, we can see that the assumption in Eq. (10.59) provides the same MMSE combining vector for all l as shown in Eq. (10.61).

In order to reduce the complexity further by avoiding the matrix inversion in Eq. (10.61), consider the eigendecomposition of $\mathbf{H}\mathbf{H}^{\mathrm{H}} = \sum_{n=1}^{N} \lambda_n \mathbf{e}_n \mathbf{e}_n^{\mathrm{H}}$, where λ_n and $\mathbf{e}_n$ are the nth eigenvalue and eigenvector of $\mathbf{H}\mathbf{H}^{\mathrm{H}}$, respectively. Then, we can show that

$$\mathbf{U}(s) = \sum_{n=1}^{N} ((1 - s)^2 \lambda_n + N_0)\mathbf{e}_n \mathbf{e}_n^{\mathrm{H}}$$

and

$$\mathbf{U}^{-1}(s)\mathbf{h}_k = \sum_{n=1}^{N} \frac{1}{(1 - s)^2 \lambda_n + N_0} \mathbf{e}_n \mathbf{e}_n^{\mathrm{H}} \mathbf{h}_k$$

$$= \mathbf{E}\mathbf{c}_k(s), \tag{10.62}$$

where $\mathbf{E} = [\mathbf{e}_1 \ \mathbf{e}_2 \ \cdots \ \mathbf{e}_N]$ and

$$[\mathbf{c}_k(s)]_n = \frac{\mathbf{e}_n^{\mathrm{H}} \mathbf{h}_k}{(1 - s)^2 \lambda_n + N_0}.$$

Consequently, we can avoid a matrix inversion for each iteration. Rather, we need to find $\mathbf{c}_k(s)$ for a given s in each iteration and then compute $\mathbf{E}\mathbf{c}_k(s)$ in Eq. (10.62) to find $\mathbf{w}_{\mathrm{sm},k}$ in Eq. (10.61). The complexity becomes $O(NK)$. The resulting MMSE-SC detector is called the simplified MMSE-SC detector.

Note that the eigendecomposition of $\mathbf{H}\mathbf{H}^{\mathrm{H}}$, which has a complexity of $O(N^3)$, is required for each EM iteration as a new estimate of $\mathbf{H}$ is available. Hence, the total complexity of detection per EM iteration becomes $O(N^3) + O(NK^2)$; it is much smaller than that obtained by the standard approach.

In order to obtain the MMSE combining vector in the simplified MMSE-SC detector, we require the correlation, s. In general, as the true value of s is not available, it has to be estimated. For each iteration, an empirical correlation can be used for an estimate of s, as follows:

$$\hat{s} = \frac{1}{KL} \sum_{k=1}^{K} \sum_{l=0}^{L-1} \text{sign}\left(\bar{b}_{k,l}^{(\text{dc})}\right) \bar{b}_{k,l}^{(\text{dc})}, \tag{10.63}$$

where $\text{sign}(\bar{b}_{k,l}^{(\text{dc})})$ replaces the true $b_{k,l}$. Note that the correlation $\hat{s}$ is zero for the initial (inner) iteration, as $\bar{b}_{k,l}^{(\text{dc})}$ is not available (or is zero).

Relation to the DFD

The decision feedback detector (DFD) for MIMO detection can also be employed to obtain the LLR of $b_{k,l}$. The DFD is employed for coded CDMA systems in Honig, Woodward, and Sun (2004). Thus, we are interested in the relationship between the DFD and the MMSE-SC detector.

Under the MMSE criterion, the feedforward combiner, $\mathbf{w}_{\text{bf},k}$, and the feedback canceler, $\mathbf{f}_{\text{bf},k}$, can be found as follows:

$$\{\mathbf{w}_{\text{df},k}, \quad \mathbf{f}_{\text{df},k}\} = \arg\min_{\mathbf{w},\mathbf{f}} E[|b_{k,l} - \mathbf{w}^{\text{H}}\mathbf{y}_l - \mathbf{f}^{\text{H}}\tilde{\mathbf{b}}_{k,l}|^2], \tag{10.64}$$

where

$$\tilde{\mathbf{b}}_{k,l} = [\tilde{b}_{1,l} \cdots \tilde{b}_{k-1,l} \ \tilde{b}_{k+1,l} \cdots \tilde{b}_{K,l}]^{\text{T}}$$

and $\tilde{b}_{k,l} = \text{sign}(\bar{b}_{k,l})$. The output of the DFD is given by

$$r_{\text{df},k,l} = \mathbf{w}_{\text{df},k,l}^{\text{H}}\mathbf{y}_l + \mathbf{f}_{\text{df},k,l}^{\text{H}}\tilde{\mathbf{b}}_{k,l}.$$

To derive the weighting vectors, $\{\mathbf{w}_{\text{df},k}, \quad \mathbf{f}_{\text{df},k}\}$, assume that the detected symbols $\tilde{b}_{k,l} = \bar{b}_{k,l}^{(\text{dc})}$ from the previous iteration are characterized as follows:

$$\tilde{b}_{k,l}^{(\text{dc})} = \begin{cases} b_{k,l}, & \text{with probability } 1 - P_{\text{b}}; \\ -b_{k,l}, & \text{with probability } P_{\text{b}}, \end{cases} \tag{10.65}$$

where P_{b} is the bit error probability. From Eq. (10.65), we can show that $E[\tilde{b}_{k,l}^{(\text{dc})} b_{k,l}] = \bar{s}$, where $\bar{s} = 1 - 2P_{\text{b}}$. The correlation $\bar{s}$ is obtained from the hard-decision, while the correlation s in Eq. (10.59) is obtained from the soft-decision. A performance analysis of a DFD for MIMO systems is carried out by using $\bar{s}$ in Liang, Sun, and Ho (2006).

To solve Eq. (10.64), we can show that

$$E\left[\begin{bmatrix} \mathbf{y}_l \\ \tilde{\mathbf{b}}_{k,l} \end{bmatrix} [\mathbf{y}_l^{\text{H}} \ \tilde{\mathbf{b}}_{k,l}^{\text{H}}]\right] = \begin{bmatrix} \mathbf{R}_{\mathbf{y}} & \mathbf{H}_k\bar{s} \\ \mathbf{H}_k^{\text{H}}\bar{s} & \mathbf{I} \end{bmatrix},$$

$$E\left[b_{k,l} \begin{bmatrix} \mathbf{y}_l \\ \tilde{\mathbf{b}}_{k,l} \end{bmatrix}\right] = \begin{bmatrix} \mathbf{h}_k \\ \mathbf{0}_{(K-1)\times 1} \end{bmatrix}. \tag{10.66}$$

Using Eqs (10.66) and applying the matrix inversion lemma, it follows that

$$
\begin{aligned}
\mathbf{w}_{\mathrm{df},k} &= \left(\mathbf{H}\mathbf{H}^{\mathrm{H}} + N_0\mathbf{I} - \bar{s}^2\mathbf{H}_k\mathbf{H}_k^{\mathrm{H}}\right)^{-1}\mathbf{h}_k \\
&= \left(\bar{s}^2\mathbf{h}_k\mathbf{h}_k^{\mathrm{H}} + \mathbf{V}(\bar{s})\right)^{-1}\mathbf{h}_k,
\end{aligned}
\tag{10.67a}
$$

$$
\begin{aligned}
\mathbf{f}_{\mathrm{df},k} &= -\bar{s}\mathbf{H}_k^{\mathrm{H}}\left(\bar{s}^2\mathbf{h}_k\mathbf{h}_k^{\mathrm{H}} + \mathbf{V}(\bar{s})\right)^{-1}\mathbf{h}_k, \\
&= -\bar{s}\mathbf{H}_k^{\mathrm{H}}\mathbf{w}_{\mathrm{df},k},
\end{aligned}
\tag{10.67b}
$$

where $\mathbf{V}(\bar{s}) = (1 - \bar{s}^2)\mathbf{H}\mathbf{H}^{\mathrm{H}} + N_0\mathbf{I}$. Hence, the output of the DFD becomes

$$
r_{\mathrm{df},k,l} = \mathbf{w}_{\mathrm{df},k,l}^{\mathrm{H}}\left(\mathbf{y}_l - \bar{s}\mathbf{H}_k\tilde{\mathbf{b}}_{k,l}\right),
$$

which is quite similar to the output of the MMSE-SC detector. The combining vector $\mathbf{w}_{\mathrm{df},k}$ can be computationally efficiently obtained by the same approach for $\mathbf{w}_{\mathrm{sm},k}$ in Eq. (10.60).

For comparison, after some manipulations, the combining vectors can be found as follows:

$$
\begin{aligned}
\mathbf{w}_{\mathrm{sm},k} &= \left(s(2-s)\mathbf{h}_k\mathbf{h}_k^{\mathrm{H}} + (1-s)^2\mathbf{H}\mathbf{H}^{\mathrm{H}} + N_0\mathbf{I}\right)^{-1}\mathbf{h}_k, \\
\mathbf{w}_{\mathrm{df},k} &= \left(\bar{s}^2\mathbf{h}_k\mathbf{h}_k^{\mathrm{H}} + (1-\bar{s}^2)\mathbf{H}\mathbf{H}^{\mathrm{H}} + N_0\mathbf{I}\right)^{-1}\mathbf{h}_k.
\end{aligned}
\tag{10.68}
$$

In addition, let $s = \bar{s}$ (it is a good approximation at a high SNR) and omit the index for symbol, l, for notational convenience. Then, we can show that

$$
\mathbf{w}_{\mathrm{sm},k} = \left(\mathbf{h}_k\mathbf{h}_k^{\mathrm{H}} + \mathbf{R}_k\right)^{-1}\mathbf{h}_k
$$

$$
\mathbf{w}_{\mathrm{df},k} = \left(\mathbf{h}_k\mathbf{h}_k^{\mathrm{H}} + \bar{\mathbf{R}}_k\right)^{-1}\mathbf{h}_k,
$$

where

$$
\bar{\mathbf{R}}_k = (1 - s^2)\mathbf{H}_k\mathbf{H}_k^{\mathrm{H}} + N_0\mathbf{I}
$$

and

$$
\mathbf{R}_k = (1 - s)^2\mathbf{H}_k\mathbf{H}_k^{\mathrm{H}} + N_0\mathbf{I}.
$$

Then, the SNRs of the MMSE-SC detector and MMSE DFD are given by

$$
\begin{aligned}
\mathrm{SNR}_{\mathrm{sm},k} &= \frac{\left|\mathbf{w}_{\mathrm{sm},k}^{\mathrm{H}}\mathbf{h}_k\right|^2}{\mathbf{w}_{\mathrm{sm},k}^{\mathrm{H}}\mathbf{R}_k\mathbf{w}_{\mathrm{sm},k}} \\
&= \mathbf{h}_k^{\mathrm{H}}\mathbf{R}_k^{-1}\mathbf{h}_k
\end{aligned}
\tag{10.69a}
$$

$$
\begin{aligned}
\mathrm{SNR}_{\mathrm{df},k} &= \frac{\left|\mathbf{w}_{\mathrm{df},k}^{\mathrm{H}}\mathbf{h}_k\right|^2}{\mathbf{w}_{\mathrm{df},k}^{\mathrm{H}}\mathbf{R}_k\mathbf{w}_{\mathrm{df},k}} \\
&= \frac{\left|\mathbf{h}_k^{\mathrm{H}}\bar{\mathbf{R}}_k^{-1}\mathbf{h}_k\right|^2}{\mathbf{h}_k^{\mathrm{H}}\bar{\mathbf{R}}_k^{-1}\mathbf{R}_k\bar{\mathbf{R}}_k^{-1}\mathbf{h}_k},
\end{aligned}
\tag{10.69b}
$$

respectively, using the matrix inversion lemma. Clearly, in the presence of soft-decision, the combining vector $\mathbf{w}_{\mathrm{sm},k}$ is optimal to maximize the (combiner's) output SNR, while the feedforward vector $\mathbf{w}_{\mathrm{df},k}$ in the DFD is not.

Consider the following two extreme cases. Because $E[|b_k|^2] = 1$, the SNR is defined by $\text{SNR} = 1/N_0$. When the SNR approaches zero, $N_0\mathbf{I}$ dominates in $\bar{\mathbf{R}}_k$ and $\mathbf{R}_k$; then we have

$$\lim_{N_0 \to \infty} \frac{\text{SNR}_{\text{sm},k}}{\text{SNR}_{\text{df},k}} = 1. \qquad (10.70)$$

This shows that no significant performance difference exists for a low SNR. For the other extreme case, where the SNR approaches infinity ($N_0 \to 0$), we can also show that

$$\lim_{N_0 \to 0} \frac{\text{SNR}_{\text{sm},k}}{\text{SNR}_{\text{df},k}} = 1. \qquad (10.71)$$

The proof is given in the appendix to this chapter. For both extreme cases, the MMSE-SC detector and the MMSE DFD have the same performance. From this, we can see that even if the MMSE DFD is not optimal in the presence of soft-decision, its performance can be close to that of the MMSE-SC detector.

10.4.4 Simulation results

For simulations, we consider a block fading MIMO channel with $K = N = \bar{N}$. Each element of the channel matrix $\mathbf{H}$ is assumed to be an independent complex Gaussian random variable with zero mean and variance $\sigma_h^2 = 1/\bar{N}$. With $\sigma_h^2 = 1/\bar{N}$, the SNR is defined by

$$\frac{E_b}{N_0} = \frac{\bar{N}}{r_c} \frac{\sigma_h^2 E[|b_{k,l}|^2]}{N_0} = \frac{1}{r_c N_0}, \qquad (10.72)$$

where E_b denotes the bit energy, r_c stands for the code rate, and $E[|b_{k,l}|^2] = 1$. For channel coding, a rate-half convolutional code (i.e. $r_c = 1/2$) is used with generating polynomial $(5, 7)$ in octal.

To obtain the ideal performance, we consider the case where the channel matrix, $\mathbf{H}$, is perfectly known (i.e. a perfect CSI is assumed) and the CAI is perfectly canceled. This leads to the MFB. Note that no iteration is required in this case. We assume that $L_1 = 16$ (the length of a training sequence) and $L_2 = 512$ bits (the length of a coded bit sequence per antenna) with a random bit interleaver. An orthogonal sequence is used for the training sequence in this case. For the double iteration, three iterations are assumed for the inner-iteration (for the iterative detector and decoder) and three iterations for the EM iteration. An initial channel estimate $\mathbf{H}^{(0)}$ is obtained by using only the training sequence based on Eq. (10.57) (which turns out to be the least square estimation). We also consider the case that the channel matrix $\mathbf{H}$ is perfectly known to see the impact of the channel estimation error with five DD iterations. Note that the outer-iteration, i.e. the EM iteration, is not required in this case.

Figure 10.9 shows the BER performance when $\bar{N} = 4$. The performance of the doubly iterative receiver can approach the performance when the channel matrix is known (it even approaches the MFB) as the SNR increases. At a (coded) BER of 10^{-3}, the difference in the required E_b/N_0 is about 1 dB, and this difference becomes smaller as the BER decreases.

As shown in Fig. 10.9, there is no significant performance difference between the MMSE-SC detector and the simplified MMSE-SC detector, while the MMSE DFD performs marginally worse in general. When the SNR is low, it is shown that the simplified MMSE-SC detector slightly outperforms the MMSE-SC detector, while the MMSE-SC

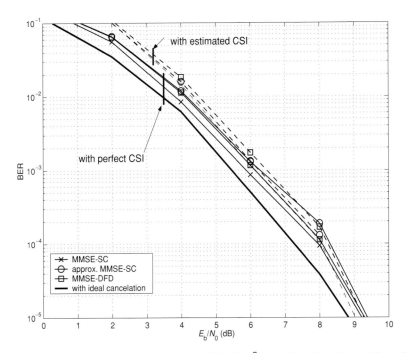

Figure 10.9. Bit error rate performance in terms of E_b/N_0 ($\bar{N} = 4$; three iterations with perfect CSI; three inner-iterations and three outer-iterations with estimated CSI). (From Choi, J. (2006). "MIMO–BICM iterative receiver with the EM based channel estimation and simplified MMSE combining with soft cancellation," *IEEE Trans. Signal Process.*, **54**, 3247–3251.)

detector outperforms the simplified MMSE-SC detector when the SNR is high. This difference results from the approach taken to determine the MMSE combining vector. In the simplified MMSE-SC detector, one global parameter, $\hat{s}$, characterizes the reliability of the extrinsic bit information to find the MMSE combining vector. At a low SNR, this global parameter can be more accurate in determining the MMSE combining vector than many local parameters (i.e. $\bar{b}_{k,l}$'s) (as in the standard MMSE-SC detector). However, when the SNR is sufficiently high, each local parameter $\bar{b}_{k,l}$ can be obtained precisely. In this case, the MMSE combining vector is individually and precisely found for each symbol vector, and it can provide a better performance than the case that one MMSE combining vector serves all symbol vectors.

Figure 10.10 shows the BER performance in terms of $\bar{N}$, where we assume that a randomly generated training sequence of length $L_1 = 4\bar{N}$ is used. As the number of antennas increases, the BER performance is improved. Since E_b/N_0 is fixed, the (average) channel gain over MIMO channels is invariant to the number of antennas. Thus, the improvement of the BER performance results from the diversity.

10.5   Summary and notes

In this chapter, we studied diversity combining techniques for the space diversity. For MIMO channels, computationally efficient MIMO detection methods were discussed. We

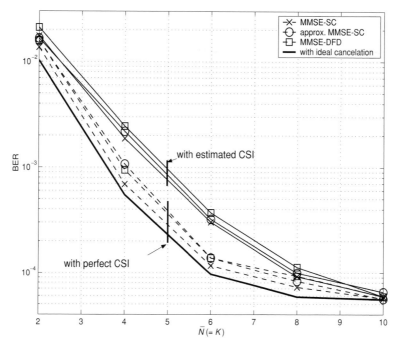

Figure 10.10. Bit error rate performance in terms of $\bar{N}$ ($E_b/N_0 = 6$ dB). (From Choi, J. (2006). "MIMO–BICM iterative receiver with the EM based channel estimation and simplified MMSE combining with soft cancellation," *IEEE Trans. Signal Process.*, **54**, 3247–3251.)

also discussed iterative receivers with computationally efficient MIMO detectors. The EM algorithm was applied to the iterative receiver for estimating the channel matrix.

Antenna or space diversity has a long history and is widely used for wireless communications. Jakes (1974) and Lee (1982) remain useful resources for the fundamental understanding of antenna diversity. The principles and fundamentals of MIMO channels can be found in Paulraj *et al.* (2003) and Tse and Viswanath (2005). Smart antennas or adaptive antennas are also important topics, although they are not included in this chapter. Adaptive antennas have a significant impact on cellular systems as the interference can be reduced effectively by beamforming; see Paulraj *et al.* (2003) for details of smart antennas and Monzingo and Miller (1980) for the fundamentals of adaptive arrays.

An excellent tutorial on iterative receivers with bit labeling and pre-coding issues can be found in Baro (2004). An overview of iterative receivers can also be found in Biglieri, Nordio, and Taricco (2004) and Haykin *et al.* (2004). Other approaches may be found, for example in Stefanov and Duman (2001) and Visoz and Berthet (2003).

10.6 Appendix to Chapter 10: Proof of Eq. (10.71)

Consider the eigendecomposition of $\mathbf{H}_k\mathbf{H}_k^{\mathrm{H}}$: $\mathbf{H}_k\mathbf{H}_k^{\mathrm{H}} = \mathbf{E}\mathbf{\Lambda}\mathbf{E}^{\mathrm{H}}$, where $\mathbf{\Lambda} = \mathrm{Diag}(\lambda_1, \lambda_2, \ldots, \lambda_N)$ (for convenience, we omit the index k). For convenience, let $\alpha = (1 - s)^2$ and $\beta = 1 - s^2$. Then, we have $\mathbf{R}_k = \sum_{n=1}^{N}(\alpha\lambda_n + N_0)\mathbf{e}_n\mathbf{e}_n^{\mathrm{H}}$ and

$\bar{\mathbf{R}}_k = \sum_{n=1}^{N}(\beta\lambda_n + N_0)\mathbf{e}_n\mathbf{e}_n^H$. In addition, it follows that

$$\mathbf{h}_k^H\mathbf{R}_k^{-1}\mathbf{h}_k = \sum_{n=1}^{N}(\alpha\lambda_n + N_0)^{-1}|g_n|^2,$$

$$\mathbf{h}_k^H\bar{\mathbf{R}}_k^{-1}\mathbf{h}_k = \sum_{n=1}^{N}(\beta\lambda_n + N_0)^{-1}|g_n|^2,$$

where $g_n = \mathbf{e}_n^H\mathbf{h}_k$. We have the following two cases.

Case I: $\mathbf{H}_k\mathbf{H}_k^H$ is full-rank ($\lambda_n > 0$ for all n). In this case, we can show that

$$\lim_{N_0 \to 0} \mathbf{h}_k^H\mathbf{R}_k^{-1}\mathbf{h}_k = \alpha^{-1}\sum_{n=1}^{N}\lambda_n^{-1}|g_n|^2, \qquad (10.73a)$$

$$\lim_{N_0 \to 0} \mathbf{h}_k^H\bar{\mathbf{R}}_k^{-1}\mathbf{h}_k = \beta^{-1}\sum_{n=1}^{N}\lambda_n^{-1}|g_n|^2, \qquad (10.73b)$$

$$\lim_{N_0 \to 0} \mathbf{h}_k^H\bar{\mathbf{R}}_k^{-1}\mathbf{R}_k\bar{\mathbf{R}}_k^{-1}\mathbf{h}_k = \frac{\alpha}{\beta^2}\sum_{n=1}^{N}\lambda_n^{-1}|g_n|^2. \qquad (10.73c)$$

Substituting Eq. (10.73) into Eq. (10.69), we obtain Eq. (10.71).

Case II: $\mathbf{H}_k\mathbf{H}_k^H$ is not full-rank. Let $V_1 = \{n|\lambda_n > 0\}$ and $V_2 = \{n|\lambda_n = 0\}$. Then, we have

$$\mathbf{h}_k^H\mathbf{R}_k^{-1}\mathbf{h}_k = \sum_{n \in V_1}(\alpha\lambda_n + N_0)^{-1}|g_n|^2 + \sum_{n \in V_2}N_0^{-1}|g_n|^2$$

$$\mathbf{h}_k^H\bar{\mathbf{R}}_k^{-1}\mathbf{h}_k = \sum_{n \in V_1}(\beta\lambda_n + N_0)^{-1}|g_n|^2 + \sum_{n \in V_2}N_0^{-1}|g_n|^2$$

$$\mathbf{h}_k^H\bar{\mathbf{R}}_k^{-1}\mathbf{R}_k\bar{\mathbf{R}}_k^{-1}\mathbf{h}_k = \sum_{n \in V_1}\frac{\alpha\lambda_n + N_0}{(\beta\lambda_n + N_0)^2}|g_n|^2 + \sum_{n \in V_2}N_0^{-1}|g_n|^2.$$

It follows that $\text{SNR}_{\text{sm},k} = N_0^{-1}(G_2 + O(N_0))$ and $\text{SNR}_{\text{df},k} = N_0^{-1}(G_2 + O(N_0))$ as $N_0 \to 0$, where $G_2 = \sum_{n \in V_2}|g_n|^2 > 0$. This completes the proof.

11 Coded OFDM and the iterative receiver

Orthogonal frequency division multiplexing (OFDM) was proposed in the 1960s (see Chang and Gibbey (1968)) and has been actively investigated since then. It can be used for both wired and wireless communications, providing several attractive features. One important feature of OFDM is that it is ISI-free. In OFDM, data symbols are transmitted by multiple orthogonal subcarriers. Each signal transmitted by a subcarrier has a narrow bandwidth and experiences flat fading without interfering with the other subcarriers' signals. From this, a simple one-tap equalizer can be used in the frequency domain to compensate for fading, while a complicated equalizer is required in a single-carrier system to overcome ISI.

It is generally known that OFDM will not outperform single-carrier systems (in terms of the average BER) when a single modulation scheme is used for all subcarriers. However, OFDM can offer a better performance if adaptive bit loading is employed. Since each subcarrier may experience different fading, the SNR varies among the subcarriers. A different number of bits per symbol can be transmitted using a different modulation scheme across subcarriers depending on the SNR for each subcarrier. For example, subcarriers with low SNR may transmit no signal or may use a lower-order modulation to stay below a certain BER ceiling, while more bits per symbol can be transmitted through subcarriers with high SNR. This approach of adaptive bit loading is used for wired communication systems (Bingham, 1990). Indeed, adaptive bit loading allows OFDM to outperform single-carrier systems. However, in some wireless communication systems, including digital terrestrial TV broadcasting, adaptive bit loading becomes impractical to implement in compensating for different fading across subcarriers. Thus, a different approach that uses channel coding is proposed (Sari, Karam, and Jeanclaude, 1995).

This chapter introduces OFDM and coded OFDM. The channel estimation problem will also be studied with an iterative receiver based on the EM algorithm.

11.1 Introduction to OFDM systems

A multicarrier system uses multiple subcarriers to transmit signals as shown in Fig. 11.1. Each subcarrier transmits signals individually. Generally, the bandwidth of each subcarrier is narrow. Hence, each subcarrier experiences flat fading and there is no ISI. This simplifies the receiver design, because a simple one-tap equalizer in the frequency domain can be employed to compensate for fading. Even though the data rate for each subcarrier is low, a

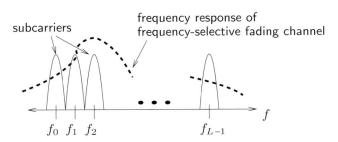

Figure 11.1. Subcarriers in the frequency domain.

total data rate, i.e. the sum of data rates of all subcarriers, can be high so that multicarrier systems are capable of transmitting multimedia information such as digital video.

In the following subsection, we introduce OFDM, which is one of the multicarrier systems. Based on the Fourier transform, orthogonal subcarriers can be generated (Weinstein and Ebert, 1971).

11.1.1 System model for OFDM

A block of multiple data symbols,

$$\mathbf{b} = [b_0 \ b_1 \cdots b_{L-1}]^\mathrm{T},$$

is transmitted, where L is the length of the OFDM block and b_l is the data symbol to be transmitted through the lth subcarrier. For simplicity, assume that $b_l \in \{-1, +1\}$. Instead of transmitting $\mathbf{b}$, the inverse discrete Fourier transform (IDFT) of $\mathbf{b}$, which is a block of modulated signals transmitted by subcarriers, is transmitted in OFDM. An L-point normalized discrete Fourier transform (DFT) may be represented as follows:

$$[\mathbf{F}]_{m,l} = f_{m,l} = \frac{1}{\sqrt{L}} e^{-j2\pi ml/L}, \quad m, l = 0, 1, \ldots, L - 1, \tag{11.1}$$

where $[\mathbf{F}]_{m,l} = f_{m,l}$ denotes the $(m + 1, l + 1)$th[†] element of $\mathbf{F}$ throughout this chapter. Then, the IDFT of $\mathbf{b}$ becomes

$$\check{\mathbf{b}} = \mathbf{F}^\mathrm{H} \mathbf{b}.$$

The normalized IDFT makes sure that the total bit energy of $\mathbf{b}$ is identical to that of $\check{\mathbf{b}}$ since $\mathbf{F}\mathbf{F}^\mathrm{H} = \mathbf{F}^\mathrm{H}\mathbf{F} = \mathbf{I}$.

Suppose that the length of CIR is P. To avoid the interblock interference,[‡] a cyclic prefix (CP) is appended to $\check{\mathbf{b}}$ as shown in Fig. 11.2. The length of CP, denoted by $\bar{P}$, should be greater than or equal to $P - 1$ (i.e. $\bar{P} \geq P - 1$). Let $\bar{L} = L + \bar{P}$. Then, the signal block

---

[†] This indexing is different from that used in the other chapters. Since m and l start from zero, this different indexing is used in this chapter.

[‡] In transmission, blocks of symbols are continuously transmitted. Without a CP or guard-interval, there would be overlapping of consecutive blocks due to a dispersive channel. This overlapping is called the interblock interference.

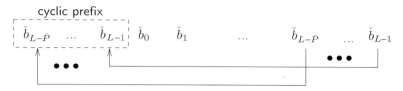

Figure 11.2. Adding the cyclic prefix.

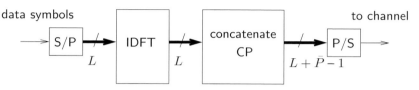

Figure 11.3. OFDM transmitter: "S/P" and "P/S" denote the serial-to-parallel and parallel-to-serial converters, respectively. A bold line slashed with a number indicates multiple lines.

with CP becomes

$$\check{\mathbf{s}} = [\begin{array}{ccccc} \check{s}_0 & \check{s}_1 & \cdots & \check{s}_{\bar{P}-1} & \check{s}_{\bar{P}} & \cdots & \check{s}_{\bar{L}-1} \end{array}]^{\mathrm{T}}$$
$$= [\begin{array}{ccccc} \check{b}_{L-\bar{P}} & \check{b}_{L-\bar{P}+1} & \cdots & \check{b}_{L-1} & \check{b}_0 & \cdots & \check{b}_{L-1} \end{array}]^{\mathrm{T}}. \tag{11.2}$$

A block diagram of the OFDM transmitter is depicted in Fig. 11.3. A block of data symbols is transformed by the IDFT. After adding the CP, signals are transmitted. Thus, the transmitted signals are different from those in the original modulation (e.g. BPSK).

Example 11.1.1 Suppose that $L = 4$ and $\bar{P} = 2$. Then, $\mathbf{F}$ is given by

$$\mathbf{F} = \frac{1}{2} \begin{bmatrix} 1 & 1 & 1 & 1 \\ 1 & -j & -1 & j \\ 1 & -1 & 1 & -1 \\ 1 & j & -1 & -j \end{bmatrix}.$$

If $\mathbf{b} = [1\ 1\ 1\ 1]^{\mathrm{T}}$, $\check{\mathbf{b}}$ and $\check{\mathbf{s}}$ become

$$\check{\mathbf{b}} = \quad\quad [2\ \ 0\ \ 0\ \ 0]^{\mathrm{T}},$$
$$\check{\mathbf{s}} = [0\ \ 0\ \ 2\ \ 0\ \ 0\ \ 0]^{\mathrm{T}}.$$

For another example, let $\mathbf{b} = [1\ -1\ 1\ 1]^{\mathrm{T}}$. Then, $\check{\mathbf{b}}$ and $\check{\mathbf{s}}$ become

$$\check{\mathbf{b}} = \quad\quad [1\ \ -j\ \ 1\ \ j]^{\mathrm{T}},$$
$$\check{\mathbf{s}} = [1\ \ j\ \ 1\ \ -j\ \ 1\ \ j]^{\mathrm{T}}.$$

Suppose that the signal is transmitted over an ISI channel. The received signal becomes

$$\check{y}_l = \sum_{p=0}^{P-1} h_p \check{s}_{l-p} + \check{n}_l, \quad l = 0, 1, \ldots, \bar{L} - 1, \tag{11.3}$$

where $\{h_p\}$ represents the CIR and $\check{n}_l$ is a circular complex Gaussian random variable with zero mean and variance $E[|\check{n}_l|^2] = N_0$. From Eq. (11.3), we can write the following:

$$\check{y}_{\bar{P}} = \sum_{p=0}^{P-1} h_p \check{s}_{\bar{P}-p} + \check{n}_{\bar{P}}$$

$$= h_0 \check{b}_0 + h_1 \check{b}_{L-1} + \cdots + h_{P-1} \check{b}_{L-P+1} + \check{n}_{\bar{P}}$$

$$\check{y}_{\bar{P}+1} = h_0 \check{b}_1 + h_1 \check{b}_0 + \cdots + h_{P-1} \check{b}_{L-P+2} + \check{n}_{\bar{P}+1}$$

$$\vdots$$

Then, the received signal vector, after deleting the first $\bar{P}$ elements corresponding to the CP part, is given by

$$\check{\mathbf{y}} = [\check{y}_{\bar{P}} \ \check{y}_{\bar{P}+1} \cdots \check{y}_{\bar{L}-1}]^{\mathrm{T}}$$
$$= \mathbf{H}\check{\mathbf{b}} + \check{\mathbf{n}}, \qquad (11.4)$$

where $\check{\mathbf{n}} = [\check{n}_{\bar{P}} \ \check{n}_{\bar{P}+1} \cdots \check{n}_{\bar{L}-1}]^{\mathrm{T}}$ and $\mathbf{H}$ becomes a circular matrix of size $L \times L$ due to the CP given by

$$\mathbf{H} = \begin{bmatrix} h_0 & h_{L-1} & \cdots & h_1 \\ h_1 & h_0 & \cdots & h_2 \\ \vdots & & \ddots & \vdots \\ h_{L-1} & h_{L-2} & \cdots & h_0 \end{bmatrix}. \qquad (11.5)$$

Here, we assume that $h_p = 0$, $P \le p \le L - 1$. Circular matrices have an important property: their eigenvectors are independently determined from the values of the coefficients $\{h_p\}$. For a circular matrix $\mathbf{H}$, we can show that

$$[\mathbf{F}\mathbf{H}\mathbf{F}^{\mathrm{H}}]_{m,l} = \sum_{p=0}^{L-1} \sum_{q=0}^{L-1} f_{m,p} h_{p,q} f_{l,q}^*$$

$$= \frac{1}{L} \sum_p e^{-j2\pi mp/L} \sum_q h_{p,q} e^{j2\pi lq/L}$$

$$= \frac{1}{L} \sum_p e^{-j2\pi mp/L} \sum_v h_v e^{j2\pi l(p-v)/L}$$

$$= \frac{1}{L} \sum_p \left(\sum_v h_v e^{-j2\pi lv/L} \right) e^{-j2\pi p(m-l)/L}$$

$$= \check{H}_l \left(\frac{1}{L} \sum_p e^{-j2\pi p(m-l)/L} \right)$$

$$= \check{H}_l \delta_{m,l}, \qquad (11.6)$$

where $h_{p,q}$ denotes the $(p+1, q+1)$th element of $\mathbf{H}$ in Eq. (11.5) and

$$\check{H}_l = \sum_{p=0}^{L-1} h_p e^{-j2\pi lp/L}$$

$$= \sum_{p=0}^{P-1} h_p e^{-j2\pi lp/L}, \quad l = 0, 1, \ldots, L - 1. \qquad (11.7)$$

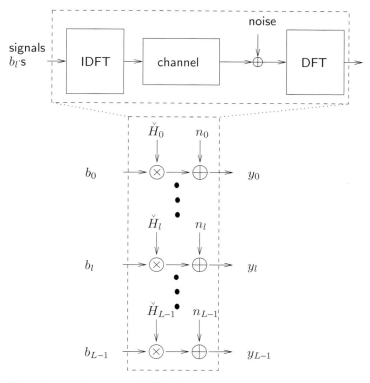

Figure 11.4. Block diagram for an overall OFDM system.

The third equality in Eq. (11.6) is derived in Section 11.5, where the final expression implies

$$\mathbf{F}\mathbf{H}\mathbf{F}^{\mathrm{H}} = \mathrm{Diag}(\check{H}_0, \check{H}_1, \ldots, \check{H}_{L-1})$$
$$\overset{\triangle}{=} \check{\mathbf{H}}. \tag{11.8}$$

Hence, the column vectors of $\mathbf{F}$ are the eigenvectors of the circular matrix $\mathbf{H}$; and the $\check{H}_l$'s are shown to be the eigenvalues of $\mathbf{H}$.

At the receiver, the signal can be recovered after the DFT as follows:

$$\mathbf{y} = \mathbf{F}\check{\mathbf{y}}$$
$$= \mathbf{F}(\mathbf{H}\check{\mathbf{b}} + \check{\mathbf{n}})$$
$$= \mathbf{F}(\mathbf{H}\mathbf{F}^{\mathrm{H}}\mathbf{b} + \check{\mathbf{n}})$$
$$= \check{\mathbf{H}}\mathbf{b} + \mathbf{n}, \tag{11.9}$$

where $\mathbf{n} = \mathbf{F}\check{\mathbf{n}}$. Let $\mathbf{y} = [y_0 \; y_1 \; \cdots \; y_{L-1}]^{\mathrm{T}}$ and $\mathbf{n} = [n_0 \; n_1 \; \cdots \; n_{L-1}]^{\mathrm{T}}$. Hence, we finally show that

$$y_l = \check{H}_l b_l + n_l, \quad l = 0, 1, \ldots, L - 1. \tag{11.10}$$

It is shown in Eq. (11.10) that b_l is transmitted through the lth subcarrier and that the signal experiences flat fading and does not interfere with the other subcarriers' signals. Figure 11.4 shows a block diagram for an overall OFDM system. In the frequency domain, L data symbols are transmitted independently.

11.1.2 Frequency-domain equalizers

At the receiver, a one-tap equalizer can be used to compensate for the fading. The tap coefficient of this ZF equalizer equals

$$\check{C}_l = \frac{1}{\check{H}_l}, \quad l = 0, 1, \ldots, L-1,$$

while the tap coefficient of the MMSE equalizer becomes

$$\check{C}_l = \frac{\check{H}_l^*}{|\check{H}_l|^2 + N_0}, \quad l = 0, 1, \ldots, L-1.$$

The equalized signal is given by

$$\hat{b}_l = \Re(\check{C}_l y_l), \quad l = 0, 1, \ldots, L-1. \tag{11.11}$$

We take the real part in Eq. (11.11) as $b_l \in \{-1, +1\}$. Note that the hard-decision after one-tap equalization is the same for both ZF and MMSE equalizers. If frequency nulls exist (i.e. some $\check{H}_l$'s are zero), the data symbols transmitted to the subcarriers at frequency nulls cannot be detected as the SNR would be low. Although there is no frequency null, the detection performance can be poor if the frequency channel gains, $|H_l|$, are low for some subcarriers. Thus, the performance of OFDM can be significantly degraded due to frequency nulls or low-frequency channel gains.

For the performance, consider the following two CIRs of length $P = 5$:

$$\text{(Channel A)} \quad \mathbf{h} = [0.227 \quad 0.460 \quad 0.688 \quad 0.460 \quad 0.227]^{\mathrm{T}},$$
$$\text{(Channel B)} \quad \mathbf{h} = [0.688 \quad 0.460 \quad 0.460 \quad 0.227 \quad 0.227]^{\mathrm{T}},$$

where $\mathbf{h} = [h_0 \ h_1 \ \cdots \ h_{P-1}]^{\mathrm{T}}$. For both Channel A and Channel B, we have $\|\mathbf{h}\|^2 = 1$. The frequency responses of Channel A and Channel B are shown in Fig. 11.5.

Simulations are carried out with $L = 1024$ and the results are shown in Fig. 11.6. A one-tap equalizer (note that both ZF and MMSE equalizers provide the same result in terms of the BER) is used at the receiver with a known CIR. The SNR is given by

$$\text{SNR} = \frac{\|\mathbf{h}\|^2}{N_0} = \frac{1}{N_0}.$$

As shown in Eq. (11.10), the total received signal power is given by $\sum_{l=0}^{L-1} |\check{H}_l|^2$ over the L subcarriers. Since

$$\|\mathbf{h}\|^2 = \frac{1}{L} \sum_{l=0}^{L-1} |\check{H}_l|^2$$

(the Parseval identity; see Roberts and Mullis (1987)), we can see that $\|\mathbf{h}\|^2$ is the average received power per subcarrier. Hence, the SNR, $\|\mathbf{h}\|^2/N_0$, becomes the average SNR of each individual data symbol.

It is clear that a better BER performance can be achieved with Channel B than with Channel A. Since Channel A has frequency nulls, data symbols transmitted through subcarriers at the frequency nulls are unlikely to be detected correctly. Hence, the average error probability over all subcarriers is raised.

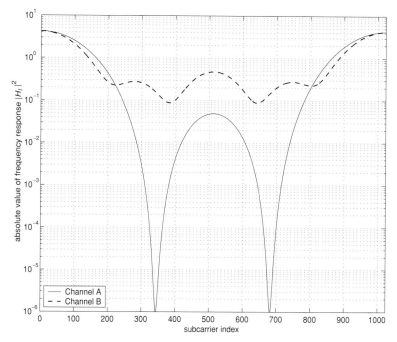

Figure 11.5. Frequency responses (power spectra) of two channels.

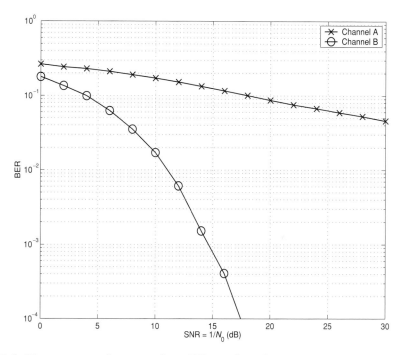

Figure 11.6. Bit error rate performance of two different channels.

To observe the performance in detail, we consider an average BER over all subcarriers. From Eq. (11.11), the BER for subcarrier l is given by

$$P_l = Q\left(\sqrt{\frac{2|\check{H}_l|^2}{N_0}}\right).$$

The average BER over all subcarriers becomes

$$P_{uc} = \frac{1}{L}\sum_{l=0}^{L-1} P_l.$$

Alternatively, it is possible to consider an empirical pdf of $|\check{H}_l|^2$, denoted by $f_{|H|^2}(\alpha)$. With the empirical cdf defined as

$$F_{|H|^2}(\alpha) = \frac{1}{L}\sum_l \mathbf{1}(|\check{H}_l|^2 \le \alpha),$$

where $\mathbf{1}(\cdot)$ is the indicator function, the empirical pdf can be derived. Then, the average BER (over subcarriers) can also be expressed as follows:

$$P_{uc} = E[P_l]$$
$$= \int Q\left(\sqrt{\frac{2\alpha}{N_0}}\right) f_{|H|^2}(\alpha)\, d\alpha. \tag{11.12}$$

Using the Chernoff bound (i.e. $Q(x) \le e^{-x^2/2}$ for $x \ge 0$), an upper bound of the average BER is given by

$$\bar{P}_{uc} = E\left[e^{-\frac{\alpha}{N_0}}\right] = \int e^{-\frac{\alpha}{N_0}} f_{|H|^2}(\alpha)\, d\alpha. \tag{11.13}$$

When the CIR is random (as in wireless communications), the BER expressions in Eqs (11.12) and (11.13) are useful. For Rayleigh fading (and independent multipath fading) as an example, we can show that

$$|\check{H}_l| = \left|\sum_{p=0}^{P-1} h_p e^{-j2\pi l p/L}\right|$$

is Rayleigh distributed. Let $\gamma = |\check{H}_l|^2/N_0$. Then, it can be shown that the pdf of γ is given by

$$f(\gamma) = \frac{1}{\bar{\gamma}} e^{-\gamma/\bar{\gamma}}, \tag{11.14}$$

where $\bar{\gamma} = E[\gamma] = E[|\check{H}_l|^2/N_0]$. From Eq. (11.13), the upper bound of the average BER over Rayleigh fading channel can be given by

$$\bar{P}_{uc} = \frac{1}{1+\bar{\gamma}}. \tag{11.15}$$

This reveals a disadvantage of OFDM. Even though there are multipaths, the diversity order becomes unity, as shown in Eq. (11.15). This would be a natural consequence as each subcarrier experiences flat fading.

If the transmitter knows the frequency responses, $\{\hat{H}_l, \forall l\}$, adaptive bit loading (Bingham, 1990) is viable. Each subcarrier has a different SNR, $|\hat{H}_l|^2 / N_0$, which determines the number of bits and corresponding modulation scheme at that subcarrier. A higher SNR allows more bits to be transmitted using a higher-order modulation scheme. If the SNR is very low, no bits would be transmitted.

Adaptive bit loading is applicable to one-to-one communications, but not broadcasting channels (which can be considered as one-to-many communications). Different receivers have different SNRs for the same subcarrier; and adaptive bit loading becomes impossible.

Another major problem of OFDM is a high peak-to-average-power ratio (PAPR). Since the IDFT of the symbol block is transmitted, the power of transmitted signals, $\breve{b}_l$, in Eq. (11.2) can vary significantly. Thus, a power amplifier is required that works over a wide range; this is usually expensive. Alternatively, clipped, the transmitted signal can be at the expense of performance degradation. The reader is referred to Bahai and Saltzberg (1999) and Heiskala and Terry (2001) for details of the PAPR problem and solutions.

11.2 Coded OFDM

To overcome the problem of frequency nulls without adaptive bit loading, a diversity technique can be employed, e.g. two antennas at the receiver, if spaced far enough, would produce different frequency responses for each channel. Even if a subcarrier's frequency response is weak at one antenna, it might be strong at the other antenna. Hence, performance can be improved after combining signals properly from the two antennas. There are other approaches, including channel coding. In this section, we introduce coded OFDM.

11.2.1 Coded OFDM with a convolutional code

In order to alleviate the adverse impact of frequency nulls in OFDM, coded sequences can be transmitted. We consider the code diversity with convolutional codes. To understand the code diversity, we need to consider error analysis of convolutional codes; see Chapter 6 for the background to convolutional codes or see Lin and Costello (1983) for a detailed account.

Suppose that $\mathbf{b}$ is the transmitted coded sequence from a convolutional encoder and that ML decoding is used to decode signals. An erroneous decision happens if

$$f(\mathbf{y}|\mathbf{b}) < f(\mathbf{y}|\mathbf{a})$$

or

$$\|\mathbf{y} - \breve{\mathbf{H}}\mathbf{b}\|^2 > \|\mathbf{y} - \breve{\mathbf{H}}\mathbf{a}\|^2,$$

where $\mathbf{a} \neq \mathbf{b}$ is another coded sequence from a convolutional code. The probability of this particular decision error is given by

$$\begin{aligned}
\Pr(\mathbf{b} \to \mathbf{a}) &= \Pr\left(\|\mathbf{y} - \breve{\mathbf{H}}\mathbf{b}\|^2 > \|\mathbf{y} - \breve{\mathbf{H}}\mathbf{a}\|^2\right) \\
&= \Pr\left(-2\Re\left(\mathbf{n}^H\breve{\mathbf{H}}(\mathbf{b} - \mathbf{a})\right) > \|\breve{\mathbf{H}}(\mathbf{b} - \mathbf{a})\|^2\right).
\end{aligned} \tag{11.16}$$

Since $\mathbf{n} = \mathbf{F}\check{\mathbf{n}}$ and $\mathbf{F}^{H}\mathbf{F} = \mathbf{I}$, $\mathbf{n}$ becomes a circular complex Gaussian random variable with $E[\mathbf{n}] = \mathbf{0}$ and $E[\mathbf{nn}^{H}] = N_0\mathbf{I}$. Thus, $\mathbf{n}^{H}\check{\mathbf{H}}(\mathbf{b} - \mathbf{a})$ is also a circular complex Gaussian random variable. Since

$$E[\mathbf{n}^{H}\check{\mathbf{H}}(\mathbf{b} - \mathbf{a})] = 0,$$
$$E[|\mathbf{n}^{H}\check{\mathbf{H}}(\mathbf{b} - \mathbf{a})|^{2}] = \text{Tr}\big(E[\check{\mathbf{H}}(\mathbf{b} - \mathbf{a})\mathbf{nn}^{H}(\mathbf{b} - \mathbf{a})^{H}\check{\mathbf{H}}^{H}]\big)$$
$$= N_0\text{Tr}\big(\check{\mathbf{H}}(\mathbf{b} - \mathbf{a})(\mathbf{b} - \mathbf{a})^{H}\check{\mathbf{H}}^{H}]\big)$$
$$= N_0\|\check{\mathbf{H}}(\mathbf{b} - \mathbf{a})\|^{2},$$

we can show that

$$\Re(\mathbf{n}^{H}\check{\mathbf{H}}(\mathbf{b} - \mathbf{a})) \sim \mathcal{N}\left(0, \frac{N_0}{2}\|\check{\mathbf{H}}(\mathbf{b} - \mathbf{a})\|^{2}\right).$$

Using the Q-function, it can be shown from Eq. (11.16) that

$$\text{Pr}(\mathbf{b} \to \mathbf{a}) = Q\left(\frac{\|\check{\mathbf{H}}(\mathbf{b} - \mathbf{a})\|}{\sqrt{2N_0}}\right). \tag{11.17}$$

The error probability depends on $\mathbf{b} - \mathbf{a}$.

With the free distance given by d_{free} (i.e. the minimum Hamming distance) of a convolutional code, we can see that there are at least d_{free} different data symbols between $\mathbf{b}$ and $\mathbf{a}$, because $\mathbf{b}$ and $\mathbf{a}$ are coded sequences. Let $\mathbf{e} = \mathbf{b} - \mathbf{a}$ and assume that the Hamming distance between $\mathbf{a}$ and $\mathbf{b}$ is d_{free} to consider the worst case. Then, the elements of $\mathbf{e}$ will be zero except for d_{free} consecutive positions. If d_{free} is greater than the number of frequency nulls, $\|\check{\mathbf{H}}\mathbf{e}\|$ is always positive for any $\mathbf{e}$. This implies that the error probability can be low even though there are frequency nulls.

To achieve a good performance, it is also important to use interleaving in the frequency domain. As shown in Fig. 11.5, the power spectrum does not vary rapidly over frequency and a group of adjacent subcarriers can experience similar severe fading. Suppose that $\mathbf{e}$ is given by

$$\{e_0, e_1, \ldots, e_{L-1}\} = \{0, 0, \ldots, 0, e_q, e_{q+1}, \ldots, e_{q+d_{\text{free}}-1}, 0, \ldots, 0\}$$

for some q. If the number of frequency nulls is larger than d_{free} and $|\hat{H}_l|^{2} \simeq 0$ for $l = q, q + 1, q + d_{\text{free}} - 1$, the error probability becomes high since $\|\check{\mathbf{H}}\mathbf{e}\|^{2} \simeq 0$. To overcome this problem, an interleaver in the frequency domain can be used to distribute clustered frequency nulls. In Fig. 11.7, randomly interleaved frequency responses of Channel A are shown. Due to interleaving, adjacent subcarriers can have different frequency responses.

It is necessary to convert the SNR into E_b/N_0 for a fair comparison with an uncoded transmission. For uncoded data symbols, the bit energy, E_b, is unity and $E_b/N_0 = \text{SNR}$ if $\|\mathbf{h}\|^{2} = 1$. However, for coded data symbols from a rate-k_c/n_c convolutional code, we have

$$k_c E_b = n_c E_{b,\text{coded}},$$

where $E_{b,\text{coded}}$ denotes the bit energy for coded symbols. For $b_l \in \{-1, +1\}$, $E_{b,\text{coded}} = 1$ and the equivalent uncoded bit energy becomes

$$E_b = \frac{n_c}{k_c}$$

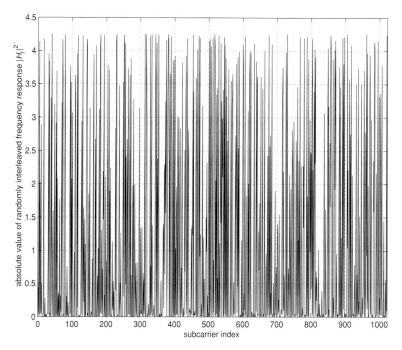

Figure 11.7. Randomly interleaved frequency responses of Channel A.

and

$$\frac{E_b}{N_0} = \frac{n_c}{k_c}\text{SNR}.$$

For a rate-half convolutional code, we have E_b/N_0 (dB) = SNR (dB) +3 (dB). That is, for the same SNR we can see that a coded symbol has 3 dB higher bit energy than an uncoded symbol in this case.

For simulations, we consider a rate-half convolutional code with generator polynomial $(7, 5)$ in octal and a random bit interleaver in the frequency domain. The coded BER performance in terms of E_b/N_0 is shown in Fig. 11.8 for both Channel A and Channel B when $L = 1024$. Due to the code diversity, coded OFDM can siginificantly outperform uncoded OFDM, as shown in Fig. 11.8. In particular, for Channel A, the coded BER can rapidly decrease with E_b/N_0, even if frequency nulls exist.

11.2.2 Performance analysis

So far, we have shown that coded OFDM can be robust against frequency nulls and provide a good performance. This performance improvement is due to the code diversity; however, this has not been discussed so far. In this subsection, we analyze the performance of coded OFDM in order to understand the code diversity.

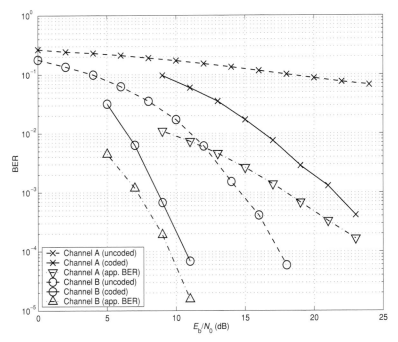

Figure 11.8. Coded and uncoded BER performance of two different channels.

If a random interleaver is used, the symbols involved in the difference, $\mathbf{e} = \mathbf{b} - \mathbf{a}$, can be randomly distributed. Hence, the average error probability is given by

$$\Pr(\mathbf{b} \to \mathbf{a}) = E\left[Q\left(\frac{\|\check{\mathbf{H}}_{\text{itl}}(\mathbf{b} - \mathbf{a})\|}{\sqrt{2N_0}}\right)\right], \tag{11.18}$$

where $\check{\mathbf{H}}_{\text{itl}}$ denotes a randomly interleaved channel response matrix. We assume that there are d_{free} different data symbols between $\mathbf{b}$ and $\mathbf{a}$. Then, the average error probability in Eq. (11.18) can be rewritten as follows:

$$E\left[Q\left(\sqrt{\frac{\|\check{\mathbf{H}}_{\text{itl}}(\mathbf{b} - \mathbf{a})\|^2}{2N_0}}\right)\right] = E\left[Q\left(\sqrt{\frac{2}{N_0}\sum_{l \in \bar{I}}|\check{H}_l|^2}\right)\right],$$

where $\bar{I}$ is a set of d_{free} random indexes for subcarriers. Let

$$X = \sum_{l \in \bar{I}}|\check{H}_l|^2. \tag{11.19}$$

If the pdf of X is available, denoted by $f_X(x)$, we can show that

$$\Pr(\mathbf{b} \to \mathbf{a}) = \int Q\left(\sqrt{\frac{2}{N_0}x}\right)f_X(x)\,\mathrm{d}x.$$

Finally, from Eq. (6.27), the coded BER can be approximated by

$$P_b \simeq \frac{1}{k_c} B_{d_{\text{free}}} \int Q\left(\sqrt{\frac{2}{N_0}}x\right) f_X(x)\,dx. \tag{11.20}$$

In general, it is not easy to find the pdf of X in Eq. (11.19) although the CIR is given. However, an approximation is available under a certain assumption. Consider an empirical pdf of $|\check{H}_l|^2$ given CIR. Assume that $|\check{H}_l|^2$ in Eq. (11.19) is iid as $|\check{H}_l|^2 \sim f_{|H|^2}(\alpha)$. Then, from Papoulis (1984) it follows that

$$f_X(\alpha) = \underbrace{f_{|H|^2}(\alpha) * f_{|H|^2}(\alpha) * \cdots * f_{|H|^2}(\alpha)}_{d_{\text{free}} \text{ times}}, \tag{11.21}$$

where $*$ stands for the convolution. The approximate BERs from Eq. (11.20) in Fig. 11.8, are shown using the empirical pdf of X from Eq. (11.21). In general, as L is finite, the assumption that $|\check{H}_l|^2$ in Eq. (11.19) is independent would not be valid. The approximate BER would also be lower than the actual BER as the independence assumption leads to an optimistic result.

The result in Eq. (11.18) depends on the CIR. To obtain a general result, we assume that $\check{H}_l$ is random. An upper bound of Eq. (11.18) is available using the Chernoff bound as follows:

$$\Pr(\mathbf{b} \to \mathbf{a}) = E\left[Q\left(\sqrt{\frac{2}{N_0}\sum_{l\in\bar{l}}|\check{H}_l|^2}\right)\right]$$

$$\leq E\left[e^{-\frac{1}{N_0}\sum_{l\in\bar{l}}|\check{H}_l|^2}\right].$$

Again, if we assume that the $\check{H}_l$'s are independent and that $|\check{H}_l|^2$ has the same pdf, $f_H(|\check{H}_l|^2)$, it follows that

$$E\left[e^{-\frac{1}{N_0}\sum_{l\in\bar{l}}|\check{H}_l|^2}\right] = \left(E\left[e^{-\frac{|\check{H}_l|^2}{N_0}}\right]\right)^{d_{\text{free}}}.$$

For Rayleigh fading, from Eq. (11.14) it can be shown that

$$\Pr(\mathbf{b} \to \mathbf{a}) \leq \left(E\left[e^{-\frac{|\check{H}_l|^2}{N_0}}\right]\right)^{d_{\text{free}}}$$

$$= \left(\frac{1}{1+\bar{\gamma}}\right)^{d_{\text{free}}}, \tag{11.22}$$

where $\bar{\gamma} = E[|\check{H}_l|^2]/N_0$. From Eq. (6.27), an upper bound of average BER of coded OFDM is given by

$$P_b \simeq \frac{1}{k_c} B_{d_{\text{free}}} \left(\frac{1}{1+\bar{\gamma}}\right)^{d_{\text{free}}}. \tag{11.23}$$

Comparing this with Eq. (11.15), we see that coded OFDM can have a diversity gain and that the diversity order is the same as the free distance. This diversity is called the code diversity. Although this result is based on several assumptions, it shows that coded OFDM can overcome frequency nulls on employing code diversity and can provide good

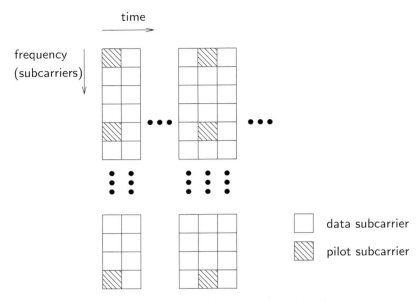

Figure 11.9. Distribution of pilot subcarriers in frequency and time domains.

performance. In addition, Eq. (11.22) suggests that a convolutional code with a large free distance is desirable to achieve a good performance.

11.3 EM-based iterative channel estimation

It is necessary to estimate the CIR at the receiver. In this section, an iterative receiver will be derived using the EM algorithm for the ML channel estimation. Since each EM iteration increases the likelihood, a better channel estimate is available and thus a better BER performance is expected.

11.3.1 Pilot subcarriers

For channel estimation, some subcarriers deliver pilot symbols and are called pilot sub-carriers. As shown in Fig. 11.9, the pilot subcarriers are typically uniformly distributed in the frequency and time domains. Channel estimation can be considered in the frequency domain or time domain. However, in this section, we focus only on channel estimation in the time domain.

11.3.2 ML channel estimation using the EM algorithm

In ML channel estimation, we estimate the CIR, $\mathbf{h}$, under the ML criterion. From Eq. (11.7), we define

$$\check{\mathbf{h}} = [\check{H}_0 \; \check{H}_1 \; \cdots \; \check{H}_{L-1}]^{\mathrm{T}}$$
$$= \mathbf{F}_{\mathrm{ch}}\mathbf{h}, \tag{11.24}$$

where the DFT matrix for the CIR, $\mathbf{F}_{\text{ch}}$, is an $L \times P$ matrix whose elements are given by

$$[\mathbf{F}_{\text{ch}}]_{l,p} = e^{-j2\pi lp/L}, \quad l = 0, 1, \ldots, L-1, \quad p = 0, 1, \ldots, P-1.$$

Given $\mathbf{y}$, the ML channel estimation is formulated as follows:

$$\hat{\mathbf{h}}_{\text{ml}} = \arg\max_{\mathbf{h}} f(\mathbf{y}|\mathbf{h}), \tag{11.25}$$

where $f(\mathbf{y}|\mathbf{h})$ is the likelihood function of $\mathbf{h}$. As this likelihood function is highly nonlinear, direct optimization is not desirable. Again, the EM algorithm can be applied for the ML channel estimation.

Let $\{\mathbf{y}, \mathbf{b}\}$ be the complete data and $\{\mathbf{y}\}$ be the incomplete data. Suppose that the index q is used for the EM iteration. Since the received signal can be rewritten as

$$\mathbf{y} = \mathbf{B}\check{\mathbf{h}} + \mathbf{n},$$

where $\mathbf{B} = \text{Diag}(b_0, b_1, \ldots, b_{L-1})$, the E-step at the $(q+1)$th iteration becomes

$$\begin{aligned} Q\left(\mathbf{h}|\hat{\mathbf{h}}^{(q)}\right) &= E\left[\log f(\mathbf{b}, \mathbf{y}|\mathbf{h})|\mathbf{y}, \hat{\mathbf{h}}^{(q)}\right] \\ &= -\frac{1}{N_0} E\left[\|\mathbf{y} - \mathbf{B}\mathbf{F}_{\text{ch}}\mathbf{h}\|^2 |\mathbf{y}, \hat{\mathbf{h}}^{(q)}\right] \\ &= -\frac{1}{N_0}\left(\|\mathbf{y}\|^2 - 2\Re\left(\mathbf{y}^H E[\mathbf{B}|\mathbf{y}, \hat{\mathbf{h}}^{(q)}]\mathbf{F}_{\text{ch}}\mathbf{h}\right) + \mathbf{h}^H \mathbf{F}_{\text{ch}}^H E[\mathbf{B}^H\mathbf{B}|\mathbf{y}, \hat{\mathbf{h}}^{(q)}]\mathbf{F}_{\text{ch}}\mathbf{h}\right), \end{aligned} \tag{11.26}$$

where the expectation is carried out with respect to $\mathbf{B}$ or $\mathbf{b}$. The M-step finds $\mathbf{h}$ that maximizes $Q(\mathbf{h}|\hat{\mathbf{h}}^{(q)})$, which becomes the $(q+1)$th estimate of $\mathbf{h}$ in the EM algorithm. Noting that

$$E\left[\mathbf{B}^H\mathbf{B}|\mathbf{y}, \hat{\mathbf{h}}^{(q)}\right] = \mathbf{I},$$

we have the M-step as follows:

$$\begin{aligned} \hat{\mathbf{h}}^{(q+1)} &= \arg\max_{\mathbf{h}} Q\left(\mathbf{h}|\hat{\mathbf{h}}^{(q)}\right) \\ &= \arg\min_{\mathbf{h}}\left\{-2\Re\left(\mathbf{y}^H\bar{\mathbf{B}}^{(q)}\mathbf{F}_{\text{ch}}\mathbf{h}\right) + \|\mathbf{F}_{\text{ch}}\mathbf{h}\|^2\right\}, \end{aligned} \tag{11.27}$$

where $\bar{\mathbf{B}}^{(q)} = E[\mathbf{B}|\mathbf{y}, \hat{\mathbf{h}}^{(q)}]$. Since $\mathbf{F}_{\text{ch}}^H\mathbf{F}_{\text{ch}} = L\mathbf{I}$, it can be shown that

$$\hat{\mathbf{h}}^{(q+1)} = \frac{1}{L}\mathbf{F}_{\text{ch}}^H\left(\bar{\mathbf{B}}^{(q)}\right)^H\mathbf{y}. \tag{11.28}$$

Using Eq. (11.24), the $(q+1)$th estimate of the frequency response of channel becomes

$$\begin{aligned} \hat{\overset{\circ}{\mathbf{h}}}^{(q+1)} &= \mathbf{F}_{\text{ch}}\hat{\mathbf{h}}^{(q+1)} \\ &= \frac{1}{L}\mathbf{F}_{\text{ch}}\mathbf{F}_{\text{ch}}^H\left(\bar{\mathbf{B}}^{(q)}\right)^H\mathbf{y}. \end{aligned} \tag{11.29}$$

In the EM algorithm, we need to find the conditional mean of b_l as follows:

$$E_{b_l}\left[b_l|\mathbf{y}, \hat{\mathbf{h}}^{(q)}\right] = \Pr\left(b_l = 1|\mathbf{y}, \hat{\mathbf{h}}^{(q)}\right) - \Pr\left(b_l = -1|\mathbf{y}, \hat{\mathbf{h}}^{(q)}\right).$$

Hence, the *a posteriori* probability of b_l is required; this can be provided by the MAP decoder.

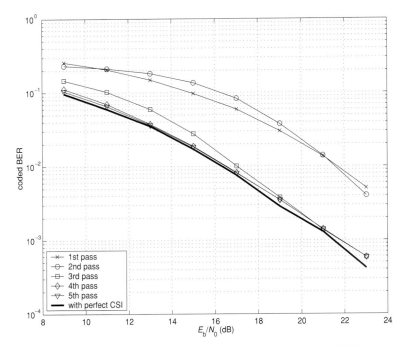

Figure 11.10. Coded BER performance of the EM-based iterative receiver with Channel A.

The iterative detector and decoder are not necessary: as no interfering signal exists, there is no need for a detector to mitigate interfering signals. With $\mathbf{y}$ and $\hat{\mathbf{h}}^{(q)}$, the MAP decoding can be directly performed without any interference mitigation. However, if there are interfering signals (caused, say, by frequency offset or time-varying CIR), a detector that is capable of mitigating interfering signals using extrinsic information would be desirable.

Figures 11.10 and 11.11 show simulation results of the EM-based iterative receiver for Channel A and Channel B, respectively. A rate-half convolutional code with generator polynomial $(7, 5)$ in octal is used with a random bit interleaver. It is assumed that $L = 1024$ and that there are 16 uniformly distributed pilot subcarriers. The simulation results show that the BER converges after four iterations for both Channel A and Channel B.

For Channel A, the performance is not improved from the first iteration to the second iteration, but a better performance is achieved thereafter. So, we see that even though the initial channel estimate is not good enough to achieve a better BER performance for the following channel decoding, the subsequent iterations can improve the performance as data subcarriers can be used as pilot subcarriers for the channel estimation.

The performance of the EM algorithm depends on the initial estimate. Since the initial channel estimate relies on pilot subcarriers, a better initial estimate is available if there are more pilot subcarriers. We show in Figs 11.12 and 11.13 simulation results with different numbers of pilot subcarriers for Channel A and Channel B, respectively. Without iterations, the performance can be improved if there are more pilot subcarriers. However, after four iterations, it is shown that the BER performance is not sensitive to the number of pilot

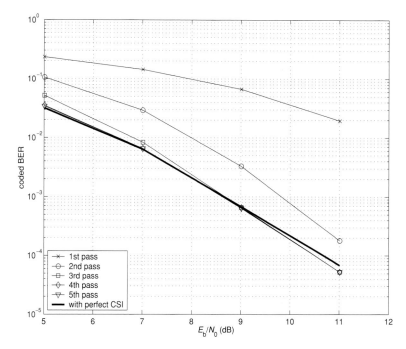

Figure 11.11. Coded BER performance of the EM-based iterative receiver with Channel B.

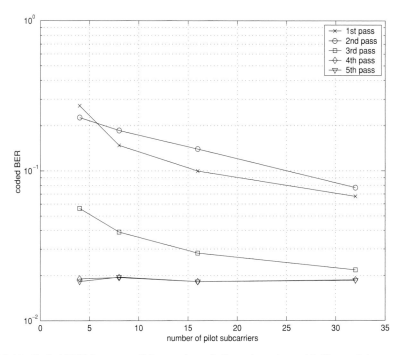

Figure 11.12. Coded BER in terms of the number of pilot subcarriers with Channel A.

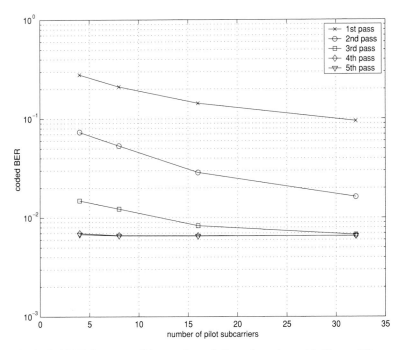

Figure 11.13. Coded BER in terms of the number of pilot subcarriers with Channel B.

subcarriers since the EM algorithm makes use of data symbols for the channel estimation. Consequently, we can see that the iterative receiver can not only improve the performance, but also increase the data throughput as a large number of pilot subcarriers is not necessary.

11.4   Summary and notes

We introduced OFDM and coded OFDM in this chapter. We also discussed the channel estimation for OFDM systems. An iterative receiver was derived using the EM algorithm for the ML channel estimation.

Since OFDM has a resistance against severe multipaths, several standards for wired and wireless communications adopt OFDM. For example, the IEEE 802.11 and 802.16 standards for wireless local area networks (LANs) and wireless metropolitan area networks (MANs), respectively, are based on OFDM (Bahai and Saltzberg, 1999; Eklund *et al.*, 2002; Nee *et al.*, 1999). Single-carrier systems with a frequency domain equalizer are investigated (Falconer *et al.*, 2002) in order to study the possibility of achieving a resistance against severe multipaths, similarly to OFDM. This approach is interesting since single-carrier systems do not suffer from a high PAPR.

The channel estimation for OFDM is well addressed in Edfors *et al.* (1998), Li, Cimini, and Sollenberger (1998), and Yang *et al.* (2001). Blind approaches are also available (Heath and Giannakis, 1999). An EM-based channel estimation without channel decoding is discussed in Ma, Kobayashi, and Schwartz (2004). For time-varying channels, an iterative receiver with channel estimation is proposed in Nguyen and Choi (2006).

11.5 Appendix to Chapter 11: Derivation of Eq. (11.6)

For the derivation from the second equality to the third equality in Eq. (11.6), we need to show that

$$\sum_{q=0}^{L-1} h_{p,q} e^{j2\pi lq/L} = \sum_{v=0}^{L-1} h_v e^{j2\pi l(p-v)/L}, \quad p = 0, 1, \ldots, L-1.$$

For a circular matrix $\mathbf{H}$, we have

$$h_{p,q} = \begin{cases} h_{p-q}, & \text{if } p \geq q; \\ h_{L+p-q}, & \text{if } p < q, \end{cases} \tag{11.30}$$

Using Eq. (11.30), it can be shown that

$$\sum_{q=0}^{L-1} h_{p,q} e^{j2\pi lq/L} = \sum_{q=0}^{p} h_{p-q} e^{j2\pi lq/L} + \sum_{q=p+1}^{L-1} h_{L+p-q} e^{j2\pi lq/L}.$$

Let $p - q = v$. Then, we have

$$\sum_{q=0}^{p} h_{p-q} e^{j2\pi lq/L} + \sum_{q=p+1}^{L-1} h_{L+p-q} e^{j2\pi lq/L}$$

$$= \sum_{v=0}^{p} h_v e^{j2\pi l(p-v)/L} + \sum_{v=p-L+1}^{-1} h_{L+v} e^{j2\pi l(p-v)/L}$$

$$= \sum_{v=0}^{p} h_v e^{j2\pi l(p-v)/L} + \sum_{\bar{v}=p+1}^{L-1} h_{L+v} e^{j2\pi l(p-\bar{v})/L} \quad (\text{letting } \bar{v} = L + v)$$

$$= \sum_{v=0}^{L-1} h_v e^{j2\pi l(p-v)/L}$$

since $e^{j2\pi l(p-\bar{v}-L)/L} = e^{j2\pi l(p-\bar{v})/L}$.

Appendix 1   Review of signal processing and the $\mathcal{Z}$-transform

A1.1   Signals and systems

We consider discrete-time signals and systems.

(i) Linear systems. Consider an operator $H[\cdot]$ which denotes a system. The output of the system $H[\cdot]$ given input x_l is denoted by $H[x_l]$. Then, a system is *linear* if

$$H[ax_l + by_l] = aH[x_l] + bH[y_l],$$

where a and b are constants and x_l and y_l are input signals.

(ii) Impulse response. The *impulse response* of a linear system is the output of the linear system when the input is a unit impulse:

$$h_l = H[\delta_l],$$

where δ_l denotes the unit impulse defined as

$$\delta_l = \begin{cases} 1, & \text{if } l = 0; \\ 0, & \text{otherwise.} \end{cases}$$

(iii) Causal. A linear system is called *causal* if

$$H[\delta_l] = 0, \quad \forall l < 0.$$

(iv) Convolution. The *convolution* of x_l and y_l is defined as follows

$$
\begin{aligned}
x_l * y_l &= \sum_m x_m y_{l-m} \\
&= \sum_m x_{l-m} y_m,
\end{aligned}
\tag{A1.1}
$$

where $*$ denotes the convolution (operation).

(v) Output of linear systems. If a linear system has impulse response $\{h_m\}$, given input x_l, the output is given by

$$H[x_l] = x_l * h_l.$$

A1.2 $\mathcal{Z}$-transform

(i) Definition. The $\mathcal{Z}$-transform of signal x_l is defined as follows:

$$X(z) = \mathcal{Z}(x_l)$$
$$= \sum_l x_l z^{-l},$$

where z is a complex variable.

(ii) Convolution property. If $y_l = x_l * h_l$, then

$$Y(z) = X(z)H(z),$$

where $Y(z)$, $X(z)$, and $H(z)$ stand for the $\mathcal{Z}$-transforms of y_l, x_l, and h_l, respectively.

(iii) Shift property. If $y_l = x_{l-m}$, then

$$Y(z) = z^{-m}X(z).$$

(iv) Parseval's theorem. Let $X(z)$ and $Y(z)$ be the $\mathcal{Z}$-transform of x_l and y_l, respectively. Then, we have

$$\sum_l x_l y_l = \frac{1}{j2\pi} \oint X(z)Y(z^{-1}) \frac{dz}{z}.$$

A1.3 Sampling theorem

An analog signal can be converted into a sampled signal (i.e. a discrete-time sequence) by the sampling process. The impulse sampling is a conceptual sampling process that can help to determine the sampling rate to recover the original analog signal from a sampled signal; see Haykin and Veen (1999) for other sampling processes.

Suppose that $x(t)$ is a bandlimited analog signal. The Fourier transform of $x(t)$, denoted by $X(f)$, is zero outside the interval $-B_W < f < B_W$, where B_W is the bandwidth. The impulse sampling uses an impulse sequence to sample analog signals as follows:

$$x_s(t) = x(t) \left(\sum_{l=-\infty}^{\infty} \delta(t - lT_s) \right)$$
$$= \sum_{l=-\infty}^{\infty} x(t)\delta(t - lT_s)$$
$$= \sum_{l=-\infty}^{\infty} x(lT_s)\delta(t - lT_s),$$

where $x_s(t)$ is the sampled signal and T_s is the sampling interval. It is important to determine the sampling interval so that the original analog signal $x(t)$ can be recovered from the sampled signal $x_s(t)$.

Using the convolution property, we can find the Fourier transform of $x_s(t)$ as follows:

$$X_s(f) = X(f) * \mathcal{F}\left(\sum_{l=-\infty}^{\infty} \delta(t - lT_s)\right)$$

$$= X(f) * \left(\frac{1}{T_s} \sum_{n=-\infty}^{\infty} \delta(f - nf_s)\right)$$

$$= \frac{1}{T_s} \sum_{n=-\infty}^{\infty} X(f - nf_s),$$

where

$$\mathcal{F}\left(\sum_{l=-\infty}^{\infty} \delta(t - lT_s)\right) = \frac{1}{T_s} \sum_{n=-\infty}^{\infty} \delta(f - nf_s)$$

and $f_s = 1/T_s$. Thus, if $f_s \geq 2B_W$, $X(f)$ can be found from $X_s(f)$ without aliasing by taking an ideal low-pass filtering. The sampling rate $f_s \geq 2B_W$ is called the Nyquist sampling rate.

Appendix 2   Important properties of matrices and vectors

A2.1   Vectors and matrices

An $N \times K$ matrix $\mathbf{A}$ is an array of numbers as follows:

$$\mathbf{A} = \begin{bmatrix} a_{1,1} & a_{1,2} & \cdots & a_{1,K} \\ a_{2,1} & a_{2,2} & \cdots & a_{2,K} \\ \vdots & \vdots & \ddots & \vdots \\ a_{N,1} & a_{N,2} & \cdots & a_{N,K} \end{bmatrix},$$

where $a_{n,k}$ denotes the (n, k)th element. If $N = K$, the matrix $\mathbf{A}$ is called square. An $N \times 1$ vector $\mathbf{a}$ is a matrix with one column:

$$\mathbf{a} = \begin{bmatrix} a_1 \\ a_2 \\ \vdots \\ a_N \end{bmatrix},$$

where a_n denotes the nth element of $\mathbf{a}$.

Basic manipulations with matrices are as follows.

(i) Addition:

$$\mathbf{C} = \mathbf{A} + \mathbf{B} \Leftrightarrow c_{n,k} = a_{n,k} + b_{n,k},$$

where the sizes of $\mathbf{A}$, $\mathbf{B}$, and $\mathbf{C}$ are the same.

(ii) Multiplications:

- For $\mathbf{A}$ and $\mathbf{B}$ of size $N \times M$ and $M \times K$, respectively:

$$\mathbf{C} = \mathbf{A}\mathbf{B} \Leftrightarrow c_{n,k} = \sum_{m=1}^{M} a_{n,m} b_{m,k},$$

where the size of $\mathbf{C}$ is $N \times K$.

- For a scalar α:

$$\mathbf{C} = \alpha \mathbf{A} \Leftrightarrow c_{n,k} = \alpha a_{n,k}.$$

(iii) Transpositions:

- The transpose of $\mathbf{A}$, denoted by $\mathbf{A}^{\mathrm{T}}$, is defined by

$$[\mathbf{A}^{\mathrm{T}}]_{n,k} = [\mathbf{A}]_{k,n},$$

where $[\mathbf{A}]_{n,k}$ stands for the (n, k)th element of $\mathbf{A}$. We also have

$$(\mathbf{A} + \mathbf{B})^{\mathrm{T}} = \mathbf{A}^{\mathrm{T}} + \mathbf{B}^{\mathrm{T}},$$
$$(\mathbf{AB})^{\mathrm{T}} = \mathbf{B}^{\mathrm{T}} \mathbf{A}^{\mathrm{T}}.$$

- The Hermitian transpose of $\mathbf{A}$, denoted by $\mathbf{A}^{\mathrm{H}}$, is defined by

$$[\mathbf{A}^{\mathrm{H}}]_{n,k} = [\mathbf{A}]_{k,n}^{*},$$

where the superscript $*$ denotes the complex conjugate. We also have

$$(\mathbf{A} + \mathbf{B})^{\mathrm{H}} = \mathbf{A}^{\mathrm{H}} + \mathbf{B}^{\mathrm{H}},$$
$$(\mathbf{AB})^{\mathrm{H}} = \mathbf{B}^{\mathrm{H}} \mathbf{A}^{\mathrm{H}}.$$

(iv) Triangular matrices: $\mathbf{A}$ is called lower triangular if

$$[\mathbf{A}]_{n,k} = \begin{cases} a_{n,k}, & \text{if } n \geq k; \\ 0, & \text{otherwise.} \end{cases}$$

If the diagonal elements of $\mathbf{A}$ are all zeros, $\mathbf{A}$ is called strictly lower triangular. $\mathbf{A}$ is called upper triangular if

$$[\mathbf{A}]_{n,k} = \begin{cases} a_{n,k}, & \text{if } n \leq k; \\ 0, & \text{otherwise.} \end{cases}$$

If the diagonal elements of $\mathbf{A}$ are all zeros, $\mathbf{A}$ is called strictly upper triangular.

(v) Symmetric: A square matrix $\mathbf{A}$ is called symmetric if

$$\mathbf{A}^{\mathrm{H}} = \mathbf{A}.$$

(vi) Identity matrix: $\mathbf{I}$ is a square matrix defined by

$$\mathbf{I} = \mathrm{Diag}(1, 1, \ldots, 1),$$

where $\mathrm{Diag}(a_1, a_2, \ldots, a_N)$ denotes a diagomal matrix given by

$$\mathrm{Diag}(a_1, a_2, \ldots, a_N) = \begin{bmatrix} a_1 & 0 & \cdots & 0 \\ 0 & a_2 & \cdots & 0 \\ \vdots & \vdots & \ddots & \vdots \\ 0 & 0 & \cdots & a_N \end{bmatrix}.$$

Hence, it follows, if multiplications can be defined, that

$$\mathbf{AI} = \mathbf{A},$$
$$\mathbf{IA} = \mathbf{A}.$$

(vii) Inverse of $\mathbf{A}$: $\mathbf{B}$ is the inverse of a square matrix $\mathbf{A}$ if

$$\mathbf{AB} = \mathbf{I};$$

$\mathbf{B}$ is denoted by $\mathbf{A}^{-1}$. We also have

$$\mathbf{A}^{-T} \triangleq (\mathbf{A}^{-1})^{\mathrm{T}} = (\mathbf{A}^{\mathrm{T}})^{-1}$$
$$\mathbf{A}^{-H} \triangleq (\mathbf{A}^{-1})^{\mathrm{H}} = (\mathbf{A}^{\mathrm{H}})^{-1}.$$

(viii) Rank. The rank of a matrix is the minimum of the numbers of linearly independent rows and columns. If the rank of an $N \times N$ matrix is N (i.e. full rank), the matrix is called nonsingular. Otherwise, the matrix is called singular.

(ix) Trace. The trace of a square matrix $\mathbf{A}$ is defined by

$$\mathrm{Tr}(\mathbf{A}) = \sum_{n=1}^{N} a_{n,n}.$$

Some properties of trace are as follows.

- $\mathrm{Tr}(\mathbf{A} + \mathbf{B}) = \mathrm{Tr}(\mathbf{A}) + \mathrm{Tr}(\mathbf{B})$;
- $\mathrm{Tr}(\mathbf{AB}) = \mathrm{Tr}(\mathbf{BA})$.

(x) Determinant. For an $N \times N$ matrix $\mathbf{A}$,

$$|\mathbf{A}| = \sum_{\text{permutation}:i_1,i_2,\ldots,i_N} \pm a_{1,i_1} a_{2,i_2} \cdots a_{N,i_N},$$

where the sum is taken over all the possible permutations and the sign is $+$ if the permutation is even and $-$ if the permutation is odd.

Some examples are as follows.

- If

$$\mathbf{A} = \begin{bmatrix} \alpha & \beta \\ \gamma & \delta \end{bmatrix},$$

$$|\mathbf{A}| = \underbrace{a_{1,1} a_{2,2}}_{\text{permutation: } 1,2} - \underbrace{a_{1,2} a_{2,1}}_{\text{permutation: } 2,1} = \alpha\delta - \beta\gamma.$$

- If $\mathbf{A}$ is square and (upper or lower) triangular, we can also show that

$$|\mathbf{A}| = \prod_{n=1}^{N} a_{n,n}.$$

Some properties of the determinant are as follows:

- $|\mathbf{A}^{\mathrm{T}}| = |\mathbf{A}|$;
- $|\mathbf{AB}| = |\mathbf{A}||\mathbf{B}|$;
- $|\mathbf{A}^{-1}| = 1/|\mathbf{A}|$.

(xi) Norms. The 2-norm of a vector $\mathbf{x}$ is defined by

$$||\mathbf{x}|| = \sqrt{\mathbf{x}^{\mathrm{T}}\mathbf{x}}.$$

The Frobenius norm of an $N \times K$ matrix $\mathbf{A}$ is defined by

$$||\mathbf{A}||_F = \sqrt{\sum_{n=1}^{N} \sum_{k=1}^{K} |a_{n,k}|^2}.$$

The 2-norm of a matrix $\mathbf{A}$ is defined by

$$||\mathbf{A}|| = \max_{\mathbf{x} \neq \mathbf{0}} \frac{||\mathbf{Ax}||}{||\mathbf{x}||}.$$

(xii) Positive definite. A square matrix $\mathbf{A}$ is called positive definite if

$$\mathbf{x}^H \mathbf{Ax} > 0 \quad \text{for any } \mathbf{x} \neq \mathbf{0}.$$

If $\mathbf{x}^H \mathbf{Ax} \geq 0$ for any $\mathbf{x} \neq \mathbf{0}$, $\mathbf{A}$ is called positive semi-definite or nonnegative definite.

(xiii) Eigenvalues and eigenvectors. A square matrix $\mathbf{A}$ has an eigenvector $\mathbf{e}$ and an eigenvalue λ if

$$\mathbf{Ae} = \lambda \mathbf{e}.$$

Generally, $\mathbf{e}$ is normalized to $||\mathbf{e}|| = 1$. An $N \times N$ symmetric matrix $\mathbf{A}$ has N real eigenvalues and is given by

$$\mathbf{A} = [\mathbf{e}_1 \ \mathbf{e}_2 \ \cdots \ \mathbf{e}_N] \text{Diag}(\lambda_1, \lambda_2, \ldots, \lambda_N)[\mathbf{e}_1 \ \mathbf{e}_2 \ \cdots \ \mathbf{e}_N]^H,$$

where λ_n denotes the nth (largest) eigenvalue and $\mathbf{e}_n$ denotes the corresponding eigenvector. In addition, $\mathbf{E} = [\mathbf{e}_1 \ \mathbf{e}_2 \ \cdots \ \mathbf{e}_N]$ has the following property:

$$\mathbf{EE}^H = \mathbf{E}^H \mathbf{E} = \mathbf{I}.$$

A2.2   Subspaces, orthogonal projection, and pseudo-inverse

(i) Subspace. A subspace is a subset of a vector space; it can be defined by a span of vectors as follows:

$$\text{Span}\{\mathbf{a}_1, \mathbf{a}_2, \ldots, \mathbf{a}_N\} = \left\{ \sum_{n=1}^{N} c_n \mathbf{a}_n \mid c_1, c_2, \ldots, c_N \text{ are complex numbers} \right\}.$$

The range of a matrix is a subspace defined by

$$\text{Range}(\mathbf{A}) = \{\mathbf{x} \mid \mathbf{x} = \mathbf{Ac} \text{ for any vector } \mathbf{c}\}$$
$$= \text{Span}\{\mathbf{a}_1, \mathbf{a}_2, \ldots, \mathbf{a}_N\}.$$

The null space of $\mathbf{A}$ is defined by

$$\text{Null}(\mathbf{A}) = \{\mathbf{x} \mid \mathbf{Ax} = \mathbf{0}\}.$$

(ii) Orthogonal projection. A matrix $\mathbf{P}$ is the orthogonal projection into the subspace $\mathcal{S}$ if

- Range($\mathbf{P}$) = $\mathcal{S}$;
- $\mathbf{P}^2 = \mathbf{PP} = \mathbf{P}$;
- $\mathbf{P}^H = \mathbf{P}$.

For an $N \times K$ matrix $\mathbf{A}$, where $N \geq K$, the orthogonal projection matrix onto the subspace Range($\mathbf{A}$) is given by

$$\mathbf{P}(\mathbf{A}) = \mathbf{A}(\mathbf{A}^H\mathbf{A})^{-1}\mathbf{A}.$$

If the inverse does not exist, the pseudo-inverse can replace the inverse. We can easily verify that

- Range($\mathbf{P}(\mathbf{A})$) $= \mathcal{S}$;
- $\mathbf{P}^2(\mathbf{A}) = \mathbf{P}(\mathbf{A})$;
- $\mathbf{P}^H(\mathbf{A}) = \mathbf{P}(\mathbf{A})$.

(iii) Pseudo-inverse. The pseudo-inverse of an $N \times M$ matrix $\mathbf{A}$, denoted by $\mathbf{A}^\dagger$, of size $M \times N$, is uniquely defined by the following equations:

$$\mathbf{A}^\dagger\mathbf{A}\mathbf{x} = \mathbf{x}, \quad \forall \mathbf{x} \in \text{Range}(\mathbf{A}^H);$$
$$\mathbf{A}^\dagger\mathbf{x} = \mathbf{0}, \quad \forall \mathbf{x} \in \text{Null}(\mathbf{A}^H).$$

If the rank of $\mathbf{A}$ is M,

$$\mathbf{A}^\dagger = (\mathbf{A}^H\mathbf{A})^{-1}\mathbf{A}^H.$$

If the rank of $\mathbf{A}$ is N,

$$\mathbf{A}^\dagger = \mathbf{A}^H(\mathbf{A}\mathbf{A}^H)^{-1}.$$

A2.3 Various forms of matrix inversion lemma

(i) A standard matrix inversion lemma is given as

$$(\mathbf{A} + \mathbf{B}\mathbf{C}\mathbf{D})^{-1} = \mathbf{A}^{-1} - \mathbf{A}^{-1}\mathbf{B}(\mathbf{C}^{-1} + \mathbf{D}\mathbf{A}^{-1}\mathbf{B})^{-1}\mathbf{D}\mathbf{A}^{-1}. \tag{A2.1}$$

(ii) Woodbury's identity:

$$(\mathbf{R} + \alpha\mathbf{u}\mathbf{u}^H)^{-1} = \mathbf{R}^{-1} - \frac{\alpha}{1 + \alpha\mathbf{u}^H\mathbf{R}^{-1}\mathbf{u}}\mathbf{R}^{-1}\mathbf{u}\mathbf{u}^H\mathbf{R}^{-1}. \tag{A2.2}$$

Note that

$$\mathbf{u}^H(\mathbf{R} + \alpha\mathbf{u}\mathbf{u}^H)^{-1}\mathbf{u} = \frac{\mathbf{u}^H\mathbf{R}^{-1}\mathbf{u}}{1 + \alpha\mathbf{u}^H\mathbf{R}^{-1}\mathbf{u}}. \tag{A2.3}$$

(iii) Sherman–Morrison formula:

$$(\mathbf{R} + \mathbf{u}\mathbf{v}^H)^{-1} = \mathbf{R}^{-1} - \frac{1}{1 + \mathbf{v}^H\mathbf{R}^{-1}\mathbf{u}}\mathbf{R}^{-1}\mathbf{u}\mathbf{v}^H\mathbf{R}^{-1}.$$

A2.4 Vectorization and Kronecker product

Consider an $N \times K$ matrix $\mathbf{A}$:

$$\mathbf{A} = [\mathbf{a}_1 \quad \mathbf{a}_2 \quad \cdots \quad \mathbf{a}_K].$$

The vectorization of $\mathbf{A}$ is given by

$$\text{vec}(\mathbf{A}) = [\mathbf{a}_1^{\mathrm{T}} \quad \mathbf{a}_2^{\mathrm{T}} \quad \cdots \quad \mathbf{a}_K^{\mathrm{T}}]^{\mathrm{T}}.$$

The Kronecker (or direct) product of two matrices, $\mathbf{A}$ and $\mathbf{B}$, is defined by

$$\mathbf{A} \otimes \mathbf{B} = \begin{bmatrix} a_{1,1}\mathbf{B} & a_{1,2}\mathbf{B} & \cdots & a_{1,K}\mathbf{B} \\ a_{2,1}\mathbf{B} & a_{2,2}\mathbf{B} & \cdots & a_{2,K}\mathbf{B} \\ & & & \\ a_{N,1}\mathbf{B} & a_{N,2}\mathbf{B} & \cdots & a_{N,K}\mathbf{B} \end{bmatrix}. \tag{A2.4}$$

Some basic properties of the Kronecker product are as follows:

(K1): $(\mathbf{A} \otimes \mathbf{B})^{\mathrm{H}} = \mathbf{A}^{\mathrm{H}} \otimes \mathbf{B}^{\mathrm{H}}$;
(K2): $\mathbf{A} \otimes (\mathbf{B} \otimes \mathbf{C}) = (\mathbf{A} \otimes \mathbf{B}) \otimes \mathbf{C}$;
(K3): $(\mathbf{A} \otimes \mathbf{B})(\mathbf{C} \otimes \mathbf{D}) = \mathbf{AC} \otimes \mathbf{BD}$, if $\mathbf{AC}$ and $\mathbf{BD}$ are defined;
(K4): $\text{Tr}(\mathbf{A} \otimes \mathbf{B}) = \text{Tr}(\mathbf{A})\text{Tr}(\mathbf{B})$, if $\mathbf{A}$ and $\mathbf{B}$ are square matrices;
(K5): $\text{vec}(\mathbf{ABC}) = (\mathbf{C}^{\mathrm{T}} \otimes \mathbf{A})\text{vec}(\mathbf{B})$.

Appendix 3 Background for probability and statistics

A3.1 Review of probability

A3.1.1 Sample space and probability measure

A sample space Ω is the set of all possible outcomes (or events) of an experiment. An outcome A is a subset of Ω. A probability measure $\Pr(\cdot)$ is a mapping from Ω to the real line with the following properties.

(i) $\Pr(A) \geq 0$, $A \in \Omega$;
(ii) $\Pr(\Omega) = 1$;
(iii) For a countable set of events, $\{A_m\}$, if $A_l \cup A_m = \emptyset$, for $l \neq m$, then

$$\Pr\left(\bigcup_{l=1}^{\infty} A_l\right) = \sum_{l=1}^{\infty} \Pr(A_l).$$

A3.1.2 Joint and conditional probability

The joint probability of two events A and B is $\Pr(A \cup B)$. The conditional probability of A given B is given by

$$\Pr(A|B) = \frac{\Pr(A \cup B)}{\Pr(B)}, \quad \Pr(B) > 0.$$

Two events A and B are independent if and only if

$$\Pr(A \cup B) = \Pr(A)\Pr(B),$$

and this implies that

$$\Pr(A|B) = \Pr(A).$$

A3.1.3 Random variables

A random variable is a mapping from an event ω in Ω to a real number, denoted by $X(\omega)$. The *cumulative distribution function* (cdf) of X is defined by

$$F_X(x) = \Pr(\{\omega | X(\omega) \leq x\})$$
$$= \Pr(X \leq x),$$

and the *probability density function* (pdf) is defined by

$$f_X(x) = \frac{d}{dx} F_X(x),$$

where the subscript X on F and f identifies the random variable. If the random variable is obvious, the subscript is often omitted.

There are some well known pdfs as follows:

(i) Gaussian pdf with mean μ and variance σ^2:

$$f(x) = \frac{1}{\sqrt{2\pi\sigma^2}} \exp\left(-\frac{1}{2}(x-\mu)^2\right);$$

(ii) expondential pdf ($a > 0$):

$$f(x) = \begin{cases} a e^{-ax}, & \text{if } x \geq 0; \\ 0, & \text{otherwise}; \end{cases}$$

(iii) Rayleigh pdf ($b > 0$):

$$f(x) = \begin{cases} \frac{x}{b} e^{-x^2/b}, & \text{if } x \geq 0; \\ 0, & \text{otherwise}; \end{cases}$$

(iv) chi-square pdf of n degrees of freedom:

$$f(x) = \begin{cases} \frac{x^{(n-2)/2} e^{-x/2}}{2^{n/2} \Gamma(n/2)}, & \text{if } x \geq 0; \\ 0, & \text{otherwise}. \end{cases}$$

A joint cdf of random variables, $X_1, X_2, \ldots, X_n$ is given by

$$F_{X_1, X_2, \ldots, X_n}(x_1, x_2, \ldots, x_n) = \Pr(X_1 \leq x_1, X_2 \leq x_2, \ldots, X_n \leq x_n).$$

The joint pdf is given by

$$f_{X_1, X_2, \ldots, X_n}(x_1, x_2, \ldots, x_n) = \frac{\partial^n}{\partial x_1 \partial x_2 \ldots \partial x_n} F_{X_1, X_2, \ldots, X_n}(x_1, x_2, \ldots, x_n).$$

The conditional pdf of X_1 given $X_2 = x_2$ is given by

$$f_{X_1|X_2}(x_1|x_2) = \frac{f_{X_1, X_2}(x_1, x_2)}{f_{X_2}(x_2)}, \quad f_{X_2}(x_2) > 0.$$

The expectation of X is given by

$$E[X] = \int x f_X(x) \, dx.$$

In addition, the expectation of $g(X)$, a function of X, is given by

$$E[g(X)] = \int g(x) f_X(x) \, dx.$$

The variance of X is given by

$$Var(X) = E[(X - E[X])^2]$$
$$= \int (x - E[X])^2 f_X(x) \, dx.$$

The conditional mean of X given $Y = y$ is given by

$$E[X|Y = y] = \int x f_{X|Y}(x|Y = y)\,dx.$$

The pdf of a real-valued joint Gaussian random vector, $\mathbf{x}$, is given by

$$f(\mathbf{x}) = \frac{1}{\sqrt{|2\pi \mathbf{C_x}|}} \exp\left(-\frac{1}{2}(\mathbf{x} - \bar{\mathbf{x}})^{\mathrm{T}} \mathbf{C_x^{-1}}(\mathbf{x} - \bar{\mathbf{x}})\right),$$

where $\bar{\mathbf{x}} = E[\mathbf{x}]$ and $\mathbf{C_x} = E[(\mathbf{x} - \bar{\mathbf{x}})(\mathbf{x} - \bar{\mathbf{x}})^{\mathrm{T}}]$. Here, $|\mathbf{A}|$ denotes the determinant of matrix $\mathbf{A}$. If $\mathbf{x}$ is a circular complex Gaussian random vector,

$$f(\mathbf{x}) = \frac{1}{|\pi \mathbf{C_x}|} \exp\left(-(\mathbf{x} - \bar{\mathbf{x}})^{\mathrm{H}} \mathbf{C_x^{-1}}(\mathbf{x} - \bar{\mathbf{x}})\right),$$

where $\bar{\mathbf{x}} = E[\mathbf{x}]$ and $\mathbf{C_x} = E[(\mathbf{x} - \bar{\mathbf{x}})(\mathbf{x} - \bar{\mathbf{x}})^{\mathrm{H}}]$.

A3.1.4 Random processes

A (discrete-time) random process is a sequence of random variables, $\{x_m\}$. The mean and autocorrelation function of $\{x_m\}$ are denoted by $E[x_m]$ and $R_x(l, m) = E[x_l x_m^*]$, respectively. A random process is called wide-sense stationary (WSS) if

$$E[x_l] = \mu, \qquad \forall l$$
$$E[x_l x_m^*] = E[x_{l+p} x_{m+p}^*], \qquad \forall l, m, p.$$

For a WSS random process, we have

$$R_x(l, m) = R_x(l - m).$$

The power spectrum of a WSS random process, $\{x_m\}$, is defined by

$$S_x(z) = \sum_{l=-\infty}^{\infty} z^{-l} R_x(l).$$

In addition, we can show that

$$R_x(l) = \frac{1}{2\pi} \int_{-\pi}^{\pi} S_x(e^{j\omega}) e^{jl\omega}\,d\omega.$$

A zero-mean WSS random process is white if

$$R_x(l) = \begin{cases} \sigma_x^2, & \text{if } l = 0; \\ 0, & \text{otherwise,} \end{cases}$$

where $\sigma_x^2 = E[|x_l|^2]$.

If x_l is an output of the linear system whose impulse response is $\{h_p\}$ with a zero-mean white random process input, n_l, the autocorrelation function of $\{x_m\}$ is given by

$$R_x(l) = \sigma_n^2 \sum_l h_l h_{l-m},$$

where $\sigma_n^2 = E[|n_l|^2]$. In addition, its power spectrum is given by

$$S_x(z) = \sigma_n^2 H(z) H(z^{-1})$$

or

$$S_x(e^{j\omega}) = \sigma_n^2 |H(e^{j\omega})|^2, \quad z = e^{j\omega}.$$

A3.2   Review of statistics

A3.2.1   Hypothesis testing

Hypothesis testing is a statistical decision process based on observations. In hypothesis testing, there are the null hypothesis denoted by H_0 and the alternative hypothesis denoted by H_1. A decision rule is a function of the observations and provides a decision as to whether H_0 is rejected or not.

There are four possible outcomes as follows:

Table A3.1. *Different types of error*

State	Choose H_0	Choose H_1
H_0 true	correct	Type I error (false alarm)
H_1 true	Type II error (missed detection)	correct

A decision rule can be designed to minimize the probability of Type II error while keeping the probability of Type I error constant.

Suppose that there are two parameters, θ_0 and θ_1. In addition, assume that the two likelihood functions given observation x, $f_X(x|\theta_0)$ and $f_X(x|\theta_1)$, are available. From observation x, the decision rule based on the likelihood ratio test (LRT) is given by

$$\phi(x) = \begin{cases} 1, & \text{if } \ell(x) > \kappa; \\ c, & \text{if } \ell(x) = \kappa; \\ 0, & \text{if } \ell(x) < \kappa, \end{cases}$$

where $\ell(x)$ is the likelihood ratio defined by

$$\ell(x) = \frac{f_X(x|\theta_1)}{f_X(x|\theta_0)},$$

where $\kappa > 0$ is a threshold and $0 \leq c \leq 1$ is a constant. If $\phi(x) = 1$, the parameter θ_1 is chosen. If $\phi(x) = 0$, the parameter θ_0 is chosen. If $\phi(x) = c$, one of θ_0 or θ_1 is chosen randomly.

In the maximum likelihood (ML) decision rule, the threshold, κ, becomes unity, since the parameter that maximizes the likelihood is chosen.

A3.2.2   Parameter estimation

Suppose that the pdf of $\mathbf{y}$ is a function of the parameter vector $\mathbf{a}$ given by

$$f(\mathbf{y}|\mathbf{a}).$$

We attempt to estimate $\mathbf{a}$ from observation $\mathbf{y}$. Generally, an estimate of $\mathbf{a}$ is a function of $\mathbf{y}$ and is given by

$$\hat{\mathbf{a}}(\mathbf{y}).$$

If $E[\hat{\mathbf{a}}(\mathbf{y})] = \mathbf{a}$, the estimate $\hat{\mathbf{a}}(\mathbf{y})$ is called an *unbiased estimate*.

Assume that the pdf is differentiable. *Fisher's information matrix* $\mathbf{J}(\mathbf{a})$ is defined as follows:

$$[\mathbf{J}(\mathbf{a})]_{n,k} = E\left[\frac{\partial \log f(\mathbf{y}|\mathbf{a})}{\partial a_n} \frac{\partial \log f(\mathbf{y}|\mathbf{a})}{\partial a_k}\right], \tag{A3.1}$$

where a_n denotes the nth element of $\mathbf{a}$. We also have

$$[\mathbf{J}(\mathbf{a})]_{n,k} = -E\left[\frac{\partial^2 \log f(\mathbf{y}|\mathbf{a})}{\partial a_n \partial a_k}\right]. \tag{A3.2}$$

To show this, it can be shown that

$$\frac{\partial^2 \log f(\mathbf{y}|\mathbf{a})}{\partial a_n \partial a_k} = -\frac{1}{f^2(\mathbf{y}|\mathbf{a})}\frac{\partial f(\mathbf{y}|\mathbf{a})}{\partial a_n}\frac{\partial f(\mathbf{y}|\mathbf{a})}{\partial a_k} + \frac{1}{f(\mathbf{y}|\mathbf{a})}\frac{\partial^2 f(\mathbf{y}|\mathbf{a})}{\partial a_n \partial a_k}.$$

Since

$$\frac{\partial \log f(\mathbf{y}|\mathbf{a})}{\partial a_n}\frac{\partial \log f(\mathbf{y}|\mathbf{a})}{\partial a_k} = \frac{1}{f^2(\mathbf{y}|\mathbf{a})}\frac{\partial f(\mathbf{y}|\mathbf{a})}{\partial a_n}\frac{\partial f(\mathbf{y}|\mathbf{a})}{\partial a_k}$$

and

$$E\left[\frac{1}{f(\mathbf{y}|\mathbf{a})}\frac{\partial^2 f(\mathbf{y}|\mathbf{a})}{\partial a_n \partial a_k}\right] = \int \frac{\partial^2 f(\mathbf{y}|\mathbf{a})}{\partial a_n \partial a_k}\, d\mathbf{y}$$

$$= \frac{\partial^2}{\partial a_n \partial a_k}\underbrace{\int f(\mathbf{y}|\mathbf{a})\, d\mathbf{y}}_{=1}$$

$$= 0,$$

Eq. (A3.2) is true.

An unbiased estimate exhibits the following inequality:

$$E[(\hat{\mathbf{a}}(\mathbf{y}) - \mathbf{a})(\hat{\mathbf{a}}(\mathbf{y}) - \mathbf{a})^{\mathsf{T}}] \geq \mathbf{J}^{-1}(\mathbf{a}).$$

This inequality is called the *Cramer–Rao* inequality or Cramer–Rao bound (CRB). To derive the CRB, we need to use the following inequality for a covariance matrix:

$$E\left[\left(\mathbf{x}_1 - \mathbf{C}_{12}\mathbf{C}_{22}^{-1}\mathbf{x}_2\right)\left(\mathbf{x}_1 - \mathbf{C}_{12}\mathbf{C}_{22}^{-1}\mathbf{x}_2\right)^{\mathsf{T}}\right] = \mathbf{C}_{11} - \mathbf{C}_{12}\mathbf{C}_{22}^{-1}\mathbf{C}_{12}^{\mathsf{T}}$$

$$\geq \mathbf{0},$$

where $\mathbf{x}_1$ and $\mathbf{x}_2$ are zero-mean random vectors and $\mathbf{C}_{nk} = E[\mathbf{x}_n \mathbf{x}_k^{\mathsf{T}}]$. The inequality is valid since a covariance matrix is positive semi-definite. In addition, the equality is achieved if

$$\mathbf{x}_1 = \mathbf{C}_{12}\mathbf{C}_{22}^{-1}\mathbf{x}_2. \tag{A3.3}$$

Let $x_1 = \hat{\mathbf{a}}(\mathbf{y}) - \mathbf{a}$ and $x_2 = \partial \log f(\mathbf{y}|\mathbf{a})/\partial \mathbf{a}$. Then, we have

$$\mathbf{C}_{11} = E[(\hat{\mathbf{a}}(\mathbf{y}) - \mathbf{a})(\hat{\mathbf{a}}(\mathbf{y}) - \mathbf{a})^{\mathrm{T}}]$$
$$\geq \mathbf{C}_{12}\mathbf{C}_{22}^{-1}\mathbf{C}_{12}^{\mathrm{T}}.$$

It can be shown that

$$\mathbf{C}_{12} = E\left[(\hat{\mathbf{a}}(\mathbf{y}) - \mathbf{a})\left(\frac{\partial \log f(\mathbf{y}|\mathbf{a})}{\partial \mathbf{a}}\right)^{\mathrm{T}}\right]$$
$$= E\left[\hat{\mathbf{a}}(\mathbf{y})\left(\frac{\partial \log f(\mathbf{y}|\mathbf{a})}{\partial \mathbf{a}}\right)^{\mathrm{T}}\right] - E\left[\mathbf{a}\left(\frac{\partial \log f(\mathbf{y}|\mathbf{a})}{\partial \mathbf{a}}\right)^{\mathrm{T}}\right].$$

Furthermore, we can show that

$$E\left[\hat{\mathbf{a}}(\mathbf{y})\left(\frac{\partial \log f(\mathbf{y}|\mathbf{a})}{\partial \mathbf{a}}\right)^{\mathrm{T}}\right] = \int \hat{\mathbf{a}}(\mathbf{y})\left(\frac{\partial f(\mathbf{y}|\mathbf{a})}{\partial \mathbf{a}}\right)^{\mathrm{T}} d\mathbf{y}$$
$$= \left[\frac{\partial}{\partial a_k}\underbrace{\int \hat{a}_n(\mathbf{y})f(\mathbf{y}|\mathbf{a})\,d\mathbf{y}}_{=a_n}\right]$$
$$= \mathbf{I}$$

and

$$E\left[\mathbf{a}\left(\frac{\partial \log f(\mathbf{y}|\mathbf{a})}{\partial \mathbf{a}}\right)^{\mathrm{T}}\right] = \mathbf{a}\int\left(\frac{\partial f(\mathbf{y}|\mathbf{a})}{\partial \mathbf{a}}\right)^{\mathrm{T}} d\mathbf{y}$$
$$= \mathbf{a}\left(\frac{\partial}{\partial \mathbf{a}}\int f(\mathbf{y}|\mathbf{a})\,d\mathbf{y}\right)^{\mathrm{T}}$$
$$= 0,$$

where $\hat{a}_n(\mathbf{y})$ denotes the nth element of $\hat{\mathbf{a}}(\mathbf{y})$. From the above, we have

$$\mathbf{C}_{12} = \mathbf{I}.$$

Since $\mathbf{C}_{22} = \mathbf{J}(\mathbf{a})$, we conclude that

$$E[(\hat{\mathbf{a}}(\mathbf{y}) - \mathbf{a})(\hat{\mathbf{a}}(\mathbf{y}) - \mathbf{a})^{\mathrm{T}}] \geq \mathbf{C}_{22}^{-1} = \mathbf{J}^{-1}(\mathbf{a}).$$

An unbiased estimate $\hat{\mathbf{a}}(\mathbf{y})$ is said to be *efficient* if it achieves the CRB. From Eq. (A3.3), an efficient estimate can be written as follows:

$$\hat{\mathbf{a}}(\mathbf{y}) = \mathbf{a} + \mathbf{J}^{-1}(\mathbf{a})\frac{\partial \log f(\mathbf{y}|\mathbf{a})}{\partial \mathbf{a}}.$$

In the above, we have ignored some mathematical steps. The reader is referred to Porat (1994) for details.

We now move on to well known estimation methods: the ML estimation and MAP estimation.

The ML estimation is widely used to estimate unknown parameters. Suppose that the pdf of the random vector $\mathbf{y}$ is given by $f(\mathbf{y}|\mathbf{a})$, where $\mathbf{a}$ is the parameter vector to be estimated.

Given realization $\mathbf{y}$, the pdf $f(\mathbf{y}|\mathbf{a})$ can be seen as a function of $\mathbf{a}$ (while $\mathbf{y}$ is serving as a parameter), called the likelihood function (of $\mathbf{a}$). The ML estimate is given by

$$\hat{\mathbf{a}}_{ml} = \arg\max_{\mathbf{a}} f(\mathbf{y}|\mathbf{a}). \qquad (A3.4)$$

We make the following remarks.

- Note that $\hat{\mathbf{a}}_{ml}$ is a function of realization or observation $\mathbf{y}$. To emphasize this, $\hat{\mathbf{a}}_{ml}(\mathbf{y})$ is used to denote $\hat{\mathbf{a}}_{ml}$.
- The log likelihood, $\log f(\mathbf{y}|\mathbf{a})$, can replace the likelihood, $f(\mathbf{y}|\mathbf{a})$. Note that since the logarithm function is strictly monotone, the logarithm does not affect the maximization, and the maximization of $f(\mathbf{y}|\mathbf{a})$ is identical to that of $\log f(\mathbf{y}|\mathbf{a})$.

The ML estimate has important properties (under certain conditions; see Porat (1994)) as follows:

(i) if $\hat{\mathbf{a}}_{ml}(\mathbf{y})$ is the ML estimate of $\mathbf{a}$, then $g(\hat{\mathbf{a}}_{ml}(\mathbf{y}))$ is the ML estimate of $g(\mathbf{a})$;
(ii) the ML estimate is an efficient estimate if there exists an efficient estimate;
(iii) the ML estimate is asymptotically normal.

If the *a priori* probability of the parameter vector $\mathbf{a}$ is available, the maximum *a posteriori* probability (MAP) estimation can be formulated as follows:

$$\hat{\mathbf{a}}_{map} = \arg\max_{\mathbf{a}} f(\mathbf{a}|\mathbf{y}),$$

where

$$f(\mathbf{a}|\mathbf{y}) = \frac{f(\mathbf{y}|\mathbf{a})f(\mathbf{a})}{f(\mathbf{y})}$$

and $f(\mathbf{a})$ is the *a priori* probability of $\mathbf{a}$.

A3.2.3 Linear estimation

Suppose that there are two *zero-mean* random vectors, $\mathbf{a}$ and $\mathbf{y}$. A linear transform can be considered to estimate $\mathbf{a}$ from $\mathbf{y}$. An estimate of $\mathbf{a}$ from a linear estimator, $\mathbf{L}$, is given by

$$\hat{\mathbf{a}} = \mathbf{L}\mathbf{y}.$$

The linear minimum mean square error (LMMSE) estimator is given by

$$\mathbf{L}_{lm} = \arg\min_{\mathbf{L}} E[||\mathbf{a} - \hat{\mathbf{a}}||^2]$$
$$= \arg\min_{\mathbf{L}} E[||\mathbf{a} - \mathbf{L}\mathbf{y}||^2].$$

The minimum can be achieved by setting the first order derivative of the MSE with respect to $\mathbf{L}$ equal to zero, because the MSE is a quadratic function of $\mathbf{L}$. Since

$$\frac{\partial}{\partial L_{n,k}} E\left[\sum_p |a_p - \sum_m L_{p,m} y_m|^2\right] = 2E\left[\left(a_n - \sum_m L_{n,m} y_m\right) y_k^*\right], \qquad (A3.5)$$

the condition to obtain the minimum can be derived by

$$E[(\mathbf{a} - \mathbf{Ly})\mathbf{y}^H] = \mathbf{0}. \tag{A3.6}$$

Let $\mathbf{e} = \mathbf{a} - \mathbf{Ly}$ be the error vector. Then, Eq. (A3.6) implies that the error vector should be uncorrelated with the observation vector $\mathbf{y}$ to achieve the minimum. This is the orthogonality principle for the LMMSE estimation.

From Eq. (A3.6), the LMMSE estimator can be written as follows:

$$\begin{aligned}
\mathbf{L}_{lm} &= E[\mathbf{ay}^H](E[\mathbf{yy}^H])^{-1} \\
&= \mathbf{R}_{ay}\mathbf{R}_{yy}^{-1},
\end{aligned}$$

where $\mathbf{R}_{ay} = E[\mathbf{ay}^H]$ and $\mathbf{R}_{yy} = E[\mathbf{yy}^H]$. The MMSE matrix is given by

$$\begin{aligned}
\text{MMSE} &= E[(\mathbf{a} - \mathbf{L}_{lm}\mathbf{y})(\mathbf{a} - \mathbf{L}_{lm}\mathbf{y})^H] \\
&= E[(\mathbf{a} - \mathbf{L}_{lm}\mathbf{y})\mathbf{a}^H] \\
&= \mathbf{R}_{aa} - \mathbf{L}_{lm}\mathbf{R}_{ya} \\
&= \mathbf{R}_{aa} - \mathbf{R}_{ay}\mathbf{R}_{yy}^{-1}\mathbf{R}_{ya},
\end{aligned}$$

where $\mathbf{R}_{aa} = E[\mathbf{aa}^H]$ and $\mathbf{R}_{ya} = E[\mathbf{ya}^H]$.

It is necessary to extend the LMMSE estimation for nonzero-mean random vectors. If $E[\mathbf{a}] = \bar{\mathbf{a}} \neq \mathbf{0}$ and $E[\mathbf{y}] = \bar{\mathbf{y}} \neq \mathbf{0}$, the LMMSE estimation problem is given by

$$\{\mathbf{L}_{lm}, \mathbf{c}_{lm}\} = \arg\min_{\mathbf{L}, \mathbf{c}} E[||\mathbf{a} - \mathbf{Ly} - \mathbf{c}||^2],$$

where $\mathbf{c}$ is a vector included to deal with nonzero-mean vectors. Let $\tilde{\mathbf{a}} = \mathbf{a} - \bar{\mathbf{a}}$ and $\tilde{\mathbf{y}} = \mathbf{y} - \bar{\mathbf{y}}$. Then, the MSE is rewritten as follows:

$$E[||\tilde{\mathbf{a}} - \mathbf{L}\tilde{\mathbf{y}} + \bar{\mathbf{a}} - \mathbf{L}\bar{\mathbf{y}} - \mathbf{c}||^2].$$

Let $\mathbf{u} = \bar{\mathbf{a}} - \mathbf{L}\bar{\mathbf{y}} - \mathbf{c}$. Then, the LMMSE estimation problem is given by

$$\{\mathbf{L}_{lm}, \mathbf{u}_{lm}\} = \arg\min_{\mathbf{L}, \mathbf{u}} E[||\tilde{\mathbf{a}} - \mathbf{L}\tilde{\mathbf{y}} + \mathbf{u}||^2].$$

The optimal vector $\mathbf{u}$ is zero to minimize the MSE, and it implies that

$$\mathbf{c}_{lm} = \bar{\mathbf{a}} - \mathbf{L}\bar{\mathbf{y}}.$$

The optimal estimator $\mathbf{L}_{lm}$ is given by

$$\begin{aligned}
\mathbf{L}_{lm} &= E[\tilde{\mathbf{a}}\tilde{\mathbf{y}}^H](E[\tilde{\mathbf{y}}\tilde{\mathbf{y}}^H])^{-1} \\
&= \mathbf{C}_{ay}\mathbf{C}_{yy}^{-1},
\end{aligned}$$

where $\mathbf{C}_{ay} = E[\tilde{\mathbf{a}}\tilde{\mathbf{y}}^H]$ and $\mathbf{C}_{yy} = E[\tilde{\mathbf{y}}\tilde{\mathbf{y}}^H]$. Note that if $\bar{\mathbf{a}} = \mathbf{0}$ and $\bar{\mathbf{y}} = \mathbf{0}$, then $\mathbf{C}_{ay} = \mathbf{R}_{ay}$ and $\mathbf{C}_{yy} = \mathbf{R}_{yy}$. The LMMSE estimate of $\mathbf{a}$ is given by

$$\begin{aligned}
\hat{\mathbf{a}}_{lm} &= \mathbf{L}_{lm}\mathbf{y} + \mathbf{c}_{lm} \\
&= \mathbf{L}_{lm}\mathbf{y} + \bar{\mathbf{a}} - \mathbf{L}_{lm}\bar{\mathbf{y}} \\
&= \bar{\mathbf{a}} + \mathbf{L}_{lm}(\mathbf{y} - \bar{\mathbf{y}}).
\end{aligned}$$

The MMSE estimate of $\mathbf{a}$ given $\mathbf{y}$ can be found as follows:

$$\hat{\mathbf{a}}_{\text{mmse}} = \arg \min_{\hat{\mathbf{a}}} E[||\mathbf{a} - \hat{\mathbf{a}}||^2 |\mathbf{y}].$$

In general, it is not necessary that $\hat{\mathbf{a}}_{\text{mmse}} = \hat{\mathbf{a}}_{\text{lm}}$. Taking the derivative with respect to $\hat{\mathbf{a}}$ and setting it equal to zero yields

$$0 = E[(\mathbf{a} - \hat{\mathbf{a}})|\mathbf{y}],$$
$$\Rightarrow \hat{\mathbf{a}}_{\text{mmse}} = E[\mathbf{a}|\mathbf{y}].$$

This shows that the conditional mean vector of $\mathbf{a}$ given $\mathbf{y}$ is the MMSE estimate of $\mathbf{a}$. Generally, the MMSE estimate of $\mathbf{a}$ given $\mathbf{y}$ is different from the LMMSE estimate $\hat{\mathbf{a}}_{\text{lm}}$, except when $\mathbf{y}$ and $\mathbf{a}$ are jointly Gaussian random vectors. If $\mathbf{y}$ and $\mathbf{a}$ are jointly Gaussian random vectors, we have

$$E[\mathbf{a}|\mathbf{y}] = \hat{\mathbf{a}}_{\text{lm}}$$
$$= \bar{\mathbf{a}} + \mathbf{L}_{\text{lm}}(\mathbf{y} - \bar{\mathbf{y}}).$$

References

Abe, T. and Matsumoto, T. (2003). "Space-time turbo equalization in frequency-selective MIMO channels," *IEEE Trans. Veh. Tech.*, **52**, 469–475.

Alamouti, S. M. (1998). "A simple transmit diversity technique for wireless communications," *IEEE J. Selec. Areas in Commun.*, **16**, 1451–1458.

Anastasopoulos, A. and Chugg, K. M. (2000). "Adaptive soft-input soft-output algorithms for iterative detection with parameter uncertainty," *IEEE Trans. Commun.*, **48**, 1638–1649.

Anderson, J. B. (1989). "Limited search trellis decoding of convolutional codes," *IEEE Trans. Inform. Theory*, **35**, 944–955.

Anderson, R. R. and Foschini, G. J. (1975). "The minimum distance for MLSE digital data systems of limited complexity," *IEEE Trans. Inform. Theory*, **21**, 544–551.

Anderson, B. D. O. and Moore, J. B. (1979). *Optimal Filtering* (Englewood Cliffs, NJ: Prentice-Hall).

Aulin, T. M. (1999). "Breadth-first maximum likelihood sequence detection: basics," *IEEE Trans. Commun.*, **47**, 208–216.

Austin, M. E. (1967). *Decision-Feedback Equalization for Digital Communication Over Dispersive Channels*," MIT Lincoln Laboratory Technical Report.

Axelsson, O. (1994). *Iterative Solution Methods* (Cambridge: Cambridge University Press).

Baccarelli, E. and Cusani, R. (1998). "Combined channel estimation and data detection using soft statistics for frequency-selective fast-fading digital links," *IEEE Trans. Commun.*, **46**, 424–427.

Bahai, A. R. S. and Saltzberg, B. R. (1999). *Multi-Carrier Digital Communications: Theory and Applications of OFDM* (New York: Kluwer Academic).

Bahl, L. R., Cocke, J., Jelinek, F., and Raviv, J. (1974). "Optimal decoding of linear codes for minimizing symbol error rate," *IEEE Trans. Inform. Theory*, **20**, 284–287.

Bapat, J. L. (1998). "Partially blind estimation: ML-based approaches and Cramer–Rao bound," *Signal Proc.*, **71**, 265–277.

Bar-Shalom, Y. and Li, X. R. (1993). *Estimation and Tracking: Principles, Techniques and Software* (Dedham, MA: Artech House).

Baro, S. (2004). "Turbo detection for MIMO systems: bit labeling and pre-coding," *Eur. Trans. Telecommun.*, **15**, 343–350.

Bello, P. A. (1963). "Characterization of randomly time-variant linear channels," *IEEE Trans. Commun. Sys.*, **11**, 360–393.

Berrou, C., Glavieux, A., and Thitimajshima, P. (1993). "Near Shannon limit error-correcting coding and decoding: turbo codes," in *IEEE Proc. ICC'93*, Geneva, Switzerland, pp. 1064–1070.

Berthet, A. O., Unal, B. S., and Visoz, R. (2001). "Iterative decoding of convolutionally encoded signals over multipath Rayleigh fading channels," *IEEE J. Select. Areas Commun.*, **19**, 1729–1743.

Bertsekas, D. P. (1987). *Dynamic Programming: Deterministic and Stochastic Models* (Upper Saddle River, NJ: Prentice-Hall).

Biglieri, E., Nordio, A., and Taricco, G. (2004). "Iterative receivers for coded MIMO signaling," *Wireless Commun. and Mobile Comp.*, **4**, 697–710.

Biglieri, E., Proakis, J., and Shamai, S. (1998). "Fading channels: information-theoretic and communications aspects," *IEEE Trans. Inform. Theory*, **44**, 2619–2692.

Bingham, J. A. C. (1990). "Multicarrier modulation for data transmission: an idea whose time has come," *IEEE Commun. Mag.*, **28**, 5–14.

Borah, D. and Hart, B. (1999). "Receiver structures for time-varing frequency-selective fading channels," *IEEE J. Selected Areas Commun.*, **17**, 1863–1875.

Boutros, J. and Caire, G. (2002). "Iterative multiuser joint decoding: unified framework and asymptotic analysis," *IEEE Trans. Inform. Theory*, **48**, 1772–1793.

Brink, S. ten (2001). "Convergence behavior of iteratively decoded parallel concatenated codes," *IEEE Trans. Commun.*, **49**, 1727–1737.

Burg, J. P., Luenberger, D. G., and Wenger, D. L. (1982). "Estimation of structured covariance matrices," *Proc. IEEE*, **70**, 963–974.

Caire, G., Taricco, G., and Biglieri, E. (1998). "Bit-interleaved coded modulation," *IEEE Trans. Inform. Theory*, **44**, 927–946.

Capon, J. (1969). "High-resolution frequency-wavenumber spectrum analysis," *Proc. IEEE*, **57**, 1408–1418.

Cavers, J. K. (1991). "An analysis of pilot symbol assisted modulation for Rayleigh fading channels," *IEEE Trans. Veh. Tech.*, **40**, 686–693.

Chang, R. W. and Gibbey, R. A. (1968). "A theoretical study of performance of an orthogonal multiplexing data transmission scheme," *IEEE Trans. Commun. Tech.*, **16**, 529–540.

Choi, J. (2006). "N/C detector and multistage receiver for coded signals over MIMO channels: performance and optimization," *IEEE Trans. Wireless Commun.*, **5**, 1207–1216.

Choi, J., Kim, S. R., and Lim, C.-C. (2004). "Receivers with chip-level decision feedback equalizer for CDMA downlink channels," *IEEE Trans. Wireless Commun.*, **3**, 300–314.

Chugg, K. M. (1998). "The condition for the applicability of the Viterbi algorithm with implications for fading channel MLSD," *IEEE Trans. Commun.*, **46**, 1112–1116.

Chugg, K. M. and Polydoros, A. (1996a). "MLSE for an unknown channel – Part I: Optimality consideration," *IEEE Trans. Commun.*, **44**, 836–846.

Chugg, K. M. and Polydoros, A. (1996b). "MLSE for an unknown channel – Part II: Tracking performance," *IEEE Trans. Commun.*, **44**, 949–958.

Cover, T. M. and Thomas, J. B. (1991). *Elements of Information Theory* (New York: Wiley).

Cozzo, C. and Hughes, B. L. (2003). "Joint channel estimation and data detection in space-time communications," *IEEE Trans. Commun.*, **51**, 1266–1270.

Dejonghe, A. and Vandendorphe, L. (2004). "Bit-interleaved turbo equalization over static frequency-selective channels: constellation mapping impact," *IEEE Trans. Commun.*, **52**, 2061–2065.

Dempster, A. P., Laird, N. M., and Rubin, D. B. (1977). "Maximum likelihood from incomplete data via the EM algorithm," *Ann. Roy. Stat. Soc.*, **39**, 1–38.

Ding, Z. and Li, G. Y. (2001). *Blind Equalization and Identification* (New york: Marcel Dekker).

Duel-Hallen, A. and Heegard, C. D. (1989). "Delayed decision feedback sequence estimation," *IEEE Trans. Commun.*, **37**, 428–436.

Duel-Hallen, A., Holtzman, J., and Zvonar, Z. (1995). "Multiuser detection for CDMA Systems," *IEEE Personal Commun.*, **2**, 46–58.

Edfors, O., Sandell, M., Beek, J.-J. van de, Wilson, S. K., and Borjesson, P. O. (1998). "OFDM channel estimation by singular value decomposition," *IEEE Trans. Commun.*, **46**, 931–939.

Eklund, C., Marks, R. B., Standwood, K. L., and Wang, S. (2002). "IEEE standard 802.16: a technical overview of the wireless MAN air interface for broadband wireless access," *IEEE Commun. Mag.*, **40**, 98–107.

Eweda, E. (1994). "Comparison of RLS, LMS, and sign algorithms for tracking randomly time-varying channels," *IEEE Trans. Signal Proc.*, **42**, 2937–2944.

Eyuboglu, M. V. and Qureshi, S. U. H. (1988). "Reduced-state sequence estimation with set partitioning and decision feedback," *IEEE Trans. Commun.*, **36**, 13–20.

Falconer, D., Ariyavisitakul, S. L., Benyamin-Seeyar, A., and Eidson, B. (2002). "Frequency domain equalization for single-carrier broadband wireless systems," *IEEE Commun. Mag.*, **40**, 58–66.

Fessler, J. A. and Hero, A. O. (1994). "Space-alternating generalized expectation-maximization algorithm," *IEEE Trans. Commun.*, **42**, 2664–2677.

Forney, G. D. Jr. (1972). "Maximum-likelihood sequence estimation of digital sequences in the presence of intersymbol interference," *IEEE Trans. Inform. Theory*, **18**, 363–378.

Forney, G. D. Jr. (1973). "The Viterbi algorithm," *Proc. IEEE*, **61**, 268–278.

Foschini, G. J. (1996). "Layered space-time architecture for wireless communication in a fading environment when using multiple-element antennas," *Bell Lab. Tech. J.*, **1**, (2), 41–59.

Foschini, G. J. and Gans, M. J. (1998). "On limits of wireless communications in a fading environment when using multiple antennas," *Wireless Personal Commun.*, **6**, 311–335.

Foschini, G. J., Chizhik, D., Gans, M. J., Papadias, C., and Valenzuela, R. A. (2003). "Analysis and performance of some basic space-time architectures," *IEEE J. Select. Areas Commun.*, **21**, 303–320.

Frost, O. L. III, (1972). "An algorithm for linearly constrained adaptive processing," *Proc. IEEE*, **60**, 926–935.

Galdino, J. F., Pinto, E. L., and de Alencar, M. S. (2004). "Analytical performance of the LMS algorithm on the estimation of wide sense stationary channels," *IEEE Trans. Commun.*, **52**, 982–991.

Garcia-Frias, J. and Villasenor, J. D. (2003). "Combined turbo detection and decoding for unknown ISI channels," *IEEE Trans. Commun.*, **51**, 79–85.

Georghiades, C. N. and Han, J. C. (1997). "Sequence estimation in the presence of random parameters via the EM algorithm," *IEEE Trans. Commun.*, **45**, 300–308.

Gitlin, R. D., Hayes, J. F., and Weinstein, S. B. (1992). *Data Communications Principles* (New York: Plenum Press).

Golub, G. H. and Van Loan, C.F. (1983). *Matrix Computations* (Baltimore, MD: The Johns Hopkins University Press).

Goodwin, G. C. and Sin, K. S. (1984). *Adaptive Filtering, Prediction, and Control* (Englewood Cliffs, NJ: Prentice-Hall).

Gray, R. M. (2006). "Toeplitz and circulant matrices: a review," *Foundations & Trends Commun. Inf. Theory*, **2**, (3), 155–239.

Hagenauer, J., Offer, E., and Papke, L. (1996). "Iterative decoding of binary block and convolutional codes," *IEEE Trans. Inform. Theory*, **42**, 429–445.

Haykin, S. (1996). *Adaptive Filter Theory*, 3rd edn (Upper Saddle River, NJ: Prentice-Hall).

Haykin, S. and Veen, B. van (1999). *Signals and Systems* (New York: John Wiley & Sons).

Haykin, S., Sayed, A. H., Zeidler, J. R., Yee, P., and Wei, P. C. (1997). "Adaptive tracking of linear time-variant systems by extended RLS algorithms," *IEEE Trans. Signal Proc.*, **45**, 1118–1128.

Haykin, S., Sellathurai, M., de Jong, Y., and Willink, T. (2004). "Turbo-MIMO for wireless communications," *IEEE Commun. Mag.*, **42**, (10), 48–53.

Heath, R. W. Jr and Giannakis, G. B. (1999). "Exploiting input cyclostationarity for blind channel identification in OFDM systems," *IEEE Trans. Signal Proc.*, **47**, 848–856.

Heiskala, J. and Terry, J. (2001). *OFDM Wireless LANs: A Theoretical and Practical Guide* (Indianapolis: Sams).

Hochwald, B. M. and Brink, S. ten (2003). "Achieving near-capacity on a multiple-antenna channel," *IEEE Trans. Commun.*, **51**, 389–399.

Honig, M. L., Woodward, G. K., and Sun, Y. (2004). "Adaptive iterative multiuser decision feedback detection," *IEEE Trans. Wireless Commun.*, **3**, 477–485.

Honig, M. L. and Tsatsanis, M. K. (2000). "Adaptive techniques for multiuser CDMA receivers," *IEEE Signal Proc. Mag.*, **17**, 49–61.

Honig, M. L., Madhow, U., and Verdu, S. (1995). "Blind adaptive multiuser detection," *IEEE Trans. Inform. Theory*, **41**, 944–960.

Honig, M. L. and Messerschmitt, D. G. (1984). *Adaptive Filters: Structures, Algorithms, and Applications* (Boston, MA: Kluwer Academic).

Iltis, R. A. (1992). "A Bayesian maximum-likelihood sequence estimation algorithm for *a priori* unknown channels and symbol timing," *IEEE J. Select. Areas Commun.*, **10**, 579–588.

Jafarkhani, H. (2005). *Space-Time Coding* (Cambridge: Cambridge University Press).

Jakes, W. C. Jr., ed. (1974). *Microwave Mobile Communications* (New York: John Wiley & Sons).

Johnson, D. H. and Dudgeon, D. E. (1993). *Array Signal Processing* (Englewood Cliffs, NJ: Prentice-Hall).

Kocian, A. and Fleury, B. H. (2003). "EM-based joint data detection and channel estimation of DS-CDMA signals," *IEEE Trans. Commun.*, **51**, 1709–1720.

Koetter, R., Singer, A. C., and Tuchler, M. (2004). "Turbo equalization," *IEEE Signal Proc. Mag.*, pp. 67–80.

Laot, C., Glavieux, A., and Labat, J. (2001). "Turbo equalization: adaptive equalization and channel decoding jointly optimized," *IEEE J. Select. Areas Commun.*, **19**, 1744–1752.

Larsson, E. G. and Stoica, P. (2003). *Space-Time Block Coding for Wireless Communications* (Cambridge: Cambridge University Press).

Lee, I. (2001). "The effect of a precoder on serially concatenated coding systems with an ISI channel," *IEEE Trans. Commun.*, **49**, 1168–1175.

Lee, W. C. Y. (1982). *Mobile Communication Engineering* (New York: McGraw-Hill).

Li, K. and Wang, X. (2005). "EXIT chart analysis of turbo multiuser detection," *IEEE Trans. Wireless Commun.*, **4**, 300–311.

Li, Y., Cimini, L. J. Jr, and Sollenberger, N. R. (1998). "Robust channel estimation for OFDM systems with rapid dispersive fading channels," *IEEE Trans. Commun.*, **46**, 902–915.

Liang, Y.-C., Sun, S., and Ho, C. K. (2006). "Block-iterative generalized decision feedback equalizers for large MIMO systems: algorithm design and asymptotic performance analysis," *IEEE Trans. Signal Proc.*, **54**, 2035–2048.

Lin, J., Proakis, J. G., Ling, F., and Lev-Ari, H. (1995). "Optimal tracking of time-varying channels: a frequency domain approach for known and new algorithms," *IEEE J. Select. Areas Commun.*, **13**, 141–154.

Lin, S. and Costello, D. J. Jr. (1983). *Error Control Coding: Fundamentals and Applications*, (Englewood Cliffs, NJ: Prentice-Hall).

Lindbom, L., Sternad, M., and Ahlen, A. (2001). "Tracking of time-varying mobile radio channels – Part I: The Wiener LMS algorithm," *IEEE Trans. Commun.*, **49**, 2207–2217.

Lindbom, L., Ahlen, A., Sternad, M., and Falkenstrom, M. (2002). "Tracking of time-varying mobile radio channels – Part II: A case study," *IEEE Trans. Commun.*, **50**, 156–167.

Lucky, R. W. (1965). "Automatic equalization for digitial communications," *Bell Syst. Tech. J.*, **44**, 547–588.

Lucky, R. W. (1966). "Techniques for adaptive equalization of digitial communications," *Bell Syst. Tech. J.*, **45**, 255–286.

Lucky, R. W., Salz, J., and Weldon, E. J. (1968). *Principles of Data Communication* (New York: McGraw-Hill).

Luo, J., Pattipati, K. R., Willett, P. K., and Hasegawa, F. (2001). "Near-optimal multiuser detection in synchronous CDMA using probabilistic data association," *IEEE Commun. Lett.*, **5**, 361–363.

Lupas, R. and Verdu, S. (1989). "Linear multiuser detectors for synchronous code-division multiple-access channels," *IEEE Trans. Inform. Theory*, **35**, 123–136.

Ma, X., Kobayashi, H., and Schwartz, S. C. (2004). "EM-based channel estimation algorithms for OFDM," *EURASIP J. Appl. Signal Proc.*, **10**, 1460–1477.

McLachlan, G. J. and Krishnan, T. (1997). *The EM Algorithm and Extensions* (New York: John Wiley & Sons).

Maybeck, P. (1979). *Stochastic Models, Estimation, and Control* (New York: Academic Press).

Monzingo, R. A. and Miller, T. W. (1980). *Introduction to Adaptive Arrays* (New York: John Wiley & Sons).

Morley, R. E. Jr and Snyder, D. L. (1979). "Maximum likelihood sequence estimation for randomly dispersive channels," *IEEE Trans. Commun.*, **27**, 833–839.

Moshavi, S. (1996). "Multiuser detection for DS-CDMA communications," *IEEE Commun. Mag.*, **34**, 124–136.

Narayanan, K. R. (2001). "Effect of precoding on the convergence of turbo equalization for partial response channels," *IEEE J. Select. Areas Commun.*, **19**, 686–698.

Nee, R. van, Awater, G., Morikura, M., Takanashi, H., Webster, M., and Halford, K. W. (1999). "New high-rate wireless LAN standards," *IEEE Commun. Mag.*, **37**, (12), 82–88.

Nefedov, N., Pukkila, M., Visoz, R., and Berthet, A. O. (2003). "Iterative data detection and channel estimation for advanced TDMA systems," *IEEE Trans. Commun.*, **51**, 141–144.

Nelson, L. B. and Poor, H. V. (1996). "Iterative multiuser receivers for CDMA channels: an EM-based approach," *IEEE Trans. Commun.*, **44**, 1700–1710.

Nguyen, H. and Choi, J. (2006). "Iterative OFDM receiver with channel estimation and frequency-offset compensation," in *Proc. IEEE Int. Conf. Communications*, Istanbul, Turkey, June 2006.

Oppenheim, A. V. and Willsky, A. S. (1996). *Signals and Systems*, 2nd edn (Upper Saddle River, NJ: Prentice-Hall).

Otnes, R. and Tuchler, M. (2004). "Iterative channel estimation for turbo equalization of time-varying frequency-selective channels," *IEEE Trans. Wireless Commun.*, **3**, 1918–1923.

Papoulis, A. (1984). *Probability, Random Variables, and Stochastic Processes*, 2nd edn (New York: McGraw-Hill).

Paulraj, A., Nabar, R., and Gore, D. (2003). *Introduction to Space-Time Wireless Communications* (Cambridge: Cambridge University Press).

Pham, D., Pattipati, K. R., Willet, P. K., and Luo, J. (2004). "A generalized probabilistic data association detector for multiple antenna systems," *IEEE Commun. Lett.*, **8**, 205–207.

Porat, B. (1994). *Digital Processing of Random Signals: Theory and Methods* (Englewood Cliffs, NJ: Prentice-Hall).

Proakis, J. G. (1995). *Digital Communications*, 3rd edn (New York: McGraw-Hill).

Raheli, R., Polydoros, A., and Tzou, C.-K. (1995). "Per-survivor processing: a general approach to MLSE in uncertain environments," *IEEE Trans. Commun.*, **43**, 354–364.

Rappaport, T. (1996). *Wireless Communications: Principles and Practice* (Piscataway, NJ: IEEE Press).

Reynolds, D. and Wang, X. (2002). "Turbo multiuser detection with unknown interferers," *IEEE Trans. Commun.*, **50**, 616–622.

Roberts, R. A. and Mullis, C. T. (1987). *Digital Signal Processing* (Reading, MA: Addison-Wesley).

Rollins, M. E. and Simmons, S. J. (1997). "Simplified per-survivor Kalman processing in fast frequency-selective fading channels," *IEEE Trans. Commun.*, **45**, 544–553.

Sari, H., Karam, G., and Jeanclaude, I. (1995). "Transmission techniques for digital terrestrial TV broadcasting," *IEEE Commun. Mag.*, **33**, (2), 100–109.

Sellami, N., Lasaulce, S., and Fijalkow, I. (2003). "Iterative channel estimation for coded DS-CDMA systems over frequency selective channels," in *Proc. 4th IEEE Workshop on Signal Processes Advances in Wireless Communication*, Rome, Italy, pp. 80–84.

Seshadri, N. (1994). "Joint data and channel estimation using blind trellis search techniques," *IEEE Trans. Commun.*, **42**, 1000–1011.

Solo, V. and Kong, X. (1995). *Adaptive Signal Processing Algorithms: Stability and Performance* (Upper Saddle River, NJ: Prentice-Hall).

Stefanov, A. and Duman, T. M. (2001). "Turbo-coded modulation for systems with transmit and receive antenna diversity over block fading channel: system model, decoding approaches, and practical considerations," *IEEE J. Select. Areas Commun.*, **19**, 958–968.

Tarokh, V., Jafarkhani, H., and Calderbank, A. R. (1999). "Space-time block codes from orthogonal designs," *IEEE Trans. Inform. Theory*, **45**, 1456–1467.

Tarokh, V., Seshadri, N., and Calderbank, A. (1998). "Space-time codes for high data rate wireless communication: performance criterion and code construction," *IEEE Trans. Inform. Theory*, **44**, 744–765.

Telatar, I. E. (1999). "Capacity of multiple-antenna Gaussian channels," *Eur. Trans. Telecommun.*, **10**, 585–595.

Tong, L., Sadler, B. M., and Dong, M. (2004). "Pilot-assisted wireless transmissions," *IEEE Signal Proc. Mag.*, **21**, 12–25.

Tse, D. and Viswanath, P. (2005). *Fundamentals of Wireless Communication* (Cambridge: Cambridge University Press).

Tuchler, M., Koetter, R., and Singer, A. (2002a). "Turbo equalization: principles and new results," *IEEE Trans. Commun.*, **50**, 754–767.

Tuchler, M., Singer, A., and Koetter, R. (2002b). "Minimum mean squared error equalization using *a priori* information," *IEEE Trans. Signal Proc.*, **50**, 673–683.

Ungerboeck, G. (1974). "Adaptive maximum-likelihood receiver for carrier-modulated data-transmission systems," *IEEE Trans. Commun.*, **22**, 624–635.

Verdu, S. (1986). "Minimum probability of error for asynchronous Gaussian multiple-access channels," *IEEE Trans. Inform. Theory*, **32**, 85–96.

Verdu, S. (1998). *Multiuser Detection* (Cambridge: Cambridge University Press).

Vikalo, H., Hassibi, B., and Kailath, T. (2004). "Iterative decoding for MIMO channels via modified sphere decoding," *IEEE Trans. Wireless Commun.*, **3**, 2299–2311.

Visoz, R. and Berthet, A. O. (2003). "Iterative decoding and channel estimation for space-time BICM over MIMO block fading multipath AWGN channel," *IEEE Trans. Commun.*, **51**, 1358–1367.

Viterbi, A. J. (1967). "Error bounds for convolutional codes and an asymptotically optimum decoding algorithm," *IEEE Trans. Inform. Theory*, **13**, 260–269.

Viterbi, A. J. (1995). *CDMA: Principles of Spread Spectrum Communications* (Reading, MA: Addison-Wesley).

Viterbo, E. and Boutros, J. (1999). "A universal lattice code decoder for fading channels," *IEEE Trans. Inform. Theory*, **45**, 1639–1642.

Wang, X. and Poor, H. V. (1999). "Iterative (turbo) soft interference cancellation and decoding for coded CDMA," *IEEE Trans. Commun.*, **47**, 1046–1061.

Weinstein, S. B. and Ebert, P. M. (1971). "Data transmission by frequency division multiplexing using the discrete Fourier transform," *IEEE Trans. Commun. Tech.*, **19**, 628–634.

Whalen, A. D. (1971). *Detection of Signals in Noise* (London: Academic Press).

Widrow, B. and Stearns, S. (1985). *Adaptive Signal Processing* (Englewood Cliffs, NJ: Prentice-Hall).

Widrow, B., McCool, J. M., Larimore, M. G., and Johnson, C. R. Jr. (1976). "Stationary and nonstationary learning characteristics of the LMS adaptive filter," *Proc. IEEE*, **64**, 1151–1162.

Wolniansky, P. W., Foschini, G. J., Golden, G. D., and Valenzuela, R. A. (1998). "V-BLAST: an architecture for realizing very high data rates over the rich-scattering wireless channel," in *Proc. ISSSE-98*, Pisa, Italy, pp. 295–300.

Worthen, A. P. and Stark, W. E. (2001). "Unified design of iterative receivers using factor graph," *IEEE Trans. Inform. Theory*, **47**, 849–853.

Xu, Z. and Tsatsanis, M. K. (2001). "Blind adaptive algorithms for minimum variance CDMA receivers," *IEEE Trans. Commun.*, **49**, 180–194.

Yang, B., Letaief, K. B., Cheng, R. S., and Cao, Z. (2001). "Channel estimation for OFDM transmission in multipath fading channels based on parametric channel modeling," *IEEE Trans. Commun.*, **49**, 467–479.

Yousef, N. R. and Sayed, A. H. (2001). "A unified approach to the steady-state and tracking analyses of adaptive filters," *IEEE Trans. Signal Proc.*, **49**, 314–324.

Zamiri-Jafarian, H. and Pasupathy, S. (1999). "EM-based recursive estimation of channel parameters," *IEEE Trans. Commun.*, **47**, 1297–1302.

Zheng, L. and Tse, D. N. C. (2003). "Diversity and multiplexing: a fundamental tradeoff in multiple-antenna channels," *IEEE Trans. Inform. Theory*, **49**, 1073–1096.

Index